Modern Group Theoretical Methods in Physics

MATHEMATICAL PHYSICS STUDIES

Series Editor:

M. FLATO, *Université de Bourgogne, Dijon, France*

VOLUME 18

Modern Group Theoretical Methods in Physics

Proceedings of the Conference in Honour of Guy Rideau

edited by

J. Bertrand

L.P.T.M.,
Université Denis Diderot,
Paris, France

M. Flato

Université de Bourgogne,
Dijon, France

J.-P. Gazeau
M. Irac-Astaud

and

D. Sternheimer

L.P.T.M.,
Université Denis Diderot,
Paris, France

KLUWER ACADEMIC PUBLISHERS

DORDRECHT / BOSTON / LONDON

A C.I.P. Catalogue record for this book is available from the Library of Congress.

ISBN 0-7923-3645-3

Published by Kluwer Academic Publishers,
P.O. Box 17, 3300 AA Dordrecht, The Netherlands.

Kluwer Academic Publishers incorporates
the publishing programmes of
D. Reidel, Martinus Nijhoff, Dr W. Junk and MTP Press.

Sold and distributed in the U.S.A. and Canada
by Kluwer Academic Publishers,
101 Philip Drive, Norwell, MA 02061, U.S.A.

In all other countries, sold and distributed
by Kluwer Academic Publishers Group,
P.O. Box 322, 3300 AH Dordrecht, The Netherlands.

Printed on acid-free paper

Printed in the Netherlands

To our friend Guy Rideau upon his retirement from CNRS

Table of Contents

GUY RIDEAU

OPENING ADDRESS

MOSHÉ FLATO
Physique Mathématique, Université de Bourgogne
B.P. 138, F-21004 Dijon Cedex, France

Ladies and Gentlemen,

Quite often a meeting which is held in honour of a retiring scientist sounds like a funeral: "He did that, he did this..." as if the scientific activity of the person stops sharply after the person retires. My few remarks will be in an opposite sense to Shakespeare's "Julius Caesar". I did not come "to bury Caesar not to praise him "but rather did I come "to praise Guy, not to bury him".

Guy Rideau for me is an old story, old and very stable friendship. I know him for *more* than 30 years (almost 32) and this explains why our friendship is stable (in contrast to the "false friendship" of Chirac and Balladur that lasted exactly 30 years).

It was one day in the autumn of 1963 that I came as a young man from Israel to the (at that time) famous "Institut Henri Poincaré" in Paris. I was going to meet Rideau's neighbour in the office and was introduced to Rideau.

Our styles were really opposite: A young Sabra interacting with a formal and well educated Frenchman could produce a rapid earthquake. And this indeed is what happened at first. But it took only six months more and there was a common article by Rideau, his neighbour and myself.

Though this is the only article we ever published together, we had since then hundreds of stimulating and creative scientific (and non-scientific) conversations. In two words: a solid and close friendship grew up and proved itself in the joys and pains that life presents to each of us.

But since I came to talk about Guy, let me try and give you some highlights of his scientific career. As a young scientist, while working on his doctoral Thesis, he was strongly involved with the mathematical aspects of quantum-field-theory. In fact in his Thesis one can find some elements of what later turned out to be constructive field theory.

As we all know physical theories have three important aspects:

1) Research of new Lagrangians which describe different phenomena. This aspect, treated mainly by theoretical physicists who are close to phenomenology, does not belong to Rideau's scientific menu.

2) Field dynamics, or equivalently algebras of observables and their automorphisms, and

3) Symmetries (and generalized symmetries) of classical and quantum dynamical systems.

During the first part of his scientific activity Rideau was engaged in point 2 of our classification. After his Thesis which touched constructive field theory, he worked for some time on quantum-field-theoretical models like the Wentzel model and the Lee model. He then got interested in the representation theory of the canonical commutation (and of course anti-commutation) relations and in the algebraic approach (Von Neumann and C^* algebras) related to them.

He was among the first (together with Kadison and Powers) to have non-trivial results on type III, later continued by Araki and culminated with the celebrated results of Alain Connes after the introduction of modular Hilbertian algebras by Tomita and Takesaki. Around the same time Rideau has shifted his activity to the study of symmetries and their dynamical consequences, namely to our point 3.

At the center of relativistic theories stands the inhomogeneous Lorentz group, often called the Poincaré group. It was the late Eugene Paul Wigner who in a pioneering paper in 1939 more or less established the unitary dual of this group. In the mid-sixties people were trying to study field theories on the manifold of the Poincaré group, and in particular to formulate the special status of massless theories in such a framework.

Rideau, after studying the reduction of tensor products of unitary irreducible representations of the Poincaré group, was able in an important paper to establish the Plancherel formula for the Poincaré group, showing among other things that the Plancherel measure of the massless part is zero, thus shooting down the claims of some authors.

The first part of the seventies was under the sign of gauge theories. Rideau then was occupied with extension theory of the (finite-dimensional) Poincaré Lie group by an (infinite-dimensional) abelian gauge-group. His paper on the Landau gauge is very well-known. This was the time when Simon and Pinczon came with the theory of extensions of representations and Rideau joined the collective effort by writing a beautiful paper on extensions of representations of the Poincaré group.

Following these lines, he also became interested in the study of non-linear representations of Lie groups and their criteria of linearizability and wrote a couple of nice papers there. This brought him straight to the subject of integrable models and non-linear covariant partial differential equations, and to nice collaboration with Bob Anderson, Pavel Winternitz and others.

In addition to those active collaborations Rideau has been quite active in some recent aspects of quantum groups.

I have no pretension to cover all of Rideau's activity in so few words. This is just a tentative to give a flavour of the problematics that characterizes Rideau's style in research.

I'll just finish by wishing our friend Guy Rideau many more healthy, happy, and very creative years to come.

Let us all applaud him very heartily!

THE PUBLICATIONS OF GUY RIDEAU

1. (1950) Sur la condensation dans la statistique de Bose-Einstein, *C.R. Acad. Sci. Paris*, **230**, p. 1036.
2. (1950) Le passage à l'espace des moments en théorie des collisions, *C.R. Acad. Sci. Paris*, **231**, p. 949.
3. (1950) La transformation de Fourier de l'équation de Dirac, *C.R. Acad. Sci. Paris*, **231**, p. 1286.
4. (1951) Méthode pour l'étude des perturbations de durée limitée, *C.R. Acad. Sci. Paris*, **232**, p. 1338.
5. (1951) Sur la conservation de l'énergie en mécanique quantique, *C.R. Acad. Sci. Paris*, **232**, p. 1409.
6. (1951) Sur la quatrième relation d'incertitude, *C.R. Acad. Sci. Paris*, **232**, p. 2007.
7. with T. Kahan, (1951) Sur un principe variationnel général en Physique Théorique, *C.R. Acad. Sci. Paris*, **233**, p. 849.
8. (1952) Sur la formulation des problèmes de diffusion, *C.R. Acad. Sci. Paris*, **234**, p. 1746.
9. (1952) Sur les méthodes de Feynman, *C.R. Acad. Sci. Paris*, **234**, p. 1852.
10. with T. Kahan, (1952) Sur la déduction de divers principes variationnels en théorie des collisions à partir d'un principe général, *Journal Phys. Radium*, **13**, p. 326.
11. (1953) Sur les principes variationnels en mécanique quantique, *C.R. Acad. Sci. Paris*, **237**, p. 1646.
12. (1954) Sur la résolution des équations de la théorie des champs I, *C.R. Acad. Sci. Paris*, **238**, p. 2057.
13. (1954) Sur la résolution des équations de la théorie des champs II, *C.R. Acad. Sci. Paris*, **238**, p. 2227.
14. (1954) Sur la résolution des équations de la théorie des champs III, *C.R. Acad. Sci. Paris*, **238**, p. 2288.

15. (1955) Sur les méthodes de résolution en théorie quantique des champs, *Thèse de Doctorat d'Etat*, Université de Paris.
16. with T. Kahan and P. Roussopoulos (1956) Les méthodes d'approximation variationnelles dans la théorie des collisions atomiques et dans la physique des piles nucléaires, *Mémorial des Sciences Mathématiques.*
17. (1958) Sur l'introduction des opérateurs habillés en théorie quantique des champs, *C.R. Acad. Sci. Paris*, **247**, p. 1098.
18. with A. Chevalier (1958) Sur la diagonalisation du modèle de Wentzel, *Nuovo Cimento*, **10**, p. 228.
19. (1960) Groupe symétrique, chap. 5 in *Théorie des groupes en physique classique et quantique*, vol.1: *Structures mathématiques et fondements quantiques*, T. Kahan, Dunod.
20. (1961) Un modèle résoluble en théorie quantique des champs : le modèle de Wentzel, *Ann. Inst. H. Poincaré*, **17**, p. 91.
21. with M. Flato and J.P. Vigier (1965) Definition of P, C, T operators for the relativistic rotator model and the Bronzan-Low symmetry, *Nucl. Phys.*, **51**, p. 250.
22. (1965) Sur la définition intrinsèque de la distribution $\Delta^{(1)}$, *C.R. Acad. Sci. Paris*, **260**, p. 2719.
23. (1965) Sur l'homomorphie entre le groupe SU(2,2) et le groupe conforme, *C.R. Acad. Sci. Paris*, **261**, p. 3297.
24. (1965) Les produits tensoriels des représentations unitaires irréductibles du groupe de Poincaré, *Ann. Inst. H. Poincaré,* **3**, p. 339.
25. (1966) Sur la réduction du produit tensoriel des représentations de la série discrète de SL(2,R), *Ann. Inst. H. Poincaré*, **4**, p. 67.
26. (1966) On the reduction of the regular representation of the Poincaré group, *Comm. Math. Phys.*, **3**, p. 218.
27. with A. Cordesse (1966) On some representations of anticommutation relations, *Nuovo Cimento,* **45A**, p. 1.
28. with A. Cordesse (1966) On some representations of anticommutation relations II, *Nuovo Cimento*, **46A**, p. 624.
29. with A. Cordesse (1967) On some representations of anticommutation relations III, *Nuovo Cimento,* **50A**, p. 244.
30. (1967) On a conjecture of D. Robinson, *Preprint Institut Henri Poincaré.*
31. (1968) On some representations of anticommutation relations, *Comm. Math. Phys.*, **9**, p. 229.
32. (1968) On some non-unitary representations of Poincaré group, *Preprint Institut Henri Poincaré.*
33. (Juin 1970) The constructive theory of fields, *Talk delivered at the Second Symposium Gunnar Källen*, Paris.
34. with J. Bertrand (1973) Non-unitary representations of the Poincaré group and Plancherel formula, *Rep. Math. Phys.* **4**, p. 47.

35. (Juin 1973) Gauges of first and second kind and Poincaré invariance, *Preprint Université Paris VII.*
36. (1974) Gauge group and extension of Poincaré group, *Physica Scripta,* **9**, p. 163.
37. (Juin 1974) Group theoretical structure of the free photon field, *Preprint Université Paris VII.*
38. (1975) On the realization of the Landau gauge, *Lett. in Math. Phys.*, **1**, p. 17.
39. (Juillet 1975) On the two-point function in Landau gauge, *Preprint Université Paris VII.*
40. (1978) Poincaré group cohomology and generalized Lorentz gauges, *J. Math. Phys.*, **19**, p. 1627.
41. with Ph. Combe, R.Rodriguez and M. Sirugue Collin (1978) On the cylindrical approximation of the Feynman path integral, *Rep. Math. Phys.*, **13**, p. 279.
42. (1978) Quantization of the Maxwell field and extension of mass zero representations of Poincaré group, *Lett. Math. Phys.*, **2**, p. 529.
43. (1979) Quantization of the Maxwell field in conformal invariant gauges, *Lett. Math. Phys.*, **3**, p. 229. cf. also Erratum and Addendum : (1979) *Lett. Math. Phys.*, **3** , p. 343.
44. (1979) On extension of mass-zero representations of Poincaré group, *Rep. Math. Phys.*, **16**, p. 251.
45. (Janvier 1979) A proof of PCT invariance for Maxwell field quantization with covariant gauge condition, *Preprint Université Paris VII.*
46. (1981) Covariant quantizations of the Maxwell field, in *Field theory, Quantization and Statistical Physics*, E.Tirapegui, ed., Reidel.
47. with J. Bertrand (1982) Stochastic jump processes in phase space representation of quantum mechanics, in "Mathematical Problems in Theoretical Physics", *Lecture Notes in Physics* , Springer-Verlag, **153**, p. 276.
48. (1982) Extension of unitary irreducible representations of Poincaré group. Application to quantum field theory, in "Mathematical Problems in Theoretical Physics", *Lecture Notes in Physics*, Springer-Verlag, **153**, p. 302.
49. with J. Bertrand (1983) An intrinsic approach to the evolution of quantum observables in terms of stochastic processes on phase space, in "Stochastic Aspects of Classical and Quantum Systems", *Lecture Notes in Mathematics*, Springer-Verlag, **1109**, p. 39.
50. with J. Bertrand (1983) Stochastic processes and the evolution of quantum observables, *Lett. Math. Phys.*, **7**, p. 327.

51. (1984) Non-trivial extensions of a representation of the Poincaré group with mass and helicity zero by its tensorial product, *Lett. Math. Phys.*, **8**, p. 421.
52. with J. Bertrand and B. Gaveau (1985), Poisson processes and quantum field theory: a model, in "Quantum probability and Applications II", *Lecture Notes in Mathematics*, Springer-Verlag, **1136**, p. 74.
53. with J. Bertrand and B. Gaveau (1985) Quantum Fields and Poisson Processes I. *Lett. Math. Phys.*, **9**, p. 73.
54. with J. Bertrand and B. Gaveau (1985), Quantum Fields and Poisson Processes II, *Lett. Math. Phys.*, **9**, p. 353.
55. (1985) An analytic non linear representation of the Poincaré group, *Lett. Math. Phys.*, **9**, p. 337.
56. with J. Bertrand (1985), An analytic non linear representation of the Poincaré group II. The case of helicities 1/2, *Lett. Math. Phys.*, **10**, p. 325.
57. (1986) Représentations non linéaires des groupes : exemples du groupe de Poincaré à 2 et 4 dimensions, *Pub. IRMA, Lille* **1**, N^0 X.
58. (1987) Non-linear representations of the Poincaré group in 3 and 4 dimensions, *XVth International Colloquium on Group Theoretical Methods in Physics*, World Scientific, p. 563.
59. with J. Bertrand (1987) Non linear representations of Poincaré group in three dimensions, *J. Math. Phys.*, **28**, p. 1972.
60. (1988) Non-trivial extensions of a massless representation of the Poincaré group in 3+1 dimensions by the tensor product of two other massless representations, *Lett. Math. Phys.*, **15**, p. 243.
61. (1989) Cohomology of extension for the Poincaré group representations: Applications to quantum mechanics with indefinite metric and to nonlinear representation theory, *Symposia Mathematica - XXXI* , p. 95.
62. (1989) Uniqueness problems in formal nonlinear representations of Poincaré group in 2+1 dimensions, *J. Math. Phys.*, **30**, p. 2411.
63. (1989) Nonlinear representations of Poincaré group in 2+1 dimensions, *XVIIth International Colloquium on Group Theoretical Methods in Physics*, World Scientific, p. 444.
64. with P. Winternitz (1990) Nonlinear equations invariant under the Poincaré similitude and conformal groups in 2-dimensional space-time, *J. Math. Phys.*, **31**, p. 1095.
65. (1990) Manifest linear covariance for non linear formal representations of Poincaré's group in 2+1 dimensions, *Lett. Math. Phys.*, **20**, p. 1.
66. with R.L. Anderson and P.W. Hebda (1991) Hamiltonian Treatment of quasi-linear systems via a Bateman Lagrangian, *Lett. Math. Phys.*, **22**, p. 335.

67. with R.L. Anderson and P.W. Hebda (1991) Canonical Quantization of Bateman-Morse-Feshbach Damped Oscillator, *Lett. Math. Phys.*, **23**, p. 79.
68. (1992) On the Representations of Quantum Oscillator Algebra, *Lett. Math. Phys.*, **24**, p. 147.
69. with M. Irac-Astaud (1992) Quantum Bihamiltonian Systems : the Example of Harmonic Oscillator, *Preprint LPTM.*
70. with P. Winternitz (1993) Evolution Equations Invariant under Two-Dimensional Space-Time Schrödinger Group, *J. Math. Phys.*, **34**, p. 558.
71. with P. Winternitz (1993) Representations of the Quantum Algebra $SU_q(2)$ on a Real Two-Dimensional Sphere, *J. Math. Phys.* , **34**, p. 6030.
72. with M. Irac-Astaud (1993) On quantum multi-hamiltonian systems, *Lett. Math. Phys.*, **29** p. 197.
73. with M. Irac-Astaud (1994) On quantum multi-hamiltonian systems, *Teoreticheskaya i Matematicheskaya Fizika* , **99**, No 3, p. 377.
74. with M. Irac-Astaud (1995) Deformed quantum harmonic oscillator, *International Journal of Modern Physics* B, **9**, to appear.
75. with J.Bertrand and G.G.Emch (1995) The cohomology of the classical and quantum Weyl CCR in curved spaces, *Lett. Math. Phys.*, to appear.

THE PROBLEM OF INVARIANTS FOR LIE-BÄCKLUND GROUPS

ROBERT L. ANDERSON
University of Georgia
Department of Physics and Astronomy
Athens, GA 30602, U.S.A.

AND

NAIL H. IBRAGIMOV
Department of Computational and Applied Mathematics
University of the Witwatersrand
Private Bag 3, P O WITS 2050
Johannesburg, South Africa

Abstract. The application of formal power series to the problem of constructing invariants for one parameter Lie-Bäcklund transformation groups [1] is discussed. Previously reported results [1] as well as an extension of these results are presented. Our method of constructing these differential invariants is built upon the fact that the only existing theory for these groups is framed in the space $[[\mathcal{A}]]$ of formal power series with coefficients in the space $\mathcal{A}$ of differential functions [2]. This theory and these spaces are discussed for completeness.

1. Introduction

The practical motivation for studying the problem of enumerating differential invariants for Lie-Bäcklund transformation groups arises because of the ubiquitous appearance of these groups as symmetry groups of nonlinear evolution equations. However there is also another unavoidable theoretical reason for studying this problem. Namely, on one hand, there is an equivalence between the generators of Lie point transformation groups and their canonical forms which are in general generators of Lie-Bäcklund transformation groups. On the other hand, a basis for differential invariants is known for the Lie point generators, whereas, in the case of Lie-Bäcklund

J. Bertrand et al. (eds.), Modern Group Theoretical Methods in Physics, 1–14.

generators we have e.g., in the one-dimensional case, no differential invariants which are locally analytic functions of the independent variable, the dependent variable and a finite number of its derivatives.

The direction we take in addressing the problem of constructing differential invariants Lie-Bäcklund transformation groups is to build on the fact that the only existing theory of these groups is framed in the space $[[\mathcal{A}]]$ of formal power series with coefficients in the space $\mathcal{A}$ of differential functions [2]. For completeness, this theory is reviewed and these spaces are defined in this paper. The question of the convergence of these series is not treated here. The reader who is interested in this aspect of the problem is referred to the classical book by Hardy [3] and for a modern approach see the program of Flato, Pinczon, and Simon [4].

2. Preliminaries

We now turn to the apparatus. The prolongation theory of Lie point and Lie contact transformation groups requires the introduction of functions depending not only on the independent variables x and dependent variables u, but also on derivatives

$$\underset{1}{u},\quad \underset{2}{u},\quad \dots$$

of finite orders. This prolongation is sufficient in the context of classical Lie theory. However, it is insufficient for the natural generalization of the classical theory given by Lie-Bäcklund transformation groups. In this generalization, one deals with transformations acting on intrinsically infinite-dimensional spaces. This new approach mandates the space $\mathcal{A}$ of differential functions as the universal space for Lie-Bäcklund transformation group analysis.

2.1. THE SPACE $\mathcal{A}$ OF DIFFERENTIAL FUNCTIONS

This space was previously discussed in reference [2] and the CCR Handbook [5]. But, for the convenience of the reader, we reproduce here the definition and necessary notation. Here, we sum over repeated indices.

Let

$$x=\{x^i\}, u=\{u^\alpha\},\ \underset{1}{u}=\{u_i^\alpha\},\ \underset{2}{u}=\{u_{ij}^\alpha\},\dots, \qquad (1)$$
$$\alpha=1,\dots,m;\quad i,j=1,\dots,n.$$

These variables are connected by the total differentiations

$$D_i=\frac{\partial}{\partial x^i}+u_i^\alpha\frac{\partial}{\partial u^\alpha}+u_{ij}^\alpha\frac{\partial}{\partial u_j^\alpha}+\dots,i=1,\dots n \qquad (2)$$

as follows:

$$u_i^\alpha = D_i(u^\alpha), u_{ij}^\alpha = D_i(u_j^\alpha) = D_jD_i(u^\alpha), \ldots \tag{3}$$

The variables x are called the independent variables, and the variables u are known as the differential variables with the successive derivatives $\underset{1}{u}$, $\underset{2}{u}$, etc.

Definition 1. A locally analytic function (i.e., locally expandable in a Taylor series with respect to all arguments) of a finite number of variables (1) is called a differential function. The highest order of derivatives appearing in the differential function is called the order of this function. The vector of all differential functions of finite order is denoted by $\mathcal{A}$.

The space $\mathcal{A}$ has the intrinsic property of being closed under the derivation given by the total derivatives D_i.

2.2. LIE-BÄCKLUND OPERATORS

Definition 2. A differential operator of the form

$$X = \xi^i \frac{\partial}{\partial x^i} + \eta^\alpha \frac{\partial}{\partial u^\alpha} + \zeta_i^\alpha \frac{\partial}{\partial u_i^\alpha} + \zeta_{i_1 i_2}^\alpha \frac{\partial}{\partial u_{i_1,i_2}^\alpha} + \ldots, \tag{4}$$

where $\xi^i, \eta^\alpha \in \mathcal{A}$, and

$$\begin{aligned} \zeta_i^\alpha &= D_i(\eta^\alpha - \xi^j u_j) + \xi^j u_{ji}, \\ \zeta_{i_1 i_2}^\alpha &= D_{i_1} D_{i_2}(\eta^\alpha - \xi^j u_j) + \xi^j u_{j i_1 i_2}, \\ &\ldots \end{aligned} \tag{5}$$

is called a Lie-Bäcklund operator. The abbreviated operator

$$X = \xi^i \frac{\partial}{\partial x^i} + \eta^\alpha \frac{\partial}{\partial u^\alpha}, \tag{6}$$

is also referred to as a Lie-Bäcklund operator.

The operator (4) is a formal sum. However, it truncates when acting on any differential function. Hence, the action of Lie-Bäcklund operators is defined on the space $\mathcal{A}$.

2.3. LIE-BÄCKLUND ALGEBRA $L_\mathcal{B}$

Let

$$X_v = \xi_v^i \frac{\partial}{\partial x^i} + \eta_v^\alpha \frac{\partial}{\partial u^\alpha}, \quad v = 1, 2, \tag{7}$$

be two Lie-Bäcklund operators (6). Their commutator (Lie bracket) is defined by the usual formula

$$[X_1, X_2] = X_1X_2 - X_2X_1. \tag{8}$$

Theorem 1. The commutator $[X_1, X_2]$ is the Lie-Bäcklund operator given by

$$[X_1, X_2] = (X_1(\xi_2^i) - X_2(\xi_1^i))\frac{\partial}{\partial x^i} + (X_1(\eta_2^\alpha) - X_2(\eta_1^\alpha))\frac{\partial}{\partial u^\alpha} + \dots, \tag{9}$$

where the terms denoted by dots are given by prolonging the coefficients of the operators $\frac{\partial}{\partial x^i}$ and $\frac{\partial}{\partial u^\alpha}$ in accordance with Equations (5).

As a consequence, the following statement is valid.

Theorem 2. The set of all Lie-Bäcklund operators is an infinite dimensional Lie algebra with respect to the Lie bracket (8). We call it the Lie-Bäcklund algebra and denote it by $L_{\mathcal{B}}$.

2.4. PROPERTIES OF $L_{\mathcal{B}}$. CANONICAL OPERATORS

The Lie-Bäcklund algebra is endowed with the following properties (see, e.g., [2])

I. The total derivation (2) is a Lie-Bäcklund operator, i.e., $D_i \in L_{\mathcal{B}}$. Moreover

$$X_* = \xi_*^i D_i \in L_{\mathcal{B}}, \tag{10}$$

for any $\xi_*^i \in \mathcal{A}$.

II. Let L_* be the set of all Lie-Bäcklund operators of the form (10). Then L_* is closed under the commutator (8), i.e., it is a subalgebra of $L_{\mathcal{B}}$. Furthermore, L_* is an ideal of $L_{\mathcal{B}}$. The last statement follows from the equation

$$[X, X_*] = (X(\xi_*^i) - X_*(\xi^i))D_i \in L_*, \tag{11}$$

where X is an arbitrary Lie-Bäcklund operator (4).

III. In accordance with property II, two operators $X_1, X_2 \in L_{\mathcal{B}}$ are said to be equivalent (i.e., $X_1 \sim X_2$) if $X_1 - X_2 \in L_*$. In particular, every operator (4) is equivalent to an operator with $\xi^i = 0$ $(i = 1, \dots, n)$, viz

$$X \sim X - \xi^i D_i \equiv (\eta^\alpha - \xi^i u_i^\alpha)\frac{\partial}{\partial u^\alpha} + \dots \tag{12}$$

Definition 3. Operators of the form

$$X = \eta^\alpha \frac{\partial}{\partial u^\alpha} + \ldots, \quad \eta^\alpha \in \mathcal{A}, \tag{13}$$

are called canonical Lie-Bäcklund operators.

It is important to note that the canonical form for the generator of a Lie point transformation group is, in general, intrinsically a Lie-Bäcklund operator. Therefore, although the canonical form was not used in classical Lie theory, for completeness, it naturally poses the question of identifying invariants for these canonical operators. Further, this equivalence naturally leads, as developed in Section 3, to a method for constructing invariants for these canonical operators.

We now turn to the notion of a formal group and show that integration of the Lie-Bäcklund equations (defined below) naturally leads to formal groups.

2.5. THE REPRESENTATION SPACE $[[\mathcal{A}]]$

For the sake of brevity, we introduce the sequence

$$z = (x, u, \underset{1}{u}, \underset{2}{u}, \ldots) \tag{14}$$

with elements $z^\nu, \nu \geq 1$. Here, all that is essential is that some ordering be established within the sets $x, u, \underset{1}{u}, \underset{2}{u}$ etc. For x and u we will use the natural ordering so that $z^i = x^i, 1 \leq i \leq n$, $z^{n+\alpha} = u^\alpha, 1 \leq \alpha \leq m$. Denote by $[z]$ any finite subsequence of z. Then elements of $\mathcal{A}$ are written as $f([z])$.

Consider formal power series in one symbol a:

$$f(z, a) = \sum_{k=0}^{\infty} f_k([z]) a^k, \quad f_k([z]) \in \mathcal{A}. \tag{15}$$

Let f and g be formal power series, where f is defined by formula (15) and g is given by a similar formula, viz

$$g(z, a) = \sum_{k=0}^{\infty} g_k([z]) a^k, \quad g_k([z]) \in \mathcal{A}. \tag{16}$$

Their linear combination $\lambda f([z]) + \mu g([z])$ with constant coefficients λ, μ and product $f([z]) \cdot g([z])$, respectively, are defined:

$$\lambda \sum_{k=0}^{\infty} f_k([z])a^k + \mu \sum_{k=0}^{\infty} g_k([z])a^k = \sum_{k=0}^{\infty}(\lambda f_k([z]) + \mu g_k([z]))a^k, \quad (17)$$

$$(\sum_{k_1=0}^{\infty} f_{k_1}([z])a^{k_1}) \cdot (\sum_{k_2=0}^{\infty} g_{k_2}([z])a^{k_2}) = \sum_{k=0}^{\infty}(\sum_{k_1+k_2=k} f_{k_1}([z])g_{k_2}([z]))a^k. \quad (18)$$

The space of all formal power series (15) endowed with the addition and multiplication operators given by Equations (17)-(18), respectively, is denoted by $[[\mathcal{A}]]$.

Lie point and Lie contact transformations, together with their prolongations of all orders, are represented by elements of the space $[[\mathcal{A}]]$. The utilization of $[[\mathcal{A}]]$ is a *fortiori* necessary in the theory of Lie-Bäcklund transformation groups. Therefore, we call $[[\mathcal{A}]]$ *the representation space of Lie-Bäcklund transformation group analysis.*

2.6. LIE-BÄCKLUND EQUATIONS

Definition 4. Given a Lie-Bäcklund operator (4), the Lie-Bäcklund equations are

$$\begin{aligned} \frac{d}{da}\overline{x}^i &= \xi^i([\overline{z}]), \quad \frac{d}{da}\overline{u}^\alpha = \eta^\alpha([\overline{z}]), \\ \frac{d}{da}\overline{u}_i^\alpha &= \zeta_i^\alpha([\overline{z}]), \quad \ldots. \end{aligned} \quad (19)$$

In the case of canonical operators (13), Equations (19) simplify to

$$\frac{d}{da}\overline{u}^\alpha = \eta^\alpha([\overline{z}]), \quad (20)$$

together with

$$\overline{u}_i^\alpha = D_i(\overline{u}^\alpha), \overline{u}_{ij}^\alpha = D_j D_i(\overline{u}^\alpha), \ldots \quad (21)$$

2.7. DEFINITION OF A FORMAL GROUP

Consider sequences of elements

$$f^\nu(z,a) \in [[\mathcal{A}]], \quad \nu \geq \infty,$$

such that

$$f^\nu(z,a) = \sum_{k=0}^{\infty} f_k^\nu([z])a^k, \quad f_0^\nu([z]) = z^\nu, \quad \nu = 1, 2, \ldots \quad (22)$$

where $f_k^\nu([z]), k = 1, 2, \ldots$ are arbitrary elements of $\mathcal{A}$.

Definition 5. Define a transformation of sequences by

$$\overline{z}^\nu = f^\nu(z, a), \quad \nu \geq 1. \tag{23}$$

Definition 6. The transformation (23) is called a formal one-parameter group if the coefficients in (22) obey the property

$$f_k^\nu([\overline{z}]) = \sum_{l=0}^{\infty} \frac{(k+l)!}{k!l!} f_{k+l}^\nu([z])a^l, \quad \nu \geq 1, \quad k = 0, 1, 2, \ldots. \tag{24}$$

Theorem 3. Equations (24) are equivalent to the usual group property

$$f^\nu(\overline{z}, b) = f^\nu(z, a+b), \quad \nu \geq 1 \tag{25}$$

written for formal power series. For the proof, see Ibragimov [2], Section 15.1.

2.8. INTEGRATION OF LIE-BÄCKLUND EQUATIONS

Consider the Lie-Bäcklund Equations (19) with the initial conditions given in Equations (22):

$$\begin{aligned} \frac{d}{da}\overline{x}^i &= \xi^i([\overline{z}]), \quad \overline{x}^i \,|_{a=0} = x^i \\ \frac{d}{da}\overline{u}^\alpha &= \eta^\alpha([\overline{z}]), \quad \overline{u}^\alpha \,|_{a=0} = u^\alpha, \\ &\ldots \end{aligned} \tag{26}$$

The formal integrability of Equations (26) is proved, e.g., in [2], Section 15.1. However for completeness, we restate the relevant theorem here.

Theorem 4. Given any Lie-Bäcklund operator (4), there exists a unique solution of the Lie-Bäcklund Equations (26) in the space $[[\mathcal{A}]]$. The solution is given by a sequence of formal power series, viz

$$\begin{aligned} \overline{x}^i &= x^i + \sum_{k=0}^{\infty} A_k^i([z])a^k, A_k^i([z]) \in \mathcal{A}, \\ \overline{u}^\alpha &= u^\alpha + \sum_{k=0}^{\infty} B_k^\alpha([z])a^k, B_k^\alpha([z]) \in \mathcal{A}, \\ &\ldots \end{aligned} \tag{27}$$

The coefficients $A^i_k([z]), B^\alpha_k([z]), \ldots$ satisfy the group property (24). The essential guiding principle is that power series representations have merit in their own right. In fact, they are necessary to develop any comprehensive, sensible theory of Lie-Bäcklund transformation groups. This situation pertains because the convergence of these formal power series cannot be universally solved and must be treated separately for each type of Lie-Bäcklund operator.

The question of the convergence of these series is not treated here. The reader who is interested in this aspect of the problem is referred to the classical book by Hardy [3] and for a modern approach see the program of Flato, Pinczon, and Simon [4].

2.9. EXPONENTIAL MAP

The question immediately arises as to how one calculates the coefficients in the series (27). The fundamental idea is provided by the exponential map in Lie group theory. For a generator X of a Lie point transformation group, the transformation is given by

$$\overline{x}^i = \exp(aX)(x^i), \quad i = 1, \ldots, n \tag{28}$$

where

$$\exp(aX) = 1 + aX + \frac{a^2}{2!}X^2 + \frac{a^3}{3!}X^3 + \ldots \,. \tag{29}$$

Similarly and more generally, as in the case of Lie theory, one can easily prove that the solution (27) of the Lie-Bäcklund Equations (26) is given by the exponential map

$$\overline{z}^\nu = \exp(aX)(z^\nu), \quad \nu \geq 1, \tag{30}$$

where $\exp(aX)$ is given by Equation (29).

As we discussed before, one can equivalently restrict one's consideration to canonical operators (13) which we shall do henceforth. The Lie-Bäcklund equations (26) coincide with Equations (20) supplemented by Equations (21). In this case the exponential map (30) reduces to the following finite number of equations:

$$\overline{u}^\alpha = \exp(aX)(u^\alpha), \quad \alpha = 1, \ldots, m, \tag{31}$$

where $\exp(aX)$ is given by Equation (29). Indeed, since the total derivatives D_i commute with canonical Lie-Bäcklund operators X, one has

$$D_i \exp(aX) = \exp(aX) D_i.$$

It follows that the transformation of all derivatives are given by (cf. Equation (21))

$$\begin{aligned} \overline{u}_i^\alpha &= \exp(aX)(u_i^\alpha), \quad \alpha = 1,\dots,m, \quad i = 1,\dots,n, \\ &\dots \end{aligned} \tag{32}$$

Finally, we remark that the origin of the fundamental difference between Lie and Lie-Bäcklund transformation groups is that the former are determined by ordinary differential equations (i.e., Lie equations) while the latter are determined by evolutionary partial differential equations (i.e., Lie-Bäcklund equations) for $\overline{u} = \overline{u}(a,x)$:

$$\frac{\partial}{\partial a}\overline{u}^\alpha = \eta^\alpha(\overline{u}, \underset{1}{\overline{u}}, \dots, \underset{k}{\overline{u}}), \alpha = 1,\dots,m. \tag{33}$$

3. Invariants

The direction we take in addressing the problem of constructing differential invariants of Lie-Bäcklund transformation groups is to build on the fact that the only existing theory for these groups is, as described in Section 2, framed in the space $[[\mathcal{A}]]$ of formal power series with coefficients in the space $\mathcal{A}$ of differential functions [2].

3.1. DIFFERENTIAL INVARIANTS IN $\mathcal{A}$

Definition 7. The differential function $F([z]) \in \mathcal{A}$ of order $k > 0$ is called a differential invariant of order k with respect to the Lie-Bäcklund transformation group (27) if

$$F([\overline{z}]) = F([z]). \tag{34}$$

This definition subsumes Lie's differential invariants of all orders and generalizes the notion to Lie-Bäcklund transformation groups.

Theorem 5. The differential function $F([z]) \in \mathcal{A}$ is an invariant of the Lie-Bäcklund group with the generator (4) if and only if

$$XF = 0. \tag{35}$$

The proof can be carried out using the exponential map (30) and is similar to the Lie case.

3.2. LACK OF DIFFERENTIAL INVARIANTS IN THE ONE-DIMENSIONAL CASE

Consider the general situation in the simplest case of Lie-Bäcklund transformation groups involving one independent variable and one dependent variable. Let x be an independent variable, D_x a derivation (2), u a dependent variable with successive derivatives $u_1, u_2, \ldots$, where $D_x(u_i) = u_{i+1}, u_0 = u$.

It is well known that local Lie groups of transformations have an infinite number of differential invariants. In the case of Lie-Bäcklund transformation groups the corresponding problem is open, if we consider invariants to be from the space $\mathcal{A}$ (Definition 7). This is because of the following situation.

Theorem 6. Any nontrivial canonical Lie-Bäcklund operator

$$X = \eta(x, u, u_1, \ldots, u_l)\frac{\partial}{\partial u} + \ldots, \tag{36}$$

has no differential invariants in the space $\mathcal{A}$ except arbitrary functions of x.

However, as we shall see, one has for Lie-Bäcklund operators the same situation as for point transformations, if one considers an approximation of invariants by divergent power series with coefficients from the space $\mathcal{A}$.

3.3. FORMAL INVARIANTS IN [[$\mathcal{A}$]]: OUTLINE OF THE METHOD

This Section is based on a recent paper by Anderson and Ibragimov [1]. Here we outline the method suggested in this paper. This is also discussed in Vol 3 of reference [5].

Definition 8. Two Lie-Bäcklund operators

$$\begin{aligned} X_1 &= \xi_1(x, u, u_1, \ldots, u_k)\frac{\partial}{\partial x} + \eta_1(x, u, u_1, \ldots, u_l)\frac{\partial}{\partial u} + \ldots \\ X_2 &= \xi_2(y, v, v_1, \ldots, v_r)\frac{\partial}{\partial y} + \eta_2(y, v, v_1, \ldots, v_s)\frac{\partial}{\partial v} + \ldots, \end{aligned}$$

are said to be similar if there exists an invertible formal transformation in [[$\mathcal{A}$]], namely given by the formal power series in one symbol ε:

$$y = x + \varepsilon Y_1(x, u, u_1, \ldots, u_{m_1}) + \varepsilon^2 Y_2(x, u, u_1, \ldots, u_{m_2}) + \ldots, Y_i \in \mathcal{A}, \tag{37}$$

$$v = u + \varepsilon V_1(x, u, u_1, \ldots, u_{n_1}) + \varepsilon^2 V_2(x, u, u_1, \ldots, u_{n_2}) + \ldots, V_i \in \mathcal{A} \tag{38}$$

such that X_1, written in the new variables y, v, coincides with X_2, i.e.

$$X_1 \equiv X_1(y)\frac{\partial}{\partial y} + X_1(v)\frac{\partial}{\partial v} = X_2. \tag{39}$$

This definition subsumes the usual definition of similarity in Lie theory.

We shall use the well known fact that the generator X of any one parameter Lie transformation group is similar to the generator of translations:

$$X_1 = \frac{\partial}{\partial x}. \tag{40}$$

Consider two Lie-Bäcklund operator: X_1 given by Equation (40) and a Lie-Bäcklund operator

$$X_2 = \eta(y, v, v_1, \ldots, v_s)\frac{\partial}{\partial v} + \ldots \tag{41}$$

The method for approximating invariants of the Lie-Bäcklund operator X_2 consists of three steps.

First, we introduce the operator

$$X_\varepsilon = (1 - \varepsilon)\frac{\partial}{\partial y} + \varepsilon X_2, \tag{42}$$

where X_2 is given by Equation (41). The operator X_ε continuously connects operators (40)-(41):

$$X_{\varepsilon=0} = X_1, X_{\varepsilon=1} = X_2.$$

Second, we establish the following result on similarity.

Theorem 7. Operator (42)

$$X_\varepsilon = (1 - \varepsilon)\frac{\partial}{\partial y} + \varepsilon X_2,$$

is similar to the operator (40)

$$X_1 = \frac{\partial}{\partial x}.$$

The proof is direct. We find the similarity transformations (37)-(38), by solving (39), in particular

$$X_1(y) = 1 - \varepsilon, \quad X_1(v) = \varepsilon \ \eta(y, v, v_1, \ldots, v_s). \tag{43}$$

Third, take the known basis of all differential invariants of the operator (40), namely,

$$u_1, u_2, u_3, \ldots, \tag{44}$$

and subject them to the similarity transformation obtained by solving Equations (43). As a result, one obtains invariants represented by elements of the space $[[\mathcal{A}]]$.

Definition 9. An element of $[[\mathcal{A}]]$, namely a formal power series in one symbol ε

$$I(y, v, v_1, \ldots; \varepsilon) = I_0 + \varepsilon I_1 + \varepsilon^2 I_2 + \ldots$$

is called an invariant of the formal group with Lie-Bäcklund operator (42) if the formal series X_ε I vanishes:

$$X_\varepsilon I = 0.$$

Here $I_0, I_1, I_2, \ldots$ depend on $y, v, v_1, \ldots$ and are elements of $\mathcal{A}$.

The following example illustrates the method.

Example. Consider the Lie-Bäcklund operator

$$X_2 = v_1 \partial/\partial v + \ldots \tag{45}$$

In this case the operator (42) becomes

$$X_\varepsilon = (1-\varepsilon)\frac{\partial}{\partial y} + \varepsilon v_1 \frac{\partial}{\partial v} + \varepsilon v_2 \frac{\partial}{\partial v_1} + \varepsilon v_3 \frac{\partial}{\partial v_2} + \ldots. \tag{46}$$

We select the simplest nontrivial solution of Equations (43) at each step and obtain the following transformation:

$$\begin{aligned} y &= (1-\varepsilon)x + \varepsilon u, \\ v &= u + \varepsilon x u_1 + \varepsilon^2[(2-u_1)xu_1 + \frac{1}{2}x^2 u_2] \\ &+ \varepsilon^3[4xu_1 + 2x^2u_2 - \frac{3}{2}x^2u_1u_2 + xu_1^3 - 4xu_1^2 + \frac{1}{3!}x^3u_3] + \ldots \end{aligned} \tag{47}$$

One can easily find the transformations of derivatives, e.g.,

$$\begin{aligned} v_1 &= u_1 + \varepsilon[2u_1 + xu_2 - u_1^2] \\ &+ \varepsilon^2[4u_1 - 4u_1^2 + 4xu_1u_2 + \frac{1}{2}x^2u_3 + u_1^3 + \ldots, \\ v_2 &= u_2 + \varepsilon[4u_2 - 3u_1u_2 + xu_3] + \ldots, \\ v_3 &= u_3 + \ldots \end{aligned} \tag{48}$$

The reason for the diminishing orders of ε with the ascending orders of the derivatives is to illustrate which powers of ε are required to invert (47)-(48) up to order $o(\varepsilon^3)$. In this process, one finds the expression

$$u = v - \varepsilon x v_1 + \frac{1}{2}\varepsilon^2 x^2 v_2 + o(\varepsilon^2), \tag{49}$$

which becomes the following transcendental expression for x:

$$y-(1-\varepsilon)x=\varepsilon v-\varepsilon^2 x v_1+\frac{1}{2}\varepsilon^3 x^2 v_2+o(\varepsilon^3). \tag{50}$$

This calculation up to order $o(\varepsilon^3)$ illustrates the method.

Moreover, in this example the following general result holds.

Theorem 8. (Anderson and Ibragimov [1]) The one-parameter family of Lie-Bäcklund operators (5)

$$X_\varepsilon=(1-\varepsilon)\frac{\partial}{\partial y}+\varepsilon\sum_{j=0}^{\infty}((D_y)^j v_1)\frac{\partial}{\partial v_j},$$

is similar to the generator (40) of the group of translations. The similarity map (37)-(38) is determined by

$$y=(1-\varepsilon)x+\varepsilon u,\quad u=\sum_{j=0}^{\infty}\frac{(-\varepsilon x)^j}{j!}v_j, \tag{51}$$

where $v=v_0$.

With this result in hand, we illustrate how to map the infinite number of differential invariants (44) of the operator (40) into equivalent ones for the operator X_ε.

Theorem 9. The Lie-Bäcklund operator (46) has a countable basis of invariants

$$u(y,v,v_1,v_2,\ldots),\quad u_1(y,v,v_1,v_2,\ldots),\quad u_2(y,v,v_1,v_2,\ldots),\ldots$$

determined recursively by Equations (51) as well as

$$D_x=D_x(y)D_y=(1-\varepsilon(1-u_1))D_y.$$

Using this theorem, one can easily find, e.g. the invariant u up to order $o(\varepsilon^2)$

$$u=v-\varepsilon y v_1+\varepsilon^2(-yv_1+\frac{1}{2}y^2 v_2+vv_1)+o(\varepsilon^2).$$

Here we shall extend these results. Theorem 8 can be improved upon in the sense that transformed variables can be obtained directly in terms of combinations of the old variables, thus avoiding the transcendental nature of the result in Theorem 8 when solving for the new variables in terms of the old variables. Specifically, we have the following new result.

Theorem 10. The one-parameter family of Lie-Bäcklund operators (46)

$$X_\varepsilon = (1-\varepsilon)\frac{\partial}{\partial y} + \varepsilon \sum_{j=0}^{\infty}((D_y)^j v_1)\frac{\partial}{\partial v_j},$$

is similar to the generator (40) of the group of translations. The similarity map (37)-(38) is determined by

$$y = (1-\varepsilon)x + \varepsilon u_1, \quad v_k = \sum_{j=0}^{\infty} \frac{(\varepsilon x)^j}{j!} c_{k+j}, \quad k = 0, 1, \ldots, \tag{52}$$

where $v = v_0$ and

$$c_k = ((1 - 2\varepsilon + \varepsilon u_2)^{-1} D_x)^k u, \quad k = 0, 1, \ldots .$$

the set $c_0 = u, u_1, c_1, c_2, \ldots$ is a basis for the set of all differential invariants for the operator (40).

References

1. Anderson, R. L. and Ibragimov, N. H., Invariants of Lie-Bäcklund transformation groups generated by formal power series, *Lie Groups and Their Applications I*, pp. 1-25 (1994).
2. Ibragimov, N.H., Transformation Groups Applied to Mathematical Physics, *Nauka*, Moscow (1983), (D. Riedel, Dordrecht (1985)).
3. Hardy, G. H., Divergent Series, *Oxford University Press*, London (1949).
4. Flato, M., Pinczon, G. and Simon, J., Nonlinear representations of Lie groups, *Ann. Scient. Ec. Norm. Sup.* **10**, pp. 405-418 (1977).
5. Anderson, R. L. and Ibragimov N.H., Lie Bäcklund symmetries: Representation by formal power series, in: *CCR Handbook of Differential Equations*, **Vol. 1**(1994), **Vol. 2-3** (to appear), Ibragimov N.H., Ed., CRC Press, Boca Raton, 1995.

GALILEAN COHERENT STATES AND WAVELETS

J-P. ANTOINE
Institut de Physique Théorique
Université Catholique de Louvain
B - 1348 Louvain-la-Neuve, Belgium

Abstract. We describe several families of coherent states and wavelets associated to the Galilei group, extended by space and time dilations. The construction follows a general method based on group representations which are square integrable only on a homogeneous space of the underlying group.

1. Motivation: Which space-time wavelets?

The wavelet transform (WT) has become by now a familiar tool in signal and image processing.[1, 2] In the latter case, it has been used mostly for the treatment of single images, either for analysis purposes (continuous WT), or for image reconstruction and data compression (discrete WT). A more challenging task is to treat image sequences, such as encountered in television or movies, in other words, time-dependent 2D signals. Clearly this requires wavelets depending both on space and time variables. How can one design them?

The answer, of course, may be found as soon as one realizes that wavelets are nothing but coherent states (CS) associated to similitude groups. And furthermore the construction of a particular type of CS may be seen as a problem in group representation theory, as formulated independently by Gilmore[3] and Perelomov[4]. In this approach, the necessary ingredients are twofold: (i) a transformation group G acting on the signals considered, and (ii) a square integrable unitary representation U of G, acting in a Hilbert space $\mathcal{H}$.

Then the corresponding CS are simply the elements of the orbit under U of a given vector $\eta \in \mathcal{H}$. In particular, the CS are called 'wavelets' whenever the underlying group G contains dilations.

J. Bertrand et al. (eds.), Modern Group Theoretical Methods in Physics, 15–26.

Adopting this point of view, one may think of several candidates for the construction of space-time wavelets.

(i) Purely kinematical wavelets:
In the simplest case, G consists of translations and dilations of space and time (independently), plus the appropriate symmetry operations: space rotations (which reduce to space reflection in one space dimension) and time reflection.[5]

(ii) Nonrelativistic wavelets:
In a genuine nonrelativistic setup, the natural kinematical group is the (extended) Galilei group, to which one adds independent space and time dilations, thus obtaining the *affine Galilei group.* An alternative solution is to restrict the dilations by the relation $a_o = a^2$, where a_o, a denote the time and space dilation, respectively. The resulting group, called the *restricted Schrödinger group* or *Galilei-Schrödinger group,* is a natural invariance group of both the Schrödinger and the heat equations.[6, 7]

(iii) Relativistic wavelets:
In a fully relativistic context, the kinematical group is the Poincaré group, and space and time must be treated in the same way, so that one requires $a_o = a$. The resulting group is called the *affine Poincaré* or *Weyl-Poincaré group.*[8]

The present paper will cover case *(ii)*. For the sake of completeness, we will also discuss the CS associated to the pure Galilei group. Before that, we will first review quickly the general construction of CS associated to a given group G, following mostly Refs. 9-10.

2. Coherent States: A General Construction

Let G be a locally compact group, with left Haar measure dg (an equivalent theory may be constructed using the right invariant measure), and U a strongly continuous, irreducible, unitary representation of G on a Hilbert space $\mathcal{H}$. We distinguish two cases.

<u>*Case 1*</u> *: U is square integrable over G*

Assume that U is *square integrable* over G, i.e. there exists a non-zero vector $\eta \in \mathcal{H}$, called an *admissible vector*, such that:

$$c(\eta, \phi) \equiv \int_G |\langle U(g)\eta|\phi\rangle|^2 \, dg < \infty, \ \ \forall \phi \in \mathcal{H} \tag{2.1}$$

or, equivalently, $c(\eta, \eta) < \infty$. It is easy to see that, if η is admissible, so is $U(g)\eta$, for any $g \in G$: the set $\mathcal{A}$ of all admissible vectors is invariant under U. Since U was assumed to be irreducible, it follows that either $\mathcal{A} = \{0\}$

or $\mathcal{A}$ is dense in $\mathcal{H}$. The former case corresponds to a representation U that is not square integrable. Finally, $\mathcal{A} = \mathcal{H}$, that is, every vector in $\mathcal{H}$ is admissible, iff G is unimodular, i.e., if its left and right invariant Haar measures coincide.

Let $\eta \in \mathcal{H}$ be a non-zero admissible vector, normalized by $c(\eta, \eta) = 1$. We denote by $\mathcal{S} = \{\eta_g = U(g)\eta \,|\, g \in G\}$ the corresponding family of coherent states. Define the linear map $W_\eta : \mathcal{H} \to L^2(G, dg)$ by

$$(W_\eta \phi)(g) = \langle \eta_g | \phi \rangle, \quad \phi \in \mathcal{H}. \tag{2.2}$$

Then:

(1) The map W_η is an isometry: $W_\eta^* W_\eta = I$; equivalently, the family $\mathcal{S}$ determines a resolution of the identity

$$\int_G |\eta_g\rangle\langle\eta_g| \, dg = I, \tag{2.3}$$

which implies that $\mathcal{S}$ is a total set (overcompleteness).

(2) The range $\mathcal{H}_\eta$ of W_η is a closed subspace of $L^2(G, dg)$ and the orthogonal projection onto $\mathcal{H}_\eta$, $P_\eta = W_\eta W_\eta^*$, is an integral operator with (reproducing) kernel $K(g, g') = \langle \eta_g | \eta_{g'} \rangle$; in other words, $\mathcal{H}_\eta$ is a *reproducing kernel Hilbert space.*

(3) The map W_η intertwines U and the left regular representation U_L of G in $L^2(G, dg)$:

$$W_\eta U(g) = U_L(g) W_\eta, \quad \forall\, g \in G, \tag{2.4}$$

where $(U_L(g)f)(g') = f(g^{-1}g'), f \in L^2(G, dg)$. It follows that U is equivalent to a subrepresentation of U_L, in other words, U belongs to the *discrete series.*

(4) From the resolution of the identity (2.3), or from the equivalent fact that W_η, being isometric, may be inverted on its range by the adjoint operator, one may write an inverse transform or *reconstruction formula.* For any $\Phi = W_\eta \phi \in \mathcal{H}_\eta$, one can recover $\phi \in \mathcal{H}$ by the relation:

$$\phi = \int_G (W_\eta \phi)(g) \eta_g dg. \tag{2.5}$$

A familiar example is the construction of one-dimensional wavelets, corresponding to the connected affine group of the line or '$ax+b$' group. Another one is that of 2D wavelets, obtained from $G = SIM(2) = \mathbb{R}^2 \circledS (\mathbb{R}_*^+ \times SO(2))$, the similitude group of the plane.

Case 2 : *U is square integrable over a homogeneous space $X = G/H$*

Let $G, U, \mathcal{H}$ be as in Case 1, H a closed subgroup of G, $X = G/H$ with an *invariant measure* ν (quasi-invariant would suffice) and $\sigma : X = G/H \to G$ a Borel section (in the principal bundle $G \to G/H$). Then we say that U is *square integrable mod*(H,σ) for the vector $\eta \in \mathcal{H}$ if the integral

$$\int_X U(\sigma(x))|\eta\rangle\langle\eta|U(\sigma(x))^* \, d\nu(x) \tag{2.6}$$

converges weakly to a bounded positive invertible operator A_σ on $\mathcal{H}$, i.e.

$$0 < c_X(\eta,\phi) \equiv \int_X |\langle U(\sigma(x))\eta|\phi\rangle|^2 \, d\nu(x) = \langle\phi|A_\sigma\phi\rangle \ < \infty, \ \ \forall\phi \in \mathcal{H}. \tag{2.7}$$

Notice that A_σ^{-1} may be unbounded. We also say that the vector η is *admissible* for (U,σ) or that the section σ is admissible for (U,η). Given a vector $\eta \in \mathcal{H}$, admissible for (U,σ) and normalized by $c_X(\eta,\eta) = 1$, we define now the family of *covariant coherent states*, indexed by points $x \in X$, as the orbit of η under G, through the representation U and the section σ:

$$\mathcal{S}_\sigma = \{\eta_{\sigma(x)} = U(\sigma(x))\eta \,|\, x \in X\}. \tag{2.8}$$

The fundamental properties of these covariant CS are a direct generalization of those described in Case 1 above:

(1) The set $\mathcal{S}_\sigma$ is total in $\mathcal{H} : (\mathcal{S}_\sigma)^\perp = \{0\}$.

(2) Define the linear map

$$W_\eta : \mathcal{H} \to L^2(X, d\nu), \quad (W_\eta\phi)(x) = \langle\eta_{\sigma(x)}|\phi\rangle. \tag{2.9}$$

Then the range $\mathcal{H}_\eta$ of W_η is complete with respect to the scalar product $\langle\Phi|\Psi\rangle_\eta \equiv \langle\Phi|W_\eta A_\sigma^{-1} W_\eta^{-1}\Psi\rangle$ and W_η is unitary from $\mathcal{H}$ onto $\mathcal{H}_\eta$. In other words, one has the resolution

$$\int_X |\eta_{\sigma(x)}\rangle\langle\eta_{\sigma(x)}| \, d\nu(x) = A_\sigma, \tag{2.10}$$

where A_σ is only required to be a bounded, positive operator with a densely defined inverse and the integral converges weakly.

(3) The orthogonal projection from $L^2(X, d\nu)$ onto $\mathcal{H}_\eta$ is an integral operator K_σ, and $\mathcal{H}_\eta$ is a reproducing kernel Hilbert space of functions:

$$\Phi(x) = \int_X K_\sigma(x,y)\Phi(y) \, d\nu(y), \ \ \forall\Phi \in \mathcal{H}_\eta. \tag{2.11}$$

The kernel is given explicitly by $K_\sigma(x,y) = \langle\eta_{\sigma(x)}|A_\sigma^{-1}\,\eta_{\sigma(y)}\rangle$, if $\eta_{\sigma(y)} \in \mathcal{D}(A_\sigma^{-1})$, $\forall y \in X$; otherwise, (2.11) holds in a distributional sense.

(4) The map W_η may be inverted on its range by the adjoint operator, $W_\eta^{-1} = W_\eta^*$ on $\mathcal{H}_\eta$, to obtain (again for $\eta_{\sigma(x)} \in \mathcal{D}(A_\sigma^{-1}), \forall x \in X$)

$$W_\eta^{-1}\Phi = \int_X \Phi(x)\, A_\sigma^{-1}\, \eta_{\sigma(x)}\, d\nu(x), \quad \Phi \in \mathcal{H}_\eta. \tag{2.12}$$

The above results simplify somewhat if A_σ^{-1} is bounded; then $\mathcal{S}_\sigma$ (and by extension A_σ itself) is called a *frame*. A further simplification occurs when $\mathcal{S}_\sigma$ is a *tight frame*, i.e., $A_\sigma = \lambda I$ (this terminology is borrowed from the theory of nonorthogonal expansions[11]).

Actually tight frames play a privileged role for the physical interpretation, in particular the formulation of the quantization process in terms of coherent states.[12] Now, if A_σ is a non-tight frame, it may be converted into a tight one. Indeed, the relation (2.10) may be formally transformed into a resolution of the identity if one multiplies it from both sides by $A_\sigma^{-1/2}$, which always exists since A_σ^{-1} is positive and self-adjoint. Then we may define the *weighted CS*, $\tilde{\eta}_x = W_\sigma(x)\eta_{\sigma(x)}$, where $W_\sigma(x)$ is the restriction of $A_\sigma^{-1/2}$ to the one-dimensional subspace generated by $\eta_{\sigma(x)}$, and obtain a resolution of the identity,

$$\int_X |\tilde{\eta}_x\rangle\langle\tilde{\eta}_x|\, d\nu(x) = I \tag{2.13}$$

(in that case, one says that the section σ is *weakly admissible* for (U,η)[13]). The corresponding frame is now tight, but the states $\tilde{\eta}_x$ are no longer genuine coherent states, i.e. elements of an orbit of η.

This general construction may be extended further on several counts: the measure ν may be taken as quasi-invariant only (appropriate cocycles have to be inserted); the rank one projection $|\eta\rangle\langle\eta|$ in the integrand of (2.6) may be replaced by an operator of rank n (this is important for treating CS of spin > 0); and the section $\sigma : X \to G$ may be combined with a smooth map $\rho : X \to X$. Further details may be found in Ref. 10.

Let us give a few examples of this construction.

(1) $H = \{e\}$: then we recover Case 1.

(2) $H = H_\eta$, the isotropy subgroup (up to a phase) of η : this is the familiar case of the Gilmore-Perelomov CS.[3, 4] Notice that now the admissibility condition (2.7) is independent of the choice of the section σ, but the CS (2.8) are *not*. Some familiar examples of groups G are,

for instance:

- Semisimple groups, such as $SU(2), SU(1,1), SO(4,2)$, etc.
- The Weyl-Heisenberg group G_{WH}, which yields the canonical CS (here H is the center of G_{WH}).
- The similitude group $SIM(n)$ of $\mathbb{R}^n$ $(n \geq 3)$: for $H = SO(n-1)$, one gets the axisymmetric n-dimensional wavelets.

(3) Relativity groups, such as the Euclidean, the Galilei or the Poincaré groups, or more generally, the semidirect product $G = V \circledS S$ of a vector space V by a semisimple group S of automorphisms of V. In that case, one usually takes for $X = G/H$ a coadjoint orbit of G, i.e. a natural phase space for the system at hand. It follows that $H \neq H_\eta$, for any η, since H contains translations. Thus it is imperative to use the general construction in order to get CS. Notice also that this provides a natural framework for a quantization process.[10, 12]

3. Galilean Coherent States

We begin with the pure Galilean case, namely the (extended) Galilei group $\widetilde{G}(1,1)$ in one space and one time dimension (the analysis is entirely similar for n space dimensions). The group elements are parametrized as: $g = (\phi, b_o, \mathbf{b}, \mathbf{v})$, where ϕ is a phase, more precisely the parameter corresponding to the central extension of mass $m > 0$, $(b_o, \mathbf{b})$ time and space translations and $\mathbf{v}$ the boost parameter. The unitary irreducible representation corresponding to spin zero is given by:

$$[U^{(m)}(g)\psi](\mathbf{k}) = \exp i(\phi + \frac{\mathbf{k}^2}{2m} b_o - \mathbf{k}\cdot\mathbf{b})\, \psi(\mathbf{k} - m\mathbf{v}), \quad \psi \in L^2(\mathbb{R}, d\mathbf{k}), \tag{3.1}$$

and it is not square integrable on $\widetilde{G}(1,1)$. Thus we have to use the general construction. The coadjoint orbit of $\widetilde{G}(1,1)$ associated to $U^{(m)}$, which may be taken as phase space, is the quotient $\Gamma = \widetilde{G}(1,1)/(\Omega \times T)$, where $\Omega \times T$ is the subgroup of $\widetilde{G}$ consisting of phase changes and time translations. It carries global coordinates $(\mathbf{q}, \mathbf{p}) \equiv (\mathbf{b} - \mathbf{v} b_o, m\mathbf{v})$ and the invariant measure is $d\mathbf{q}\, d\mathbf{p}$.

We consider first the Galilean section $\sigma_o : \Gamma \to \widetilde{G}(1,1)$ given by

$$\sigma_o(\mathbf{q}, \mathbf{p}) = (0, 0, \mathbf{q}, \frac{\mathbf{p}}{m}). \tag{3.2}$$

Then an affine section is an arbitrary section of the form:

$$\begin{aligned} \sigma(\mathbf{q}, \mathbf{p}) &= \sigma_o(\mathbf{q}, \mathbf{p})\,(0, \mathbf{q}\cdot\boldsymbol{\vartheta}(\mathbf{p}), \mathbf{0}, \mathbf{0}) \\ &= (\mathbf{q}\cdot\boldsymbol{\vartheta}(\mathbf{p})\frac{p^2}{2m}, \mathbf{q}\cdot\boldsymbol{\vartheta}(\mathbf{p}), \mathbf{q} + \frac{\mathbf{p}}{m}\mathbf{q}\cdot\boldsymbol{\vartheta}(\mathbf{p}), \frac{\mathbf{p}}{m}), \end{aligned} \tag{3.3}$$

with $\boldsymbol{\vartheta}$ a real continuous function that indexes the section.

The admissibility condition can be calculated explicitly for the class of affine sections. The vector $\eta \in L^2(\mathbb{R}, d\mathbf{k})$ is admissible mod$(\Omega \times T, \sigma)$ if it satisfies the following two conditions:

(i) $\forall\, \mathbf{k} \in \mathbb{R},\ \mathbf{p} \cdot \boldsymbol{\vartheta}(\mathbf{k}-\mathbf{p}) \geq m$ implies $\eta(\mathbf{p}) = 0$.
(ii) The following integral converges:

$$\mathcal{A}_\sigma^\eta(\mathbf{k}) \equiv 2\pi \int d\mathbf{p}\ \left(1 - \frac{\mathbf{p}}{m} \cdot \boldsymbol{\vartheta}(\mathbf{k}-\mathbf{p})\right)^{-1} |\eta(\mathbf{p})|^2. \tag{3.4}$$

Thus, contrary to the Poincaré case,[9] the admissibility of η depends here explicitly on the particular section σ, and a support restriction must be imposed on admissible vectors η (except for the Galilean section σ_o, corresponding to $\boldsymbol{\vartheta} = 0$, for which condition (i) is satisfied identically.

Given a vector $\eta \in L^2(\mathbb{R}, d\mathbf{k})$ admissible mod$(\Omega \times T, \sigma)$, we obtain Galilean coherent states, namely

$$\begin{aligned} \eta_{\sigma(\mathbf{q},\mathbf{p})}(\mathbf{k}) &= [U^{(m)}(\sigma(\mathbf{q},\mathbf{p}))\eta](\mathbf{k}) \\ &= \exp i \left(\frac{|\mathbf{k}-\mathbf{p}|^2}{2m} \mathbf{q} \cdot \boldsymbol{\vartheta}(\mathbf{p}) - \mathbf{k} \cdot \mathbf{q}\right) \eta(\mathbf{k}-\mathbf{p}), \end{aligned} \tag{3.5}$$

and these vectors constitute a frame of CS indexed by points in phase space. The corresponding resolution generator A_σ^η is a positive, bounded, invertible multiplication operator :

$$(A_\sigma^\eta \psi)(\mathbf{k}) = \mathcal{A}_\sigma^\eta(\mathbf{k})\psi(\mathbf{k}),\ \psi \in L^2(\mathbb{R}, d\mathbf{k}), \tag{3.6}$$

with kernel $\mathcal{A}_\sigma^\eta(\mathbf{k})$ given in (3.4).

We conclude with two examples of admissible vectors:

- Gaussian vector: $\eta_G(\mathbf{k}) \sim \exp(-\mathbf{k}^2/2mu)$.
- Binomial vector: $\eta_\alpha(\mathbf{k}) \sim (1 + \mathbf{k}^2/2mu)^{-\alpha/2}$, $\alpha > 1/2$.

Here $u > 0$ is a normalization factor, which may be interpreted as a kind of internal energy. It is interesting to note that these vectors are the limits of analogous admissible vectors for the Poincaré group $\mathcal{P}_+^\uparrow(1,1)$ under the contraction $c \to \infty$. Furthermore the corresponding sets of CS go into one another in the same limit.

4. Galilean Wavelets

We turn now to the affine Galilei group in n space dimensions,[14] which is obtained by combining:

(i) The Galilei group $G(1,n)$, with generic element $(b_o, \mathbf{b}, \mathbf{v}, R)$, where $b_o \in \mathbb{R}$ and $\mathbf{b} \in \mathbb{R}^n$ are the time and space translations, respectively, $\mathbf{v} \in \mathbb{R}^n$ is the boost parameter and $R \in SO(n)$ is a rotation.

(ii) The group of space and time dilations, $a \in \mathbb{R}_*^+$, resp. $a_o \in \mathbb{R}_*^+$.

(iii) The central extension by $\Omega \sim \mathbb{R}$ (phase changes), which is necessary to get genuine representations, not merely projective ones.

The resulting group G, the extended affine Galilei group, with mass $m > 0$, has the following structure :

$$G = (\mathbb{R} \times \mathbb{R} \times \mathbb{R}^n) \circledS [\mathbb{R}^n \circledS (\mathbb{R}_*^+ \times \mathbb{R}_*^+ \times SO(n))], \tag{4.1}$$

where $\circledS$ denotes again a semidirect product and $\times$ a direct product (there are several such semidirect product decompositions). Correspondingly, and in the same order, a generic element is written

$$g = (\phi, b_o, \mathbf{b}; \mathbf{v}; a_o, a, R), \tag{4.2}$$

where $\phi \in \Omega$ is the extension parameter. The action of g on space-time is then

$$\left\{ \begin{array}{lll} \mathbf{x} & \mapsto & a(R\mathbf{x} + t\mathbf{v} + \mathbf{b}), \\ t & \mapsto & a_o t + b_o. \end{array} \right. \tag{4.3}$$

The group law reads:

$$\begin{aligned} g.g' &= (\phi + \phi' + \mathbf{v}.(aR\mathbf{b}' + \frac{1}{2}a_o b_o' \mathbf{v}), a_o b_o' + b_o, \\ &\qquad aR\mathbf{b}' + a_o b_o' \mathbf{v} + \mathbf{b}; aR\mathbf{v}' + \mathbf{v}; a_o a_o', aa', RR'), \end{aligned} \tag{4.4}$$

and the inverse element is given by

$$\begin{aligned} g^{-1} &= (-\phi + \mathbf{v}.(\mathbf{b} - \frac{1}{2}b_o \mathbf{v}), -a_o^{-1} b_o, a^{-1}R^{-1}(b_o \mathbf{v} - \mathbf{b}); \\ &\qquad - a^{-1}R^{-1}\mathbf{v}; a_o^{-1}, a^{-1}, R^{-1}). \end{aligned} \tag{4.5}$$

The center of G is the subgroup $\Omega \sim \mathbb{R}$ of phase changes. In addition, G contains as subgroup the restricted Schrödinger group or Galilei-Schrödinger group, namely $S(n) = \{g \in G : \phi = 0, a_o = a^2\}$, which is a natural kinematical invariance group of the free Schrödinger equation and of the heat equation. (The maximal invariance group is the full Schrödinger group, which contains in addition the so-called expansions, nonrelativistic analogues of conformal transformations[6, 7]).

In order to build space-time wavelets associated to G, we consider a unitary irreducible representation of spin zero, which may be obtained by the orbit method.[14] The Hilbert space is $\mathcal{H} = L^2(\mathbb{R}^n \times \mathbb{R}, d\mathbf{k}d\omega)$, where the manifold $\mathbb{R}^n \times \mathbb{R}$ may be identified with the usual Minkowski space, and the representation reads:

$$[U(g)\psi](\mathbf{k},\omega) = \sqrt{a_o a^n}\ \exp i(\phi + \mathbf{k}\cdot\mathbf{b} - \omega b_o)\psi(\mathbf{k}',\omega') \tag{4.6}$$

with

$$\mathbf{k}' = aR^{-1}(\mathbf{k} + m\mathbf{v}), \tag{4.7}$$

$$\omega' = a_o(\omega - \mathbf{k}\cdot\mathbf{v} - \frac{1}{2}m\mathbf{v}^2) \tag{4.8}$$

(the mass m enters as a coordinate in the dual of the Lie algebra of Ω). As expected, this representation is not square integrable over G, so that we are again forced to look for appropriate quotients. We will consider three cases $X_i = G/H_i$, $i = 1, 2, 3$.

(A) $H_1 = \{$ *phase changes* ϕ *and time dilations* $a_o\}$

The space $X_1 = G/H_1$ is parametrized by points $x_1 = (b_o, \mathbf{b}; \mathbf{v}; a, R)$. As before, we consider first the basic section $\sigma_1 : X_1 \to G$:

$$\sigma_1(x_1) = (0, b_o, \mathbf{b}; \mathbf{v}; 1, a, R). \tag{4.9}$$

A straightforward calculation shows that a vector $\eta \in \mathcal{H}$ is admissible mod(H_1, σ_1) iff the following integral converges:

$$\mathcal{A}_1^\eta(\mathbf{k},\omega) \equiv c_n \iint d\mathbf{k}' d\omega'\ \frac{[\mathbf{k}^2 + 2m(\omega - \omega')]^{n-1}}{|\mathbf{k}'|^{2n}} |\eta(\mathbf{k}',\omega)|^2 < \infty. \tag{4.10}$$

Then the resolution operator A_1^η is the operator of multiplication by the function $\mathcal{A}_1^\eta(\mathbf{k},\omega)$, and one obtains a frame. The latter is in general not tight, but may be converted into a tight frame of weighted CS, as explained in Section 2 above.

We consider now a general section:

$$\sigma_{\beta_1}(x_1) = \sigma_1(x_1)\,(0, 0, \mathbf{0}; \mathbf{0}; \beta_1(x_1), 1, I), \tag{4.11}$$

where $\beta_1 : X_1 \to \mathbb{R}_*^+$ is a Borel function and $\beta_1(x_1) \in \mathbb{R}_*^+$ represents a time dilation. The corresponding set of Galilean wavelets is given as

$$\eta_{\beta_1}(\mathbf{k},\omega) = \sqrt{a^n \beta_1(x_1)}\ e^{i(\mathbf{k}\cdot\mathbf{b} - \omega b_o)}\ \eta(\mathbf{k}',\omega'), \tag{4.12}$$

with

$$\mathbf{k}' = aR^{-1}(\mathbf{k} + m\mathbf{v}), \tag{4.13}$$

$$\omega' = \beta_1(x_1)\,(\omega - \mathbf{k}\cdot\mathbf{v} - \frac{1}{2}m\mathbf{v}^2). \tag{4.14}$$

In the simple case where $\beta_1(x_1) \equiv \beta_1(a, \mathbf{v})$, the admissibility condition for η may be resolved explicitly, in the sense that one may find conditions (actually a system of partial differential equations) on the function β_1 such that the corresponding resolution operator $A^{\eta}_{\beta_1}$ is again a multiplication operator. In particular, if we assume that β_1 depends only on a, we obtain in this way a two-parameter family of (weakly admissible) sections characterized by the functions β_1:

$$\beta_1(x_1) \equiv \beta_1(a) = \frac{2a^2}{\lambda_o + 2\mu_o a^2}, \quad \lambda_o, \mu_o \in \mathbb{R}. \tag{4.15}$$

For each pair (λ_o, μ_o) the vector η generates a frame of Galilean wavelets, which is non-tight in general, but may be converted into a tight frame of weighted CS. This family contains two simpler cases, namely:

. for $\lambda_o = 0, \mu_o = 1$, we recover the basic section $\beta_1(x_1) \equiv 1$, $\sigma_{\beta_1} = \sigma_1$;
. for $\lambda_o = 2, \mu_o = 0$, we get $\beta_1(x_1) = a^2$, that is, the Schrödinger case.

(B) $H_2 = \{$ phase changes ϕ and space dilations $a\}$

The space $X_2 = G/H_2$ is parametrized by points $x_2 = (b_o, \mathbf{b}; \mathbf{v}; a_o, R)$. It possesses a natural measure, which is both left and right G-invariant, namely $d\nu(x_2) = a_o^{-2} da_o\, db_o\, d\mathbf{b}\, d\mathbf{v}\, d\mu(R)$, where $d\mu(R)$ is the Haar measure on $SO(n)$. The basic section $\sigma_2 : X_2 \to G$ reads

$$\sigma_2(x_2) = (0, b_o, \mathbf{b}; \mathbf{v}; a_o, 1, R), \tag{4.16}$$

and a general section:

$$\sigma_{\beta_2}(x_2) = \sigma_2(x_2)\, (0, 0, \mathbf{0}; \mathbf{0}; 1, \beta_2(x_2), I), \tag{4.17}$$

where $\beta_2(x_2) \in \mathbb{R}^+_*$ is now a space dilation. As in the previous case, one may take $\beta_2(x_2) \equiv \beta_2(a_o, \mathbf{v})$. If β_2 satisfies the following system of differential equations:

$$\beta_2^{-3} \frac{\partial \beta_2}{\partial a_o} = 2\lambda_o, \qquad \beta_2^{-2}\, R\, \frac{\partial \beta_2}{\partial \mathbf{v}} = \mathbf{v_o}, \tag{4.18}$$

where $\lambda_o \in \mathbb{R}$ and $\mathbf{v_o} \in \mathbb{R}^3$ are two constants, then there is a dense set of vectors $\eta \in \mathcal{H}$ admissible mod(H_2, σ_{β_2}). In addition, the corresponding resolution operator $A^{\eta}_{\beta_2}$ is now automatically a multiple of the identity, that is, η generates a tight frame.

For instance, if we take $\beta_2(x_2) \equiv \beta_2(a_o)$, the system (4.18) yields again a two-parameter family of functions β_2:

$$\beta_2(x_2) \equiv \beta_2(a_o) = (\mu_o + \lambda_o a_o)^{1/2}, \quad \lambda_o, \mu_o \in \mathbb{R}. \tag{4.19}$$

For each of these, any admissible vector η generates a tight frame of Galilean wavelets

$$\eta_{\beta_2(x_2)}(\mathbf{k},\omega) = \sqrt{a_o(\mu_o + \lambda_o a_o)^{n/2}}\; e^{i(\mathbf{k}\cdot\mathbf{b}-\omega b_o)}\, \eta(\mathbf{k}',\omega'), \tag{4.20}$$

with $(\mathbf{k}',\omega')$ given by

$$\mathbf{k}' = (\mu_o + \lambda_o a_o)^{1/2} R^{-1}(\mathbf{k} + m\mathbf{v}), \tag{4.21}$$

$$\omega' = a_o\,(\omega - \mathbf{k}\cdot\mathbf{v} - \frac{1}{2}m\mathbf{v}^2), \tag{4.22}$$

and

$$\iint_{X_2} |\eta_{\beta_2(x_2)}\rangle\langle\eta_{\beta_2(x_2)}|\; a_o^{-2} da_o\, db_o\, d\mathbf{b}\, d\mathbf{v}\, d\mu(R) = c(\eta) I, \tag{4.23}$$

$$x_2 \equiv (b_o, \mathbf{b}; \mathbf{v}; a_o, R).$$

Notice that here too we recover the Schrödinger case $a_o = a^2$, for $\mu_o = 0$.

(C) $H_3 = \{$ *phase changes* ϕ *and boosts* $\mathbf{v}\}$

This case yields the same results as (A), the main difference being that the function fixing the section is now vector-valued, $\boldsymbol{\beta}_3 : X_3 \to \mathbb{R}^3$, since $\boldsymbol{\beta}_3(x_3)$ represents a boost. This makes the calculation much more difficult, except for the basic section $\boldsymbol{\beta}_3 = \mathbf{0}$, which yields a tight frame.

Notice that this choice of subgroup leads to the purely kinematical wavelets introduced by Duval-Destin and Murenzi,[5] and which correspond to the similitude group *SIM*(n).

In addition to these three cases, one may also consider directly the Galilei-Schrödinger group $S(n)$,[15] that is, impose from the beginning the relation $a_o = a^2$. It turns out that the restriction to $S(n)$ of the Stone-von Neumann representation (4.6) is square integrable on $S(n)/\Omega$, where Ω is the center of $S(n)$. Thus, in this case, one may return to the usual formalism of Gilmore[3] and Perelomov.[4]

Typical wavelets that seem useful are, for instance:

- the Schrödinger-Marr wavelet:

$$\psi_{SM}(t,\mathbf{x}) = (i\partial_t + \Delta)\, e^{-(x^2+t^2)/2};$$

- the Schrödinger-Cauchy wavelet:

$$\psi_{SC}(t,\mathbf{x}) = (i\partial_t + \Delta)\, \frac{1}{(t+i)\prod_{j=1}^{n}(x_j+i)}.$$

5. Outcome

The conclusion is that Galilean coherent states and wavelets may be obtained with the general method of construction of coherent states developed in Refs. 9-10. For the pure Galilei group, the novel phenomenon is the appearance of restrictions on the support of admissible vectors. As for wavelets, several different families may be constructed, corresponding to various subgroups to be quotiented out, and among them one recovers the kinematical wavelets[5] and the Galilei-Schrödinger wavelets.[15]

The next step is to implement these wavelets and apply them to physical problems. A natural candidate, for $n = 2$, is the analysis of image sequences. Another one, for $n = 3$, is the description of certain processes in atomic physics, in which the quasiclassical behavior is dominant, such as the Rydberg states of an atom. Work is in progress in these two directions.

References

1. Y. Meyer (ed.), *Wavelets and their Applications (Proc. Marseille 1989)*, Masson, Paris and Springer-Verlag, Berlin 1991
2. Y. Meyer and S. Roques (eds.), *Progress in Wavelet Analysis and Applications (Proc. Toulouse 1992)*, Ed. Frontières, Gif-sur-Yvette 1993
3. R. Gilmore, "Geometry of symmetrized states", *Ann. Phys. (NY)* **74** (1972) 391-463; "On properties of coherent states", *Rev. Mex. Fis.* **23** (1974) 143-187
4. A. Perelomov, "Coherent states for arbitrary Lie group", *Commun. Math. Phys.* **26** (1972) 222-236; *Generalized Coherent States and their Applications*, Springer-Verlag, Berlin 1986
5. M. Duval-Destin and R. Murenzi, "Spatio-temporal wavelets: Application to the analysis of moving patterns", in Ref.2, pp. 399-408
6. U. Niederer, "The maximal kinematical invariance group of the free Schrödinger equation", *Helv. Phys. Acta* **45** (1972) 803
7. M. Perroud, "Projective representations of the Schrödinger group", *Helv. Phys. Acta* **50** (1977) 233
8. J. Bertrand and P. Bertrand, "Reflectivity study of time-varying targets", in *Signal Processing VI: Theory and Applications (Proc.EUSIPCO 92, Brussels)*, pp. 1813-1816; J. Vandewalle et al. (eds.), Elsevier, Amsterdam 1992
9. S.T. Ali, J-P. Antoine and J-P. Gazeau, "Square integrability of group representations on homogeneous spaces. I. Reproducing triples and frames. II. Coherent and quasi-coherent states. The case of the Poincaré group", *Ann. Inst. H. Poincaré* **55** (1991) 829-855; 857-890
10. S.T. Ali, J-P. Antoine, J-P. Gazeau and U.A. Mueller, "Coherent states and their generalizations – A mathematical overview", *Reviews Math. Phys.* (to appear)
11. I. Daubechies, A. Grossmann and Y. Meyer, "Painless nonorthogonal expansions", *J. Math. Phys.* **27**(1986) 1271-1283
12. S.T. Ali, J-P. Antoine and J-P. Gazeau, "Relativistic quantum frames", *Ann. Phys.(NY)* **222** (1993) 38-88
13. C. Kalisa and B. Torrésani, "N-dimensional affine Weyl-Heisenberg wavelets", *Ann. Inst. H. Poincaré* **59** (1993) 201-236
14. J-P. Antoine, N.M. Hounkonnou and C. Kalisa, "Coherent states for the affine Galilei group", (in preparation)
15. J-P. Antoine, A. Haffad and R. Murenzi, "A wavelet transform associated to the Galilei-Schrödinger group", (in preparation).

ON THE USE OF QUANTUM ALGEBRAS IN ROTATION-VIBRATION SPECTROSCOPY

R. BARBIER AND M. KIBLER
Institut de Physique Nucléaire de Lyon
IN2P3-CNRS et Université Claude Bernard
43 Boulevard du 11 Novembre 1918
F-69622 Villeurbanne Cedex, France

Abstract. A two-parameter deformation of the Lie algebra u_2 is used, in conjunction with the rotor system and the oscillator system, to generate a model for rotation-vibration spectroscopy of molecules and nuclei.

1. Introduction

Quantum algebras and quantum groups [1] are nowadays widely used in physics. It is a point of fact that most of the physical applications of quantum algebras are based on one-parameter deformations. Nevertheless, during the last years, multiparameter (mainly two-parameter) quantum algebras have been constructed [2-13].

In nuclear and molecular physics, one-parameter quantum algebras have been used for describing rotation spectra of nuclei [14-16] and rotation-vibration spectra of diatomic molecules [17-22]. In more details, we can distinguish three kinds of models. First, models describing only rotation spectra in terms of the $U_q(su_2)$ quantum algebra [14, 15, 17] or the q-Poincaré algebra [16]. Second, models describing only vibration spectra of diatomic molecules and based on the q-deformation of the $u_2 \supset o_2$ chain [18], or on the quantum algebra $U_q(su_{1,1})$ [19] or on the q-deformation of the Heisenberg algebra h_4 [20]. Third, models describing rotation-vibration spectra of diatomic molecules in terms of a $q(j)$-deformation of the Heisenberg algebra h_4 [21] or a q-deformation of the $u_4 \supset so_4 \supset so_3$ chain [22]. Recently, a two-parameter Hopf algebra [10] has been applied to the description of rotational spectra of superdeformed nuclei [23-25].

J. Bertrand et al. (eds.), Modern Group Theoretical Methods in Physics, 27–36.

The work presented in this communication constitutes a first step towards a model for describing simultaneously rotation and vibration spectra of molecules and nuclei. This model (to be developed in section 3) relies on the use of two copies of the two-parameter quantum algebra $U_{qp}(\mathrm{u}_2)$ (described in section 2). The approach is entirely of a phenomenological nature although it is implicitly based on the using of a non-rigid rotor system and an anharmonic oscillator system.

2. The Quantum Algebra $U_{qp}(\mathrm{u}_2)$

We begin with a brief description of the quantum algebra $U_{qp}(\mathrm{u}_2)$ restricted to those aspects necessary for the model to be investigated here. Two-parameter deformations of su_2 and u_2 have been considered by several authors. We follow here the presentation of Refs. [10, 12, 13]. According to Ref. [10], the quantum algebra $U_{qp}(\mathrm{u}_2)$ is introduced by the following commutation relations

$$[J_0, J_\alpha] = 0 \qquad [J_3, J_\pm] = \pm J_\pm \qquad [J_+, J_-] = (qp)^{J_0 - J_3}\, [[2J_3]]_{qp} \tag{1}$$

between four operators J_α ($\alpha = 0, 3, +, -$). In this work, we use the notations

$$[[X]]_{qp} := \frac{q^X - p^X}{q - p} \qquad [X]_q := [[X]]_{qq^{-1}} = \frac{q^X - q^{-X}}{q - q^{-1}} \tag{2}$$

for the qp- and the q-deformations of a real number or an operator X.

The generators J_α verify the co-product rules

$$\begin{aligned} \Delta(J_0) &= J_0 \otimes 1 + 1 \otimes J_0 && (3)\\ \Delta(J_3) &= J_3 \otimes 1 + 1 \otimes J_3 && (4)\\ \Delta(J_+) &= J_+ \otimes (qp)^{J_0} (qp^{-1})^{+J_3} + 1 \otimes J_+ && (5)\\ \Delta(J_-) &= J_- \otimes 1 + (qp)^{J_0} (qp^{-1})^{-J_3} \otimes J_- && (6) \end{aligned}$$

Note that (5) and (6) depends on the two independent parameters q and p. The universal $\mathcal{R}$-matrix associated to the co-product Δ reads

$$\mathcal{R}_{qp} = \begin{pmatrix} p & 0 & 0 & 0 \\ 0 & 1 & 0 & 0 \\ 0 & p-q & pq & 0 \\ 0 & 0 & 0 & p \end{pmatrix} \tag{7}$$

in the representation $\frac{1}{2} \otimes \frac{1}{2}$.

The operator [10]

$$C_2(U_{qp}(\mathrm{u}_2)) = \frac{1}{2}(J_+J_- + J_-J_+) + \frac{1}{2}\, [[2]]_{qp}\, (qp)^{J_0 - J_3}\, ([[J_3]]_{qp})^2 \tag{8}$$

is invariant under the generators of the quantum algebra $U_{qp}(\mathrm{u}_2)$. It is characterized by the judicious introduction of the element J_0 which is itself a linear invariant of $U_{qp}(\mathrm{u}_2)$.

An irreducible representation of $U_{qp}(\mathrm{u}_2)$, in which the operators J_0 and $C_2(U_{qp}(\mathrm{u}_2))$ are simultaneously diagonal (in the case where neither q nor p are roots of unity), is given by a Young pattern $[\varphi_1; \varphi_2]$ with $\varphi_1 - \varphi_2 = 2j$ and $\varphi_1 + \varphi_2 = 2j_0$ where $2j_0$ and $2j$ are non-negative integers. We introduce the vectors $|j_0 jm)$ (with $m = -j, -j+1, \cdots, +j$) as the basis vectors for the representation $[j_0 + j \; ; \; j_0 - j]$. The vector $|j_0 jm)$ is connected to the highest weight vector $|j_0 jj)$ by the relation [12]

$$|j_0 jm) \;=\; (qp)^{-\frac{1}{2}(j_0-j)(j-m)} \sqrt{\frac{[[j+m]]_{qp}!}{[[j-m]]_{qp}! \; [[2j]]_{qp}!}} \; (J_-)^{j-m} \; |j_0 jj) \qquad (9)$$

Note that the unusual introduction of the central element J_0 in (9) comes from the commutation relations (1) and will play a crucial role in what follows. By using (9), we obtain the following action of the generators J_α on the state vector $|j_0 jm)$

$$\begin{aligned} J_0 \; |j_0 jm) &= j_0 \; |j_0 jm) \\ J_3 \; |j_0 jm) &= m \; |j_0 jm) \\ J_\pm \; |j_0 jm) &= (qp)^{\frac{1}{2}(j_0-j)} \sqrt{[[j \mp m]]_{qp} [[j \pm m + 1]]_{qp}} \; |j_0 jm \pm 1) \end{aligned} \qquad (10)$$

Note that Eq. (10) gives back the Jimbo [1] representation of $U_q(\mathrm{su}_2)$ when $p = q^{-1}$.

With the help of (10), the eigenvalues of (8) in the irreducible representation $[j_0 + j \; ; \; j_0 - j]$ become

$$E(j_0, j)_{qp} \;=\; (qp)^{j_0-j} \; [[j]]_{qp} \; [[j+1]]_{qp} \qquad (11)$$

A factorisation of the eigenvalues (11) into a j_0-dependent and j-dependent part is given by

$$E(j_0, j)_{qp} \;=\; P^{2j_0-1} \; [j]_Q \; [j+1]_Q \qquad (12)$$

where we use the notation (2) with the parameters

$$Q = (qp^{-1})^{\frac{1}{2}} \qquad P = (qp)^{\frac{1}{2}} \qquad (13)$$

as an alternative to q and p.

The calculation of the Clebsch-Gordan coefficients (CGc's) corresponding to the algebra homomorphism $\Delta : U_{qp}(\mathrm{u}_2) \longrightarrow U_{qp}(\mathrm{u}_2) \otimes U_{qp}(\mathrm{u}_2)$ defined by Eqs. (3-6) is in progress. These coupling coefficients depend on the two parameters q and p. It is to be emphasized that it seems not possible to express them in terms of only one of the two parameters Q or P given by Eq. (13). In this respect, it is perhaps interesting to note that, should we have chosen, instead of Δ, the co-product Δ_{qp} defined by

$$\begin{aligned}\Delta_{qp}(J_0) &= J_0 \otimes 1 + 1 \otimes J_0 \\ \Delta_{qp}(J_3) &= J_3 \otimes 1 + 1 \otimes J_3 \\ \Delta_{qp}(J_\pm) &= J_\pm \otimes (qp)^{\frac{1}{2}J_0}(qp^{-1})^{+\frac{1}{2}J_3} + (qp)^{\frac{1}{2}J_0}(qp^{-1})^{-\frac{1}{2}J_3} \otimes J_\pm \end{aligned} \quad (14)$$

we would have obtained CGc's depending only on the parameter Q (see Appendix).

All that precedes is of pivotal importance for application to rotational spectroscopy. We now continue with some aspects of $U_{qp}(\mathrm{u}_2)$ useful in vibrational spectroscopy.

The quantum algebra $U_{qp}(\mathrm{u}_2)$ can be constructed from two pairs, say $\{a_+^+, a_+\}$ and $\{a_-^+, a_-\}$, of qp-deformed (creation and annihilation) boson operators. The action of these qp-bosons on a non-deformed two-particle Fock space $\{|n_+ n_-\rangle : n_+ \in \mathrm{N},\ n_- \in \mathrm{N}\}$ may be chosen to be controlled by

$$\begin{aligned} a_+^+ \, |n_+ n_-\rangle &= \sqrt{[[n_+ + \frac{1}{2} + \frac{1}{2}]]_{qp}} \, |n_+ + 1 \; n_-\rangle \\ a_+ \, |n_+ n_-\rangle &= \sqrt{[[n_+ + \frac{1}{2} - \frac{1}{2}]]_{qp}} \, |n_+ - 1 \; n_-\rangle \\ a_-^+ \, |n_+ n_-\rangle &= \sqrt{[[n_- + \frac{1}{2} + \frac{1}{2}]]_{qp}} \, |n_+ \; n_- + 1\rangle \\ a_- \, |n_+ n_-\rangle &= \sqrt{[[n_- + \frac{1}{2} - \frac{1}{2}]]_{qp}} \, |n_+ \; n_- - 1\rangle \end{aligned} \quad (15)$$

The vectors $|n_+ n_-\rangle$ are generated from

$$|n_+ n_-\rangle = \frac{1}{\sqrt{[[n_+]]_{qp}! [[n_-]]_{qp}!}} (a_+^+)^{n_+} (a_-^+)^{n_-} \, |00\rangle \quad (16)$$

The two pairs $\{a_+^+, a_+\}$ and $\{a_-^+, a_-\}$ of qp-bosons commute and satisfy [8, 26]

$$a_\pm a_\pm^+ = [[N_\pm + 1]]_{qp} \qquad a_\pm^+ a_\pm = [[N_\pm]]_{qp} \quad (17)$$

where N_+ and N_- are the usual number operators with

$$N_\pm \, |n_+ n_-\rangle = n_\pm \, |n_+ n_-\rangle \quad (18)$$

Of course, the qp-bosons $a_\pm^+$ and $a_\pm$ reduce to ordinary bosons in the limiting situation where $p = q^{-1} \to 1$.

The passage from the (harmonic oscillator) state vectors $|n_+n_-\rangle$ to angular momentum state vectors $|jm)$ is achieved through the relations

$$j := \frac{1}{2}(n_+ + n_-) \qquad m := \frac{1}{2}(n_+ - n_-) \tag{19}$$

and

$$|jm) \equiv |j+m\ j-m\rangle = |n_+n_-\rangle \tag{20}$$

Equations (15) may thus be rewritten as

$$\begin{aligned} a_\pm^+ \ |jm) &= \sqrt{[[j\pm m + \frac{1}{2} + \frac{1}{2}]]_{qp}}\ |j+\frac{1}{2}\ m\pm\frac{1}{2}) \\ a_\pm \ |jm) &= \sqrt{[[j\pm m + \frac{1}{2} - \frac{1}{2}]]_{qp}}\ |j-\frac{1}{2}\ m\mp\frac{1}{2}) \end{aligned} \tag{21}$$

so that the qp-bosons behave as ladder operators for the quantum numbers j and m (with $|m| \leq j$).

Contact with the quantum algebra $U_{qp}(\mathrm{u}_2)$ may now be established. Let us define the four operators $\tilde{J}_\alpha$ $(\alpha = 0, 3, +, -)$ as

$$\tilde{J}_0 := \frac{1}{2}(N_+ + N_-) \quad \tilde{J}_3 := \frac{1}{2}(N_+ - N_-) \quad \tilde{J}_+ := a_+^+ a_- \quad \tilde{J}_- := a_-^+ a_+ \tag{22}$$

A simple calculation shows that the operators $\tilde{J}_\alpha$ satisfy the same commutation relations as the operators J_α [see Eq. (1)]. Consequently, Eq. (22) provides us with a boson realization of $U_{qp}(\mathrm{u}_2)$. We may ask what kind of representation we thus obtain ? Indeed, repeated application of Eqs. (21) yields

$$\begin{aligned} \tilde{J}_0 \ |jm) &= j \ |jm) \\ \tilde{J}_3 \ |jm) &= m \ |jm) \\ \tilde{J}_\pm \ |jm) &= \sqrt{[[j \mp m]]_{qp}[[j \pm m + 1]]_{qp}}\ |jm \pm 1) \end{aligned} \tag{23}$$

so that the boson realization (22) concerns the irreducible representation $[2j\ ;\ 0]$ for which $j_0 = j$ [cf. Eq. (10)].

At this stage, it is useful to mention the range of variation of the parameters q and p of interest for practical applications. As a matter of fact, from Hermitean conjugation requirements, the values of the parameters q and p must be restricted to some domains that can be classified as follows: (i) $q \in \mathbf{R}$ and $p \in \mathbf{R}$, (ii) $q \in \mathbf{C}$ and $p \in \mathbf{C}$ with the constraint $p = q^*$, and (iii) $q = p^{-1} = \mathrm{e}^{\mathrm{i}\beta}$ with $0 \leq \beta < 2\pi$.

3. A Model for Rotation-Vibration Spectroscopy

We are now in position to present a model for describing rotation-vibration spectra of diatomic molecules and nuclei. This model relies on two dynamical systems, viz., the non-rigid rotor system with the dynamical symmetry $U_{qp}(\mathrm{u}_2)$ and the anharmonic oscillator system with the dynamical symmetry $U_{sr}(\mathrm{u}_2) \supset \mathrm{u}_1$. One of the cornerstones of this model is the Hamiltonian

$$H = E_{\text{rovib}}\, C_2(U_{qp}(\mathrm{u}_2)) \;+\; E_{\text{vib}}\, C(U_{sr}(\mathrm{u}_2) \supset \mathrm{u}_1) \;+\; E_0 \tag{24}$$

where E_{rovib} and E_{vib} are two constants and E_0 the zero energy term. The first term $C_2(U_{qp}(\mathrm{u}_2))$ is the invariant (8) of the quantum algebra $U_{qp}(\mathrm{u}_2)$ and thus is appropriate for the rotational part. We shall see how to make this term vibration-dependent too by introducing a convenient coupling with the second term $C(U_{sr}(\mathrm{u}_2) \supset \mathrm{u}_1)$. The latter term describes the vibrational part. We take it in the form

$$C(U_{sr}(\mathrm{u}_2) \supset \mathrm{u}_1) \;=\; \left([[2\tilde{J}_0]]_{sr}\right)^2 - (sr)^{2(\tilde{J}_0 - \tilde{J}_3)} \left([[2\tilde{J}_3]]_{sr}\right)^2 \tag{25}$$

where $\tilde{J}_0$ and $\tilde{J}_3$ are defined by (22) and refer to a second quantum algebra $U_{sr}(\mathrm{u}_2)$. The operator $C(U_{sr}(\mathrm{u}_2) \supset \mathrm{u}_1)$ is invariant under the generators $\tilde{J}_0$ and $\tilde{J}_3$ of the chain $U_{sr}(\mathrm{u}_2) \supset \mathrm{u}_1$. The Hamiltonian H clearly exhibits the $U_{qp}(\mathrm{u}_2) \times (U_{sr}(\mathrm{u}_2) \supset \mathrm{u}_1)$ dynamical symmetry.

The passage from the sr-boson state vectors $|n_+ n_-\rangle$ [see Eq. (16)] to the vibration state vectors $|nv)$ is accomplished by means of

$$|nv) \;\equiv\; |n - v, v\rangle \tag{26}$$

In other words, we put

$$n = n_+ + n_- \qquad n - 2v = n_+ - n_- \tag{27}$$

The total number n of sr-bosons is connected to the maximal vibration quantum number $v_{\max}$ by $v_{\max} = n/2$ or $(n-1)/2$ depending on whether n is even or odd. On the other hand, the rotational state vectors are chosen to be $|j_0 jm)$ [see Eq. (9)]. The coupling between the rotational part and the vibrational part is achieved by assuming that $j_0 = v$. This leads to the rotation-vibration state vectors

$$|nvjm) = |vjm) \otimes |nv) \tag{28}$$

where v and j are the vibrational and rotational quantum numbers, respectively. The diagonalisation of H on the subspace spanned by the vectors $|nvjm)$ leads to the eigenvalues

$$E(n,v,j) = E_{\text{rovib}}(qp)^{v-j}[[j]]_{qp}[[j+1]]_{qp} + E_{\text{vib}}[[2v]]_{sr}[[2(n-v)]]_{sr} + E_0 \tag{29}$$

Alternatively, Eq. (29) can be rewritten as

$$E(n,v,j) = E_{\mathrm{rovib}} P^{2v-1}[j]_Q[j+1]_Q + E_{\mathrm{vib}} R^{2(n-1)}[2v]_S[2(n-v)]_S + E_0 \quad (30)$$

where we have introduced the parameters

$$S = (sr^{-1})^{\frac{1}{2}} \qquad R = (sr)^{\frac{1}{2}} \quad (31)$$

to be compared with the parameters Q and P of Eq. (13).

The model described by Eq. (29) depends on four quantum algebra parameters (namely, q, p, s and r). In the most general situation, we thus have eight real parameters. This number may be reduced following the discussion at the end of section 2. Here, we shall limit ourselves to the case (ii) of the latter discussion for the pairs (q,p) and (s,r). This yields four real parameters. Furthermore, the consideration of limiting processes leads us to take

$$q = p^* = \mathrm{e}^{\beta\cos\gamma}\,\mathrm{e}^{+\mathrm{i}\beta\sin\gamma} \qquad s = r^* = \mathrm{e}^{\tau\cos\gamma}\,\mathrm{e}^{+\mathrm{i}\tau\sin\gamma} \quad (32)$$

so that we end up with three real parameters β, γ and τ. Then, Eq. (29) becomes

$$\begin{aligned} E(n,v,j) = {} & E_{\mathrm{rovib}}\,\mathrm{e}^{(2v-1)\beta\cos\gamma}\,\frac{\sin(j\beta\sin\gamma)\,\sin[(j+1)\beta\sin\gamma]}{\sin^2(\beta\sin\gamma)} \\ & + E_{\mathrm{vib}}\,\mathrm{e}^{(2n-2)\tau\cos\gamma}\,\frac{\sin(2v\tau\sin\gamma)\,\sin[2(n-v)\tau\sin\gamma]}{\sin^2(\tau\sin\gamma)} + E_0 \quad (33) \end{aligned}$$

Equation (33) can be developed as

$$E(n,v,j) \;=\; \sum_{l,k} Y_{lk}\,(v+\frac{1}{2})^l\,[j(j+1)]^k \quad (34)$$

which resembles the Dunham [27] expansion obtained from the Morse oscillator system. The coefficients Y_{lk} in Eq. (34) depends on the parameters β, γ and τ (and also n which can be employed to characterize the anharmonicity constant of the oscillator). They shall be reported elsewhere. It should be noted that the limiting case $\gamma = \frac{\pi}{2}$ corresponds to a model where the rotational and vibrational parts are decoupled; in this case, the only non-vanishing coefficients Y_{lk} in Eq. (34) are of the type Y_{0k} and Y_{l0}.

4. Closing Remarks

We have concentrated in this work on an $U_{qp}(\mathrm{u}_2) \times (U_{sr}(\mathrm{u}_2) \supset \mathrm{u}_1)$ model that unifies and extends various models developed in recent years. The limiting cases ($E_{\text{rovib}} = 0$, $q = p^{-1} = s = r^{-1} = \mathrm{e}^{\mathrm{i}\tau}$) and ($E_{\text{rovib}} = 0$, $q = p^{-1} = s = r^{-1} = 1$) correspond to the $U_q(\mathrm{u}_2) \supset \mathrm{o}_2$ model and $\mathrm{u}_2 \supset \mathrm{o}_2$ model worked out in Ref. [18] and Ref. [28], respectively, for vibrational spectra of molecules, while the limiting case ($E_{\text{vib}} = 0$, $q = p^{-1} = \mathrm{e}^{\mathrm{i}\beta}$) corresponds to the $U_q(\mathrm{su}_2)$ model introduced [15] for rotational spectroscopy of nuclei. The particular case ($E_{\text{vib}} = 0$, $v = j_0 = j$) is nothing but the $U_{qp}(\mathrm{u}_2)$ rotor model successfully applied to rotational bands of superdeformed nuclei [23, 25].

A second important step towards the understanding of the dynamics inherent to the $U_{qp}(\mathrm{u}_2) \times (U_{sr}(\mathrm{u}_2) \supset \mathrm{u}_1)$ model remains to be made. This represents a difficult task involving some difference equations and quantum inverse scattering methods. Another appealing project should be to replace the Fock oscillator states corresponding to Eqs. (15-18) by the states (which do not have classical limit) recently introduced by Rideau [29]. We hope to return on these matters in the future.

One of the authors (M.K.) would like to express his sincere gratitude to Guy Rideau for his help, friendly advices and interesting discussions on various occasions during the last ten years. Thanks are due to Guy Rideau and Pavel Winternitz for useful comments at several stages of this work. Finally, M.K. is indebted to Moshe Flato for friendly and constructive criticism.

A. Clebsch-Gordan Coefficients

The CGc's

$$(j_{01} j_{02} j_1 j_2 m_1 m_2 | j_0 j m)_{qp} \equiv (m_1 m_2 | m)_{qp} \tag{35}$$

corresponding to the co-product Δ_{qp} defined through Eq. (14) can be seen to satisfy the following three-term recursion relations [12]

$$\begin{aligned} &\sqrt{[j \mp m]_Q \, [j \pm m + 1)]_Q} \; (m_1 m_2 | m \pm 1)_{qp} \\ &\quad = \; Q^{+m_2} \sqrt{[j_1 \pm m_1]_Q \, [j_1 \mp m_1 + 1]_Q} \; (m_1 \mp 1, m_2 | m)_{qp} \\ &\quad + \; Q^{-m_1} \sqrt{[j_2 \pm m_2]_Q \, [j_2 \mp m_2 + 1]_Q} \; (m_1, m_2 \mp 1 | m)_{qp} \end{aligned} \tag{36}$$

which are identical to the ones [30] satisfied by the CGc's $(j_1 j_2 m_1 m_2 | j m)_Q$ of the quantum algebra $U_Q(\mathrm{su}_2)$. Therefore, there exists a proportionality constant between the qp-CGc's and the Q-CGc's. For q and p real, reality and normalization conditions can be used to justify that the proportionality constant is taken to be equal to 1. In fact, this may be checked by direct calculation: By adapting the method of projection operators used for

$U_q(\mathrm{su}_2)$ in Ref. [31], we can show that

$$(j_{01} j_{02} j_1 j_2 m_1 m_2 | j_0 j m)_{qp} = \delta(j_0, j_{01} + j_{02}) \, (j_1 j_2 m_1 m_2 | j m)_Q \tag{37}$$

with

$$(j_1 j_2 m_1 m_2 | j m)_Q = (-1)^{j_1+j_2-j} Q^{-\frac{1}{2}(j_1+j_2-j)(j_1+j_2+j+1)+j_1 m_2 - j_2 m_1}$$
$$\left(\frac{[j_1 - j_2 + j]![j_1 + j_2 - j]![j_1 + j_2 + j + 1]![j_2 - m_2]![j + m]!}{[j - j_1 + j_2]![j_2 + m_2]![j - m]![j_1 - m_1]![j_1 + m_1]!} \right)^{\frac{1}{2}}$$
$$[2j+1]^{\frac{1}{2}} \sum_z \frac{(-1)^z Q^{z(j_1+m_1)} [2j_2 - z]![j_1 + j_2 - m - z]!}{[z]![j_1 + j_2 - j - z]![j_2 - m_2 - z]![j_1 + j_2 + j + 1 - z]!} \tag{38}$$

where we have used the abbreviation $[x]_Q \equiv [x]$.

The results (37) and (38) can be also justified as follows. The universal R-matrix associated to the co-product Δ_{qp} reads

$$\mathcal{R}_{pq} = \begin{pmatrix} p & 0 & 0 & 0 \\ 0 & (pq)^{\frac{1}{2}} & 0 & 0 \\ 0 & p - q & (pq)^{\frac{1}{2}} & 0 \\ 0 & 0 & 0 & p \end{pmatrix} \tag{39}$$

which can be factorised in terms of Q- and P-depending parts. Such a factorisation corresponds to the decomposition

$$U_{qp}(\mathrm{u}_2) = \mathrm{u}_1 \otimes U_Q(\mathrm{su}_2) \tag{40}$$

where $U_Q(\mathrm{su}_2)$ is spanned by

$$A_0 := J_0 \qquad A_3 := J_3 \qquad A_\pm := (qp)^{-\frac{1}{2}(J_0 - \frac{1}{2})} \, J_\pm \tag{41}$$

with the following commutation relations

$$[A_3, A_\pm] = \pm A_\pm \qquad [A_+, A_-] = [2A_3]_Q \tag{42}$$

In terms of co-product, we have

$$\Delta_{qp}(J_\pm) = P^{\Delta_Q(A_0) - \frac{1}{2}} \, \Delta_Q(A_\pm) \tag{43}$$

where the co-product Δ_Q is given via

$$\begin{aligned} \Delta_Q(A_0) &= A_0 \otimes 1 + 1 \otimes A_0 \\ \Delta_Q(A_3) &= A_3 \otimes 1 + 1 \otimes A_3 \\ \Delta_Q(A_\pm) &= A_\pm \otimes Q^{+A_3} + Q^{-A_3} \otimes A_\pm \end{aligned} \tag{44}$$

which relations correspond to the CGc's $(j_1 j_2 m_1 m_2|jm)_Q$.

References

1. Kulish, P.P. and Reshetikhin, N.Yu. (1981) *Zap. Sem. LOMI* **101**, 101 [(1983) *J. Soviet. Math.* **23**, 2435]; Sklyanin, E.K. (1982) *Funkt. Anal. Pril.* **16**, 27 [(1982) *Funct. Anal. Appl.* **16**, 262]; Drinfeld, V.G. (1985) *Soviet. Math. Dokl.* **32**, 254; (1986) in: *Proc. Int. Congr. Math.*, Ed., A.M. Gleason (AMS, Providence, RI) pp. 798; Jimbo, M. (1985) *Lett. Math. Phys.* **10**, 63; (1986) *Commun. Math. Phys.* **102**, 537; Woronowiccz, S.L. (1987) *Publ. RIMS-Kyoto* **23**, 117; (1987) *Commun. Math. Phys.* **111**, 613
2. Sudbery, A. (1990) *J. Phys. A: Math. Gen.* **23**, L697
3. Demidov, E.E., Manin, Yu.I., Mukhin, E.E. and Zhdanovich, D.V. (1990) *Prog. Theo. Phys. Suppl.* **102**, 203
4. Reshetikhin, N. (1990) *Lett. Math. Phys.* **20**, 331
5. Fairlie, D.B. and Zachos, C.K. (1991) *Phys. Lett. B* **256**, 43
6. Schirrmacher, A., Wess, J. and Zumino, B. (1991) *Z. Phys. C* **49**, 317
7. Vokos, S.T. (1991) *J. Math. Phys.* **32**, 2979
8. Chakrabarti, R. and Jagannathan, R. (1991) *J. Phys. A: Math. Gen.* **24**, L711
9. Dobrev, V.K. (1992) *J. Math. Phys.* **33**, 3419
10. Kibler, M.R. (1993) in: *Symmetry and Structural Properties of Condensed Matter*, Eds., W. Florek, D. Lipiński and T. Lulek (World Scientific, Singapore) pp. 445
11. Chakrabarti, R. and Jagannathan, R. (1994) *J. Phys. A: Math. Gen.* **27**, 2023
12. Kibler, M.R., Asherova, R.M. and Smirnov, Yu.F. (1994) preprint LYCEN 9439
13. Jagannathan, R. and Van der Jeugt, J. (1994) preprint hep-th/9411200
14. Iwao, S. (1990) *Prog. Theor. Phys.* **83**, 363
15. Raychev, P.P., Roussev, R.P. and Smirnov, Yu.F. (1990) *J. Phys. G: Nucl. Phys.* **16**, L137
16. Celeghini, E., Giachetti, R., Sorace, E. and Tarlini, M. (1992) *Phys. Lett. B* **280**, 180
17. Bonatsos, D., Raychev, P.P., Roussev, R.P. and Smirnov, Yu.F. (1990) *Chem. Phys. Lett.* **175**, 300
18. Bonatsos, D., Raychev, P.P. and Faessler, A. (1991) *Chem. Phys. Lett.* **178**, 221
19. Bonatsos, D., Argyres, E.N. and Raychev, P.P. (1991) *J. Phys. A: Math. Gen.* **24**, L403
20. Chang, Z., Guo, H.Y. and Yan, H. (1991) *Phys. Lett. A* **156**, 192
21. Chang, Z. and Yan, H. (1991) *Phys. Lett. A* **158**, 242
22. Pan, F.Z. (1993) *J. Phys. B: At. Mol. Opt. Phys.* **26**, L47
23. Barbier, R., Meyer, J. and Kibler, M. (1994) *J. Phys. G: Nucl. Phys.* **17**, L67
24. Kibler, M. (1994) in: *Generalized Symmetries in Physics*, Eds., H.-D. Doebner, V.K. Dobrev and A.G. Ushveridze (World Scientific, Singapore) pp. 55
25. Barbier, R., Meyer, J. and Kibler, M. (1994) preprint LYCEN 9437
26. Katriel, J. and Kibler, M. (1992) *J. Phys. A: Math. Gen.* **25**, 2683
27. Dunham, J.L. (1932) *Phys. Rev.* **41**, 721
28. Van Roosmaleen, O.S., Levine, R.D. and Dieperink, A.E. (1983) *Chem. Phys. Lett.* **101**, 512
29. Rideau, G. (1992) *Lett. Math. Phys.* **24**, 147
30. Kibler, M., Campigotto, C. and Smirnov, Yu.F. (1994) in: *Proceedings of the International Workshop "Symmetry Methods in Physics, in Memory of Professor Ya.A. Smorodinsky"*, Eds., A.N. Sissakian, G.S. Pogosyan and S.I. Vinitsky, (JINR, Dubna, Russia) pp. 246
31. Smirnov, Yu.F., Tolstoǐ, V.N. and Kharitonov, Yu.I. (1991) *Sov. J. Nucl. Phys.* **53**, 593

INVARIANT DIFFERENTIAL CALCULUS ON A DEFORMATION OF THE WEYL-HEISENBERG ALGEBRA

J. BERTRAND AND M. IRAC-ASTAUD
LPTM, Université Paris VII, 75251 Paris Cedex 05, France

A Guy Rideau, en témoignage d'amitié.

Abstract. A quantum hyperplane is defined in terms of three generators satisfying quadratic commutation relations that can be seen as defining a deformation of the Weyl-Heisenberg algebra. The corresponding quantum invariance is then investigated and conditions for the existence of an invariant differential calculus are obtained. A quantum group is constructed that cannot be obtained as a special case of any known deformed $GL(3)$.

1. Introduction

General procedures to set up a differential calculus on quantum groups or quantum hyperplanes have been widely studied (see for example [1, 3] [4]). But the specific examples of quantum hyperplanes that appear in the literature are usually of the special form:

$$x^i x^j = q^{ij} x^j x^i, \quad q^{ij} = \frac{1}{q^{ji}}, \quad q_{ii} = 1 \tag{1}$$

where x^i, $i = 1, ...n$, are the coordinates on the quantum space and q are deformation parameters. The differential calculus is then set up subject to an external constraint of invariance either by the dilation group [5] or a quantum group $GL_q(n)$ [6, 7]. In the limit $q^{ij} \to 1$, such a scheme goes over to the usual differential calculus on space $\mathbb{R}^n$, invariant by group $GL(n)$. Thus, by construction, the physically important case of the Weyl-Heisenberg algebra cannot be recovered.

The present paper investigates the possible deformations of the Weyl-Heisenberg algebra $\mathcal{H}$ on which an invariant differential calculus can be set

J. Bertrand et al. (eds.), Modern Group Theoretical Methods in Physics, 37–49.

up. The starting point is the usual $\hat{a}$, $\hat{a}^\dagger$ commutation relations written as:

$$(R) \begin{cases} x^1x^2 - x^2x^1 = (x^3)^2 \\ x^1x^3 = x^3x^1 \\ x^2x^3 = x^3x^2 \end{cases} \tag{2}$$

where x^3 is a number and:

$$\hat{a} = \frac{x^1}{x^3} \quad \hat{a}^\dagger = \frac{x^2}{x^3} \tag{3}$$

These relations are invariant if the vector of components x^i is transformed by a matrix T of the form:

$$T = (T^i_j) = \begin{pmatrix} a & b & c \\ g & d & l \\ 0 & 0 & m \end{pmatrix} \tag{4}$$

provided the elements of T are such that:

$$ad - bg = m^2 \tag{5}$$

The problem now is to define a quantum space by deforming relations (1) and to find a relevant quantum group performing linear transformations simultaneously on that space and on the associated space of forms. This is carried out in sec.1 where a two-parameter quantum group is obtained as a deformation of the group of matrices (4). The invariant differential calculus is then set up in sec.2 by usual methods [1, 9]. Finally, in sec.3, a realization of the quantum space is shown and compared to known results.

2. Construction of the invariance quantum group

2.1. TRANSFORMATION LAW OF THE QUANTUM HYPERPLANE

The quantum space X considered here is defined as the quotient algebra $\mathbb{C} < x^1, x^2, x^3 > / R_{xx}$, where:

- $\mathbb{C} < x^1, x^2, x^3 >$ is the free associative algebra generated by the elements x^i, $i = 1,2,3$ (it can be viewed as the space of formal power series in x^i),

- R_{xx} are deformed Weyl-Heisenberg relations given by:

$$(R_{xx}) \quad \begin{cases} x^1x^2 - q\,x^2x^1 - s(x^3)^2 &= 0 \\ x^1x^3 - u\,x^3x^1 &= 0 \\ x^2x^3 - v\,x^3x^2 &= 0 \end{cases} \tag{6}$$

with arbitrary non-zero parameters q, s, u, v belonging to $\mathbb{C}$. This is the most general deformation of relations (2) which preserves their quadratic character and does not introduce any new term. It is easily seen that the parameters cannot be independent. Indeed, multiplying the first relation (6) by x^3, we obtain the consistency condition:

$$uv = 1 \tag{7}$$

To give a meaning to the invariance requirement, a relevant space of quantum matrices has to be introduced and its action on X defined. Let $\mathbb{C} < T^i_j >$, $i, j = 1, 2, 3$ denote the associative algebra generated by the elements T^i_j of the matrix T defined by (4). This algebra will be factorized by commutation relations between the elements T^i_j. Such relations can be written in terms of a 9×9 matrix $\widehat{R}$ as:

$$\widehat{R}^{ji}_{kl}\, T^k_m T^l_n = T^j_l T^i_k\, \widehat{R}^{lk}_{mn} \tag{8}$$

or more intrinsically as:

$$\widehat{R}\,(T \otimes T) = (T \otimes T)\,\widehat{R} \tag{9}$$

where $\otimes$ denotes the Kronecker product of matrices.

The $\widehat{R}$-matrix and hence relations (9) are unknown so far and will be determined by requiring invariance of the quantum hyperplane.

In the present context, the action of matrix T on the quantum space $\mathbb{C} < x^1, x^2, x^3 >$ is defined as a homomorphism [8]:

$$\delta : \mathbb{C} < x^i > \longrightarrow \mathbb{C} < T^i_j > \otimes\, \mathbb{C} < x^i > \tag{10}$$

given explicitly by:

$$\delta(x^i) = \sum_{j=1}^{3} T^i_j \otimes x^j \tag{11}$$

This has the form of an ordinary matrix multiplication rule with the tensor product in place of the ordinary product. The tensorial sign $\otimes$ will be dropped in the following.

Now the requirement that relations R_{xx} be invariant by the action (11) produces the following commutation relations between the elements of matrix T:

$$\begin{aligned} am &= ma & gm &= u^{-2}\, mg \\ bm &= u^2\, mb & dm &= md \\ cm &= u\, mc & lm &= u^{-1}\, ml \\ ag &= q\, ga & bd &= q\, db \end{aligned} \tag{12}$$

$$\begin{aligned} u\,al - q\,la &= qu\,gc - cg && (13)\\ bl - qu\,lb &= q\,dc - u\,cd && (14)\\ ad - da &= q\,gb - q^{-1}\,bg && (15)\\ (ad - q\,gb)s &= s\,m^2 - cl + q\,lc && (16) \end{aligned}$$

As a consequence, the action δ becomes defined as a mapping on X itself.

To define an inverse matrix T^{-1}, we try and set:

$$T^{-1} = \begin{pmatrix} dm & \alpha\,bm & \beta_1\,bl + \beta_2\,lb + \gamma\,cd \\ \xi\,gm & \theta\,am & \delta_1\,al + \delta_2\,la + \epsilon\,cg \\ 0 & 0 & \zeta_1\,ad + \zeta_2\,da + \eta\,bg \end{pmatrix} \times D^{-1} \qquad (17)$$

where the determinant D and the parameters $\alpha, ..., \xi$ have to be determined. Writing that $TT^{-1}\,D = D$ gives three equivalent expressions for the determinant:

$$D \equiv (ad + \xi\,bg)m = (\alpha\,gb + \theta\,da)m = (\zeta_1\,ad + \zeta_2\,da + \eta\,bg)m \qquad (18)$$

Several other relations are also obtained but for the moment, we need only write the simplest ones:

$$ab = -\frac{\theta}{\alpha}\,ba \qquad gd = -\xi\,dg \qquad (19)$$

2.2. INVARIANT QUANTUM EXTERIOR ALGEBRA

Since the ultimate goal of the study is the construction of a differential calculus on X, a space of forms will have to be introduced. Let Y^1 be the vector space generated by the elements ξ^i, $i = 1, 2, 3$, and denote its exterior products by $Y^2 \equiv \wedge^2 Y^1$ and $Y^3 \equiv \wedge^3 Y^1$. The ξ^i are chosen such that:

$$\xi^i \wedge \xi^i = 0 \qquad (20)$$

for all values of i. The wedges will be dropped in the following.

The action of T on ξ^i is defined to be the same as (11). The constraint of invariance of relations (20) by this action can then be written down.The use of relations (12) and (19) leads to the following commutation relations:

$$\begin{aligned} \xi^2\xi^1 &= \xi\,\xi^1\xi^2 \\ \xi^3\xi^1 &= \lambda\,\xi^1\xi^3 \\ \xi^3\xi^2 &= \mu\,\xi^2\xi^3 \end{aligned} \qquad (21)$$

where λ and μ are arbitrary constants. In addition, the value of α is given by:

$$\alpha = \theta/\xi \tag{22}$$

The requirement that the action of T upon $\wedge^3 Y$ be such that:

$$(T\xi)^1(T\xi)^2(T\xi)^3 = D\,\xi^1\xi^2\xi^3 \tag{23}$$

leads to one more relation between ad and da:

$$da - ad = \xi\, bg - \xi^{-1}\, gb \tag{24}$$

and the comparison with relations (15) and (18) yields necessarily:

$$\theta = 1 \quad gb = -(\xi/q)\, bg \tag{25}$$

In addition, the $\xi^i\xi^j$ commutation relations obtained in (21) must also be conserved when T is applied. This requirement leads to a condition on the parameters:

$$\mu = u^2\,\lambda \tag{26}$$

and to still more relations between the T^i_j:

$$\begin{array}{ll} ac + \lambda\, ca = 0 & bc + \lambda u^2\, cb = 0 \\ gl + \lambda\, lg = 0 & dl + \lambda u^2\, ld = 0 \end{array} \tag{27}$$

$$\xi\, al - \lambda\, la = \xi\lambda\, cg - gc \tag{28}$$

$$\xi\, bl - \lambda u^2\, lb = \xi\lambda u^2\, cd - dc \tag{29}$$

The consistency of the various conditions imposed so far to the elements of matrix T has to be verified. This amounts to a comparison of the four-term relations (13-28) and (14-29) which are found compatible provided:

$$\lambda u = q\xi \tag{30}$$

and

$$dc = -u\xi\, cd \qquad -\xi\, cg = u\, gc \tag{31}$$

Finally, to eliminate all relations involving products of three elements of T, the commutation of relation (16) with d and g is performed and yields:

$$\xi = -q^{-1} \qquad q = u^2 \tag{32}$$

It must be stressed that these relations which constrain the algebra X itself cannot be avoided by changing condition (16) to a set of two independent relations:

$$cl = q\, lc, \quad ad - q\, gb = m^2 \tag{33}$$

The inverse matrix can now be completely determined and is given by :

$$T^{-1} = \begin{pmatrix} dm & -u^2\, bm & bl - u\, cd \\ -u^{-2}\, gm & am & -u^{-2}\, al + u^{-3}\, cg \\ 0 & 0 & ad - u^{-2}\, bg \end{pmatrix} \times D^{-1} \tag{34}$$

where the determinant D can be written as :

$$D \equiv (ad - u^{-2}\, bg)m \tag{35}$$

Remark that the determinant is not central and its inverse D^{-1} must be added to the set of generators of the algebra. Its commutation relations are easily deduced from those of D.

2.3. RESULT: THE HOPF ALGEBRA $\mathcal{M}$

The requirement of covariance has restricted the possible choice of relations for both quantum spaces (functions and forms) while defining precisely the corresponding algebra of quantum matrices $\mathcal{M}$. Gathering the above results, we find that the relevant algebra $\mathcal{M}$ is defined as the quotient of $\mathbb{C} < T^i_j, D^{-1} >$ by two sets of relations, namely those involving only T^i_j:

$$\begin{array}{llll} ab = u^{-2}\, ba & ac = u^{-1}\, ca & ad = da & ag = u^2\, ga \\ al = u\, la & am = ma & bc = u\, cb & bd = u^2\, db \\ bg = u^4\, gb & bl = u^3\, lb & bm = u^2\, mb & cd = u\, dc \\ cg = u^3\, gc & cm = u\, mc & dg = u^2\, gd & dl = u\, ld \\ dm = md & gl = u^{-1}\, lg & gm = u^{-2}\, mg & lm = u^{-1}\, ml \\ \multicolumn{4}{l}{cl - u^2\, lc + s(ad - u^2\, gb - m^2) = 0} \end{array} \tag{36}$$

and those involving D^{-1}:

$$\begin{array}{lll} D^{-1}a = aD^{-1} & u^{-6}\, D^{-1}b = bD^{-1} & u^{-3}\, D^{-1}c = cD^{-1} \\ D^{-1}d = dD^{-1} & D^{-1}g = u^{-6}\, gD^{-1} & D^{-1}l = u^{-3}\, lD^{-1} \\ D^{-1}m = mD^{-1} & & \end{array} \tag{37}$$

The most general $\widehat{R}$-matrix reproducing relations (36) when substituted in (11) is found to be:

$$\widehat{R} = r_1\, I - r_2 \Gamma \tag{38}$$

where r_1 and r_2 are arbitrary constants and where the matrix Γ is defined by:

$$\Gamma = \begin{pmatrix} 1 & 0 & 0 & 0 & 0 & 0 & 0 & 0 & 0 \\ 0 & 0 & 0 & u^2 & 0 & 0 & 0 & 0 & s \\ 0 & 0 & 0 & 0 & 0 & 0 & u & 0 & 0 \\ 0 & u^{-2} & 0 & 0 & 0 & 0 & 0 & 0 & -s/u^2 \\ 0 & 0 & 0 & 0 & 1 & 0 & 0 & 0 & 0 \\ 0 & 0 & 0 & 0 & 0 & 0 & 0 & 1/u & 0 \\ 0 & 0 & 1/u & 0 & 0 & 0 & 0 & 0 & 0 \\ 0 & 0 & 0 & 0 & 0 & u & 0 & 0 & 0 \\ 0 & 0 & 0 & 0 & 0 & 0 & 0 & 0 & 1 \end{pmatrix} \tag{39}$$

It has the property that it is equal to its inverse:

$$\Gamma = \Gamma^{-1} \tag{40}$$

In addition, matrix Γ can be shown to verify the Yang-Baxter equation written in the form [8]:

$$\Gamma_{12}\Gamma_{23}\Gamma_{12} = \Gamma_{23}\Gamma_{12}\Gamma_{23} \tag{41}$$

where the usual notation has been used:

$$\Gamma_{12} \equiv \Gamma \otimes I, \quad \Gamma_{23} \equiv I \otimes \Gamma \quad I = \text{unit matrix} \tag{42}$$

It may be noticed that only the Γ-part, i.e. the part of $\widehat{R}$ satisfying the Yang-Baxter equation, contributes to the establishment of relations R_{xx} and $R_{\xi\xi}$ defining the quantum spaces.

From the way it was constructed, the algebra $\mathcal{M}$ can easily be given a quantum group structure. The co-product Δ, co-unit γ and antipode S are defined by:

$$\Delta(T^i_j) \equiv \sum_{k=1}^{3} T^i_k \otimes T^k_j, \quad \Delta(D^{-1}) \equiv D^{-1} \otimes D^{-1} \tag{43}$$

$$\gamma(T, D^{-1}) \equiv (I, 1), \quad S(T) \equiv T^{-1}, \quad S(D) \equiv D^{-1} \tag{44}$$

With these definitions, it may be verified that $\mathcal{M}$ is a Hopf algebra.

3. Invariant Differential Calculus

3.1. THE GRADED ALGEBRA OF FORMS

Thus far, two spaces invariant by action of the matrix T have been obtained. The first space $C < x^i > / R_{xx}$ is defined by relations R_{xx} given by (6), (7) and (32):

$$R_{xx}: \quad \left\{ \begin{array}{rcl} x^1x^2 - u^2\, x^2x^1 - s(x^3)^2 & = & 0 \\ x^1x^3 - u\, x^3x^1 & = & 0 \\ x^2x^3 - u^{-1}\, x^3x^2 & = & 0 \end{array} \right. \tag{45}$$

The second space $\oplus_{i=1}^3 Y^i / R_{\xi\xi}$, which can be viewed as a deformation of a Grassmann algebra, is defined by relations:

$$R_{\xi\xi}: \quad \left\{ \begin{array}{rclrcl} (\xi^i)^2 & = & 0, & \xi^1\xi^2 & = & -u^2\, \xi^2\xi^1 \\ \xi^1\xi^3 & = & -u\, \xi^3\xi^1, & \xi^2\xi^3 & = & -u^{-1}\, \xi^3\xi^2 \end{array} \right. \tag{46}$$

The variables ξ^i will now be identified with the differentials of x^i. Let $\mathcal{A}$ be the associative graded algebra defined by:

$$\mathcal{A} = \oplus_{i=1}^3 \mathcal{A}_i$$

where $\mathcal{A}_0 \equiv X$ and $\mathcal{A}_k$ is the space of k-forms on X. The exterior differential operator d is defined in $\mathcal{A}$ by the properties:

- $d(x^i) = \xi^i$ (the complementary coordinates ξ are identified with the differentials of the x)
- d is linear
- $d^2 = 0$
- d satisfies the graded Leibniz rule :

$$d(fg) = (df)g + (-1)^k f(dg) \tag{47}$$

where $f, g \in \mathcal{A}$ and f is of degree k.

For a complete characterization of $\mathcal{A}$, the commutation relations between x^i and ξ^i must be established. Let us assume that they are of the form

$$x^k\xi^l = C^{kl}_{mn}\xi^m x^n \tag{48}$$

A constraint on the coefficients C is obtained by applying operator d on the left and requiring that the result written as:

$$\xi^k\xi^l = -C^{kl}_{mn}\xi^m\xi^n \tag{49}$$

be identical to the defining relations $R_{\xi\xi}$. Moreover, relations (48) must be invariant by action of T. These two constraints are sufficient to determine the coefficients C^{ij}_{kl}.

An example of the procedure is given by writing (48) for $k = 2$, $l = 3$ and using relations (46) to obtain:

$$\left\{ \begin{array}{rcl} C^{23}_{32} & = & (1 + C^{23}_{23})u^{-1} \\ C^{23}_{ij} & = & 0 \quad \text{for } (i,j) \neq \{(1,1),(2,2),(3,3)\} \end{array} \right. \tag{50}$$

The invariance by transformation T gives

$$C_{11}^{12} = C_{22}^{12} = C_{33}^{12} = C_{23}^{12} = 0 \tag{51}$$

and the following commutation relation results :

$$x^2\xi^3 = u^{-1}\xi^3 x^2 \tag{52}$$

In that manner, it is shown that the coefficients C_{mn}^{kl} are given by the elements Γ_{mn}^{kl} of the matrix Γ defined by (39). Hence the commutation relations (48) can be written explicitly as:

$$R_{x\xi}\left\{\begin{array}{lll} x^1\xi^1 = \xi^1 x^1, & x^1\xi^3 = u\xi^3 x^1, & x^1\xi^2 = u^2\xi^2 x^1 + s\xi^3 x^3 \\ x^2\xi^2 = \xi^2 x^2, & x^2\xi^3 = u^{-1}\xi^3 x^2, & x^2\xi^1 = u^{-2}\xi^1 x^2 - su^{-2}\xi^3 x^3 \\ x^3\xi^3 = \xi^3 x^3, & x^3\xi^2 = u\xi^2 x^3, & x^3\xi^1 = u^{-1}\xi^1 x^3 \end{array}\right. \tag{53}$$

The algebra $\mathcal{A}$ quotiented by the two-sided ideal generated by the relations $R_{xx}, R_{x\xi}, R_{\xi\xi}$ is now well-defined.

3.2. PARTIAL DERIVATIVES

To introduce the partial derivatives ∂_i, we set :

$$d \equiv \xi^i\, \partial_i \tag{54}$$

and require d to be invariant under the action of the quantum group defined by $T \otimes (T^{-1})^t$, where t denotes the transposition. The above definition implies:

$$\partial_l(x^k) = \delta_l^k \tag{55}$$

The next task is to find the commutation relations between the partial derivatives ∂_i and the elements x^i, ξ^i and ∂_i.

Applying the Leibniz rule on $x^k f$, $f \in \mathcal{A}$, and taking into account relations (48) with $C = \Gamma$, we obtain the commutation relations between ∂_i and x^k [1]:

$$\partial_l x^k = \delta_l^k + \Gamma_{ln}^{km} x^n \partial_m \tag{56}$$

or, explicitly:

$$R_{x\partial}\left\{\begin{array}{lll} \partial_1 x^1 = 1 + x^1\partial_1, & \partial_1 x^2 = u^{-2}x^2\partial_1, & \partial_1 x^3 = u^{-1}x^3\partial_1 \\ \partial_2 x^2 = 1 + x^2\partial_2, & \partial_2 x^3 = ux^3\partial_2, & \partial_3 x^1 = ux^1\partial_3 + sx^3\partial_2 \\ \partial_3 x^3 = 1 + x^3\partial_3, & \partial_2 x^1 = u^2 x^1\partial_2, & \partial_3 x^2 = u^{-1}x^2\partial_3 - su^{-2}x^3\partial_1 \end{array}\right. \tag{57}$$

Assuming that they are quadratic, the commutation relations between the partial derivatives ∂_k and the forms ξ^l can be written as:

$$\partial_k\xi^l = D_{kn}^{lm}\xi^n\partial_m \tag{58}$$

Multiplying this equation by x^r on the right and commuting x^r to the left we obtain a polynomial in ξ and $x\partial\xi$ whose coefficients must vanish identically. This yields two equations for D:

$$D = \Gamma^{-1}(=\Gamma) \tag{59}$$
$$D^{ir}_{mn}\Gamma^{ml}_{jk}\Gamma^{nu}_{lv} = D^{ik}_{jl}\Gamma^{ru}_{km}\Gamma^{lm}_{kv} \tag{60}$$

Thus D is determined and the second equation gives a constraint on Γ that can be rewritten using (40) and (42) as:

$$\Gamma_{12}\Gamma_{23}\Gamma_{12} = \Gamma_{23}\Gamma_{12}\Gamma_{23} \tag{61}$$

This is just the Yang-Baxter equation (41) that is indeed satisfied by matrix Γ. Thus the commutation relations between ∂ and ξ can be written explicitly as:

$$R_{\xi\partial}\quad\left\{\begin{array}{ll}\partial_1\xi^1 = \xi^1\partial_1, & \partial_2\xi^2 = \xi^2\partial_2\\ \partial_3\xi^3 = \xi^3\partial_3 & \\ \partial_1\xi^2 = u^{-2}\xi^2\partial_1, & \partial_1\xi^3 = u^{-1}\xi^3\partial_1\\ \partial_2\xi^1 = u^2\xi^1\partial_2 & \partial_2\xi^3 = u\xi^3\partial_2\\ \partial_3\xi^1 = u\xi^1\partial_3 + s\xi^3\partial_2 & \partial_3\xi^2 = u^{-1}\xi^2\partial_3 - su^{-2}\xi^3\partial_1\end{array}\right. \tag{62}$$

Still unknown are the commutation relations between the partial derivatives ∂ themselves. Since they are invariant by action of $(T^{-1})^t$, these relations can be determined by general methods [2] [8]. First the "R-matrix" corresponding to $(T^{-1})^t$ will be found. Substituting (38) and (39) in (9), we have:

$$\Gamma^{ij}_{kl}T^k_mT^l_n = T^i_kT^j_l\Gamma^{kl}_{mn} \tag{63}$$

Multiplying this relation on the right by $(T^{-1})^n_{n'}(T^{-1})^m_{m'}$ and on the left by $(T^{-1})^{j'}_j(T^{-1})^{i'}_i$ we get :

$$(T^{-1})^{j'}_j(T^{-1})^{i'}_i\Gamma^{ij}_{m'n'} = \Gamma^{i'j'}_{mn}(T^{-1})^n_{n'}(T^{-1})^m_{m'} \tag{64}$$

or, in a more compact form:

$$\widetilde{\Gamma^t}((T^{-1})^t\otimes(T^{-1})^t) = ((T^{-1})^t\otimes(T^{-1})^t)\widetilde{\Gamma^t} \tag{65}$$

where $\widetilde{\Gamma^t}$ denotes the matrix defined by $\widetilde{\Gamma^t}^{ji}_{nm} = \Gamma^{t\,ij}_{mn}$. Since the matrix $\widetilde{\Gamma^t}$ is such that:

$$(\widetilde{\Gamma^t})^2 - I = 0 \tag{66}$$

It will lead to two different quantum vector spaces invariant by $(T^{-1})^t$. The relevant relations for the partial derivatives are found to be:

$$(\widetilde{\Gamma^t} - 1)^{kl}_{ij}\partial_k\partial_l = 0 \tag{67}$$

or explicitly :

$$R_{\partial\partial} \quad \begin{cases} \partial_1\partial_2 - u^2\partial_2\partial_1 &= 0 \\ \partial_1\partial_3 - u\partial_3\partial_1 &= 0 \\ \partial_2\partial_3 - u^{-1}\partial_3\partial_2 &= 0 \end{cases} \tag{68}$$

The second invariant algebra is generated by the variables ξ_1, ξ_2, ξ_3 and the relations:

$$(\widetilde{\Gamma^t} + 1)_{ij}^{kl}\xi_k\xi_l = 0 \tag{69}$$

These variables satisfy the same quadratic relations as ξ^1, ξ^2, ξ^3 and can be identified with the covariant differential forms.

4. Realization of the quantum space.

4.1. THE WEYL-HEISENBERG ALGEBRA

In the particular case where u and s are equal to 1, relations R_{xx} as given in (45) coincide with those of the Weyl-Heisenberg algebra $\mathcal{H}$ whose realization is well-known. However, the invariance quantum group defined by relations (36) is not necessarily reduced to the original group defined by (4) and (5); It is still possible to have non-commuting elements l and c such that:

$$cl - lc + ad - gb - m^2 = 0 \tag{70}$$

while the five other elements generate the center of the corresponding quantum group.

In any case, the results of the preceding section show that the invariant differential calculus involves the element x^3 of the center in a non trivial manner. However, this variable can be ignored by setting $x^3 = 1$ and hence $\xi^3 = dx^3 = 0$. As ∂_3 does not appear anymore in the expression (54) of d, all the equations involving it have to be dropped. We then get the invariant calculus on the algebra defined by two generators x^1 and x^2.

4.2. THE DEFORMED WEYL-HEISENBERG ALGEBRA

A possible realization of the quantum space defined by relations R_{xx} given in (45) is obtained by setting:

$$x^1 = u^{\alpha N}\,\hat{a}, \quad x^2 = u^{\beta N}\,\hat{a}^\dagger, \quad x^3 = u^\gamma N \tag{71}$$

with α, β and γ arbitrary complex constants. The operators $\hat{a}$ and $\hat{a}^\dagger$ are supposed to be mutually adjoint and to verify undeformed commutation relations with N:

$$[\hat{a}, N] = \hat{a}, \quad [\hat{a}^\dagger, N] = -\,\hat{a}^\dagger \tag{72}$$

Relations (45) then imply that $\gamma = 1$ and that $\hat{a}$, $\hat{a}^\dagger$ verify:

$$\hat{a}\,\hat{a}^\dagger - u^{2-(\alpha+\beta)}\,\hat{a}^\dagger\,\hat{a} = s u^{-\beta}\,u^{(2-(\alpha+\beta))N} \tag{73}$$

The action (11) of T on x^i induces a transformation on $\hat{a}$, $\hat{a}^\dagger$ and N which must respect the property that $\hat{a}^\dagger$ is the adjoint of $\hat{a}$. This requirement puts restrictions on the values of the parameters and leads to define a star operation on the quantum group. To proceed, it is convenient to define an element λ in the quantum group by $m \equiv u^\lambda$. Two cases must be considered:

(i) $u \notin \mathbb{R}$

Then necessarily $\alpha = \beta = 1$ and $u^{-1}s$ is real (it can be chosen equal to one without loss of generality). The conjugation (denoted $\dagger$) on the quantum group is defined by:

$$\lambda^\dagger = \lambda,\; a^\dagger = (\bar{u}/u)^\lambda\, d,\; b^\dagger = \bar{u}^{-2}(\bar{u}/u)^\lambda\, g,\; c^\dagger = \bar{u}^{-1}(\bar{u}/u)^\lambda\, l \tag{74}$$

These relations are compatible with (36) and a $*$-bialgebra is obtained in this way. Thus, in this case, the quantum space is the oscillator algebra generated by the elements $\hat{a}$, $\hat{a}^\dagger$ and N but the corresponding quantum group, far from being trivialized, has acquired a new structure.

(ii) $u \in \mathbb{R}$

Now, we must have:

$$\alpha = \bar{\beta}, \quad s\,u^{-\beta} \in \mathbb{R} \tag{75}$$

The value of s will be chosen equal to u^β for convenience. The star operation on the quantum group $\mathcal{M}$ is necessarily defined by:

$$\lambda^\dagger = \lambda,\; a^\dagger = d,\; b^\dagger = u^{-(\beta+\bar{\beta})}\, g,\; c^\dagger = u^{-\beta}\, l,\; m^\dagger = m \tag{76}$$

The relations are compatible with (36) and allow to define a Hopf $\star$-algebra.

A representation of relations (72) and (73) is easily exhibited for case (ii). Assume that there exists a vector $|0>$ verifying $N\,|0>= 0$ and define a basis $|n>$ according to:

$$|n>= ([n]!)^{-1/2}\,\hat{a}^{\dagger n}\,|0> \quad \text{where } [n] \equiv n\,u^{(2-(\beta+\bar{\beta}))(n-1)},\; n \in Z \tag{77}$$

The action of the operators on this basis is found to be:

$$\begin{cases} N\,|n>= n\,|n> \\ \hat{a}^\dagger\,|n>= [n+1]^{1/2}\,|n+1> \\ \hat{a}\,|n>= [n]^{1/2}\,|n-1> \end{cases} \tag{78}$$

Remark that the algebra (73) is essentially different from the so-called quantum oscillator algebra (see [10] [11]) which would correspond to the choice $q = u^{-2}$ in relations (6). Such a choice has been eliminated in the present case by the requirement of covariance of the companion space of forms.

5. Conclusion

We have been able to deform simultaneously the Weyl-Heisenberg algebra and its group of invariance (a subgroup of $GL(3)$). In addition, an invariant differential calculus has been set up on the resulting quantum space. However, it must be stressed that the whole construction cannot be carried out for arbitrary values of the deformation parameters and that the final result depends only on one complex number u.

The construction has yielded a quantum group $\mathcal{M}$ that is original and cannot be obtained as a special case of known deformations of $GL(3)$ [7]. Moreover, if the invariant quantum space generated by the elements x^i, $i = 1, 2, 3$ satisfying relations (45) is realized in terms of canonical operators $\hat{a}$, $\hat{a}^\dagger$ and N, an extra structure may be defined on $\mathcal{M}$ and, in some cases, a Hopf $\star$-algebra is obtained.

References

1. Wess, J. and Zumino, B. (1990) Covariant Differential Calculus on the Quantum Hyperplane *Nucl. Phys. (Proc. Suppl.)*, **B 18**, pp. 301-313.
2. Zumino, G. (1992) Differential calculus on quantum spaces and quantum groups, in *Group Theoretical Methods in Physics*, M.A. del Olmo, M Santander and J.Mateos Guilarte (Eds.), Anales de Física Monografias, CIEMAT (Spain), pp. 41-59.
3. Schupp, P., Watts, P. and Zumino, B. (1992) Differential geometry on linear quantum groups *Lett. Math. Phys.* , **25**, pp. 139-147.
4. Manin, Yu. I. (1988) Quantum groups and non-commutative geometry, (Publications du CRM, Université de Montréal, Montréal).
5. Brzeziński, T., Dabrowski, H. and Rembieliński (1992) On the quantum differential calculus and the quantum holomorphicity, *J. Math. Phys.*, **33**, pp. 19-24.
6. Schirrmacher, A., Wess, J. and Zumino, B. (1991) The two-parameter deformation of GL(2), its differential calculus, and Lie algebra, *Z. Phys. C)*, **49**, pp. 317-324.
7. Schirrmacher, A. (1991) The Multiparametric Deformation of GL(n) and the Covariant Differential calculus on the quantum vector space, *Z. Phys. C*, **50**, pp 321-327.
8. Reshetikhin, N.Yu., Takhtadzhyan, L.A. and Faddeev, L.D. (1990) Quantization of Lie groups and Lie algebras, *Leningrad Math. J.*, **1**, pp. 193-225.
9. Sudbery, A. (1993) The algebra of differential forms on a full matric bialgebra, *Math. Proc. Camb. Phil. Soc.*, **114**, pp. 111-130.
10. Mac Farlane, A.J. (1989) On q-analogues of the quantum harmonic oscillator and the quantum group $SU(2)_q$, *J.Phys. A*, **22**, p. L 4581.
11. Biedenharn, L.C. (1989) The quantum group $SU_q(2)$ and a q-analogue of the boson operators *J.Phys. A*, **22**, p. L 873.

NONCOMPACT TOPOLOGICAL QUANTUM GROUPS

FRÉDÉRIC BIDEGAIN
Université de Bourgogne : Laboratoire d'Algèbre et d'Analyse : Théorie des Représentations – Physique-Mathématique BP 138 - F-21 004 DIJON Cedex - FRANCE

Abstract. A star-product construction of quantum semisimple real Lie groups is performed for the noncompact case.

To Guy Rideau, with my best wishes for his retirement.

Introduction

The usual formulation of the quantum mechanics is obtained by a correspondence principle which gives a representation of the observables by operators. However, a different approach (called *star-product* theory) was proposed about twenty years ago, using deformations theory (see for instance [1]). In this formulation, the algebra of observables is presented as a parameter family of algebras underlain by the vector space of classical observables (functions on the phase space). For a more general survey of the star-products and its applications, see D. Sternheimer's talk.

On the other hand, during the seventies, L. D. Faddeev and the Leningrad school developed the (classical and quantum) inverse scattering method ([24]) to study integrable systems. A new object appeared: the quantized algebra of $\mathbf{sl}(2)$ (*in* [20]). Built up by V. G. Drinfel'd, the *quantum group* theory is the issue of this development and is now a mathematical theory of the most prolific ones (*e. g.* see [7]). There are many points of view: the most important ones are V. G. Drinfel'd's ([9], [10], [11]), M. Jimbo's ([19]), L. D. Faddeev, N. Y. Reshetikhin and L. A. Takhtajan's ([12]), S. L. Woronowicz's ([27]).

The connection between star-product and quantum group was proposed by M. Flato and D. Sternheimer ([13]) and recently developed in order to

J. Bertrand et al. (eds.), Modern Group Theoretical Methods in Physics, 51–62.

unify the previously quoted models by P. Bonneau, M. Flato, M. Gerstenhaber and G. Pinczon ([6], [5] and [4]).

The aim of this talk is to describe a star-product approach of (real) quantum groups. The present model is founded on the duality "functions–distributions" (in the sense of topological Hopf algebras) and on the fact that the vector space of compactly-supported distributions on a real Lie group is topologically generated by the group. I give here a fast survey of some interesting results about semi-simple real Lie group and I shall succinctly give an example of *noncompact quantum semigroup*, constructed from the discrete series of $\mathbf{SL}(2,\mathbb{R})$.

For the cohomological aspects of the theory, I refer to the P. Bonneau's talk or to the articles [15], [25], [16], [14], [5]...

1. Hopf algebras of functions and distributions associated to a Lie group

Let $\mathbf{G}$ be a real Lie group countable at infinity, $\mathfrak{g}$ its complexified Lie algebra and $\mathfrak{U}$ the universal enveloping algebra of $\mathfrak{g}$.

The (complex) vector space $\mathcal{E}(\mathbf{G})$ of the infinitely differentiable (complex-valued) functions on $\mathbf{G}$, with its usual (nuclear and Fréchet) topology, is endowed with a natural topological Hopf algebra structure (*i. e.* the product, the unity, the coproduct, the counity and the antipode are continuous maps) inherited from the Lie group $\mathbf{G}$:

$$\begin{cases} f \times g : x \longmapsto f(x)g(x), \\ \mathbb{1} : x \longmapsto 1, \\ \delta f : (x,y) \longmapsto f(xy), \\ \partial_e f = f(e), \\ s(f) : x \longmapsto f(x^{-1}) \end{cases}$$

for all functions f and g in $\mathcal{E}(\mathbf{G})$ (e denotes the neutral element of $\mathbf{G}$).

The space $\mathcal{E}'(\mathbf{G})$ of compactly-supported distributions possesses a structure of topological Hopf algebra:

$$\begin{cases} \mathrm{S} * \mathrm{T} : f \longmapsto \int_{\mathbf{G}} \int_{\mathbf{G}} f(xy) d\mathrm{S}_x d\mathrm{T}_y, \\ \partial_e : f \longmapsto f(e), \\ \Delta(\mathrm{S}) : (f, g) \longmapsto \int_{\mathbf{G}} f(x) g(x) d\mathrm{S}_x, \\ \varepsilon(\mathrm{S}) = \int_{\mathbf{G}} d\mathrm{S}_x, \\ \sigma(\mathrm{S}) : f \longmapsto \int_{\mathbf{G}} f(x^{-1}) d\mathrm{S}_x \end{cases}$$

where S and T are compactly-supported distributions (the product is exactly the usual convolution product).

The topological Hopf algebra $\mathcal{E}'(\mathbf{G})$ is dual to the topological Hopf algebra $\mathcal{E}(\mathbf{G})$ in the following sense:

- the topological vector space $\mathcal{E}'(\mathbf{G})$ is the strong dual of $\mathcal{E}(\mathbf{G})$,
- the product, the unity, the coproduct, the counity and the antipode of $\mathcal{E}'(\mathbf{G})$ are transpositions respectively of the coproduct, the counity, the product, the unity and the antipode of $\mathcal{E}(\mathbf{G})$.

Conversely, $\mathcal{E}'(\mathbf{G})$ is the dual topological Hopf algebra of $\mathcal{E}(\mathbf{G})$. The duality is given by:

$$\begin{cases} < f \mid \mathrm{S} * \mathrm{T} > &= & < \delta(f) \mid \mathrm{S} \otimes \mathrm{T} >, \\ < f \mid \partial_e > &= & \partial_e f, \\ < f \otimes g \mid \Delta(\mathrm{S}) > &= & < f \times g \mid \mathrm{S} >, \\ \varepsilon(\mathrm{S}) &= & < \mathbb{1} \mid \mathrm{S} >, \\ < f \mid \sigma(\mathrm{S}) > &= & < s(f) \mid \mathrm{S} > \end{cases}$$

for all functions f and g in $\mathcal{E}(\mathbf{G})$ and all distributions S and T in $\mathcal{E}'(\mathbf{G})$.

The choice of the spaces $\mathcal{E}(\mathbf{G})$ and $\mathcal{E}'(\mathbf{G})$ is supported by the following remarks:

- Any spaces of functions (even classical) are not topological Hopf algebras (for the natural structure): for instance, the coproduct of a compactly-supported (infinitely differentiable) function is not necessarily compactly-supported! By transposition, this means that the convolution product of two distributions is not necessarily defined.
- let ∂_x be the Dirac measure on x ($\partial_x(f) = f(x)$): the Lie group $\mathbf{G}$ is topologically embedded in $\mathcal{E}'(\mathbf{G})$ by the map $\partial : \mathbf{G} \longrightarrow \partial_{\mathbf{G}} : x \longmapsto \partial_x$

which is a morphism of semigroups:

$$\partial_x * \partial_y = \partial_{xy},$$

for each x and y in $\mathbf{G}$. More precisely, the group $\mathbf{G}$ is isomorphic to the topological group of the group-like elements of $\mathcal{E}'(\mathbf{G})$

$$\mathbf{G} \simeq \{\mathrm{S} \in \mathcal{E}'(\mathbf{G}) \mid \Delta(\mathrm{S}) = \mathrm{S} \otimes \mathrm{S} \text{ and } \varepsilon(\mathrm{S}) = 1\}$$

and spanned a dense subspace of $\mathcal{E}'(\mathbf{G})$.

Let $\mathcal{E}(\mathbf{G})[[\mathrm{h}]]$ be the space of formal series with coefficients in $\mathcal{E}(\mathbf{G})$, endowed with the product topology:

$$\mathcal{E}(\mathbf{G})[[\mathrm{h}]] \simeq \prod_k \mathcal{E}(\mathbf{G})\mathrm{h}^k \simeq \mathcal{E}(\mathbf{G})\hat{\otimes}\mathbb{C}[[\mathrm{h}]].$$

The complex topological Hopf algebra structure of $\mathcal{E}(\mathbf{G})$ obviously extend to $\mathcal{E}(\mathbf{G})[[\mathrm{h}]]$ as a topological Hopf algebra structure on the (topological) ring $\mathbb{C}[[\mathrm{h}]]$ but it is not the only one. Each topological Hopf algebra structure on the $\mathbb{C}[[\mathrm{h}]]$-module $\mathcal{E}(\mathbf{G})[[\mathrm{h}]]$ verifying the following isomorphism of Hopf topological spaces:

$$\mathcal{E}(\mathbf{G})[[\mathrm{h}]] / \mathrm{h}\mathcal{E}(\mathbf{G})[[\mathrm{h}]] \simeq \mathcal{E}(\mathbf{G}).$$

is called a formal deformation of $\mathcal{E}(\mathbf{G})$. In other terms, a formal deformation of $\mathcal{E}(\mathbf{G})[[\mathrm{h}]]$ is given by a product, an unity, a coproduct, a counity and an antipode, all $\mathbb{C}[[\mathrm{h}]]$-linear and continuous, in the following form:

$$\begin{cases} f \times_{\mathrm{h}} g &= f \times g + \sum_{k \geqslant 1} p_{\mathrm{h}}^k(f, g)\mathrm{h}^k, \\ \mathbb{1}_{\mathrm{h}} &= \mathbb{1} + \sum_{k \geqslant 1} \mathbb{1}_{\mathrm{h}}^k \mathrm{h}^k, \\ \delta_{\mathrm{h}}(f) &= \delta(f) + \sum_{k \geqslant 1} \delta_{\mathrm{h}}^k(f)\mathrm{h}^k, \\ \partial_{\mathrm{h}}(f) &= \partial_e(f) + \sum_{k \geqslant 1} \partial_{\mathrm{h}}^k(f)\mathrm{h}^k, \\ s_{\mathrm{h}}(f) &= s(f) + \sum_{k \geqslant 1} s_{\mathrm{h}}^k(f)\mathrm{h}^k. \end{cases}$$

By transposition, a formal deformation of $\mathcal{E}(\mathbf{G})$ induces a formal deformation of $\mathcal{E}'(\mathbf{G})$:

$$\left\{\begin{array}{rcl} < f \mid \mathrm{S} *_{\mathrm{h}} \mathrm{T} > & = & < \delta_{\mathrm{h}}(f) \mid \mathrm{S} \otimes \mathrm{T} >, \\ < f \mid \partial_{\mathrm{h}} > & = & \partial_{\mathrm{h}} f, \\ < f \otimes g \mid \Delta_{\mathrm{h}}(\mathrm{S}) > & = & < f \times_{\mathrm{h}} g \mid \mathrm{S} >, \\ \varepsilon_{\mathrm{h}}(\mathrm{S}) & = & < \mathbb{1}_{\mathrm{h}} \mid \mathrm{S} >, \\ < f \mid \sigma_{\mathrm{h}}(\mathrm{S}) > & = & < s_{\mathrm{h}}(f) \mid \mathrm{S} > . \end{array}\right.$$

Conversely, each formal deformation of $\mathcal{E}'(\mathbf{G})$ gives an formal deformation of $\mathcal{E}(\mathbf{G})$.

More generally, any (smooth) action of $\mathbf{G}$ on a real manifold $\mathbf{M}$:

$$\begin{array}{rcl} \boldsymbol{.} : \mathbf{G} \times \mathbf{M} & \longrightarrow & \mathbf{M} \\ (x, \mathbf{m}) & \longmapsto & x \boldsymbol{.} \mathbf{m} \end{array}$$

extends in a single way to a continuous linear action of $\mathcal{E}'(\mathbf{G})$ on the (topological) coalgebra $\mathcal{E}'(\mathbf{M})$ of the compactly-supported distributions on $\mathbf{M}$:

$$\begin{array}{rl} \boldsymbol{.} : \mathcal{E}'(\mathbf{G}) \times \mathcal{E}'(\mathbf{M}) \longrightarrow & \mathcal{E}'(\mathbf{M}) \\ (\mathrm{S}, \Sigma) \longmapsto & \left(\mathrm{S} \boldsymbol{.} \Sigma : \varphi \longmapsto \int_{\mathbf{G}} \int_{\mathbf{M}} \varphi(x \boldsymbol{.} \mathbf{m}) d\mathrm{S}_x d\Sigma_{\mathbf{m}}\right). \end{array}$$

Similarly, any representation ρ of $\mathbf{G}$ on a locally convex Hausdorff vector space $\mathbf{V}$ induces a representation of $\mathcal{E}'(\mathbf{G})$ on the space of differentiable vectors of $\mathbf{V}$ (endowed with its natural topology) by:

$$\rho(\mathrm{S})\mathbf{v} = \int_{\mathbf{G}} \rho(x) \mathbf{v} d\mathrm{S}_x.$$

The (left and right) regular representation and the coadjoint one on the space $\mathcal{E}(\mathbf{G})$ are typical examples of such representations.

As we just have seen, the space $\mathcal{E}'(\mathbf{G})$ is closely connected with the group $\mathbf{G}$. Roughly speaking, $\mathbf{G}$ is used for $\mathcal{E}'(\mathbf{G})$: "group deformation" of $\mathbf{G}$ means Hopf deformation of $\mathcal{E}'(\mathbf{G})$ or, dually, Hopf deformation of $\mathcal{E}(\mathbf{G})$ (*idem* for action, representation, *etc.*).

2. Hopf deformations of $\mathfrak{U}$ and star-product

Let us say that a Hopf formal deformation of $\mathfrak{U}$ is *twisted* if the product, the unity and the counity are the initial ones and if there exists a *twist*, *i. e.* an invertible element P of $(\mathfrak{U} \otimes \mathfrak{U})[[\mathrm{h}]]$ such that:

$$\left\{\begin{array}{c} \Delta_{\mathrm{h}} = P \Delta_0 P^{-1}; \\ P = 1 \otimes 1 + P_1 \mathrm{h} + O(\mathrm{h}^2) \text{ with } P_1 \in \mathfrak{g} \otimes \mathfrak{g}. \end{array}\right.$$

From quantum group theory, the following proposition is very important (See [11] and [5]):

Theorem (Drinfel'd): *Each Hopf formal deformation of the universal enveloping algebra $\mathfrak{U}$ of a semi-simple algebra is formally equivalent to a twisted deformation.*

Henceforth, the deformations of $\mathfrak{U}$ are considered twisted.

The main result about the present model is:

Theorem: *Each twisted Hopf formal deformation of the universal enveloping algebra $\mathfrak{U}$ of a semi-simple Lie algebra $\mathfrak{g}$ extends to a Hopf formal deformation of the space $\mathcal{E}'(\mathbf{G})$ with the initial convolution product, for all connected semi-simple real Lie group $\mathbf{G}$ with complexified Lie algebra $\mathfrak{g}$.*

and its

Dual version: *Each twisted Hopf formal deformation of the universal enveloping algebra $\mathfrak{U}$ of a semi-simple Lie algebra $\mathfrak{g}$ induces a preferred (ı. e. with initial coproduct and counity) Hopf formal deformation of the space $\mathcal{E}(\mathbf{G})$, for all connected semi-simple real Lie group $\mathbf{G}$ with complexified Lie algebra $\mathfrak{g}$, such that the map*

$$\begin{array}{ccc} \mathcal{E}(\mathbf{G})\widehat{\otimes}\mathcal{E}(\mathbf{G}) & \longrightarrow & \mathcal{E}(\mathbf{G}) \\ f\otimes g & \longmapsto & \left.\dfrac{f\times_{\mathrm{h}} g - g\times_{\mathrm{h}} f}{\mathrm{h}}\right|_{\mathrm{h}=0} \end{array}$$

is a Poisson bracket and a first-order bidifferential operator.

The proof of those results is in [5] for compact semi-simple group and in [3] for general semi-simple group. It is based on the construction of a pair of dual topological Hopf algebras $(\mathcal{H}(\mathbf{G}), \mathcal{A}(\mathbf{G}))$ from finite-dimensional representations.

Let (ρ, V_ρ) be a finite-dimensional linear representation of $\mathbf{G}$ and M a linear map of V_ρ. The coefficient $c_\rho^M : \mathbf{G} \longrightarrow \mathbb{C}$ is defined as the map $x \longmapsto \mathrm{tr}(M\rho(x))$. Let Π be a complete set of irreducible finite-dimensional representations, two by two non equivalent, and let assume that Π contains the non-zero one-dimensional representation and is globally stable by contraposition. The coefficients of Π's representations verify:

$$\begin{cases} c_\rho^M \times c_{\rho'}^{M'} &= c_{\rho\otimes\rho'}^{M\otimes M'}, \\ \mathbb{1} &= c_{\rho_0}^{(1)}, \\ \delta(c_\rho^{e_i^*\otimes e_j}) &= \sum_k c_\rho^{e_i^*\otimes e_k} \otimes c_\rho^{e_k^*\otimes e_j}, \\ \varepsilon(c_\rho^M) &= c_\rho^M(e), \\ s(c_\rho^M) &= c_{\check{\rho}}^{\,{}^tM}, \end{cases}$$

where ρ et ρ' belong to Π, M and M' are matrices respectively on V_ρ and $V_{\rho'}$, $\{e_i\}_i$ is a basis of V_ρ, $\{e_i^*\}_i$ its dual basis, where ${}^tM$ is the transposed matrice of M and $\check{\rho}$ the contragredient representation of ρ.

Let us denote $\mathcal{H}(\mathbf{G})$ the vector space generated by the coefficients of the finite-dimensional representations, endowed with the finest locally convex topology: $(\mathcal{H}(\mathbf{G}), \times, \mathbb{1}, \delta, \varepsilon, s)$ is a topological Hopf algebra. About the algebraic structure of this space, we have:

Proposition 1: *The (topological) Hopf algebra $\mathcal{H}(\mathbf{G})$ is finitely generated.*

The space $\mathcal{H}(\mathbf{G})$ is a very useful tool in the compact case (see [5] and [6]) and remains important in a larger framework, owing by

Proposition 2: *The space $\mathcal{H}(\mathbf{G})$ is dense in $\mathcal{E}'(\mathbf{G})$ if and only if the Lie group $\mathbf{G}$ has a faithful finite-dimensional representation (or equivalently if $\mathbf{G}$ is linear* i. e. *if $\mathbf{G}$ is isomorphic to a closed subgroup of a group $\mathbf{GL}(n)$ - see* [26]).

Let $\mathcal{A}(\mathbf{G})$ be the dual topological Hopf algebra of $\mathcal{H}(\mathrm{G})$. Its elements, called here *generalized distributions*, are series of compactly-supported distributions non necessarily convergent in $\mathcal{E}'(\mathbf{G})$. If $\mathbf{G}$ is linear, the universal enveloping algebra $\mathfrak{U}$ is dense in $\mathcal{A}(\mathbf{G})$. This property is very important and often used. Unfortunately, for a non-linear group, other methods of demonstration are required. For instance, in the proof of the before stated theorem, a point of the most delicate ones is to prove coassociativity of the $\mathfrak{U}$'s coproduct extension to $\mathcal{E}'(\mathbf{G})$. By the following inclusion sequence:

$$\mathfrak{U} \subset \mathcal{E}'(\mathbf{G}) \subset \mathcal{A}(\mathbf{G}),$$

using the coassociativity of the deformed coproduct on $\mathfrak{U}$ and density of $\mathfrak{U}$ in $\mathcal{A}(\mathbf{G})$ if $\mathbf{G}$ is linear, conclusion is direct. If $\mathbf{G}$ is not linear, another proof is necessary: for instance, it is possible to prove that the coassociativity of the deformed coproduct on $\mathfrak{U}$ first conveys on the group $\mathbf{G}$ (identified with the group-like elements set), then, by density of the subspace spanned by $\mathbf{G}$, on $\mathcal{E}'(\mathbf{G})$ (for details, see [3]).

To cut the matter short, here are some remarks about the three various cases:

- if the Lie group $\mathbf{G}$ is compact, the coefficients of any self-dual faithful finite-dimension representation generate the space $\mathcal{H}(\mathbf{G})$,
- if $\mathbf{G}$ is linear but noncompact, the coefficients of any self-dual faithful finite-dimensional representation generate a subspace of $\mathcal{H}(\mathbf{G})$ *dense in* $\mathcal{E}'(\mathbf{G})$ (It is a generalization of Stone-Weierstraß theorem!) but there exists such a representation which generates the whole space $\mathcal{H}(\mathbf{G})$,
- if the Lie group is not linear (hence noncompact), there is no faithful finite-dimensional representation.

3. A candidate for a noncompact quantum group

The previous construction ($\mathcal{A}$ and $\mathcal{H}$) can be generalized to other families of representation if the tensor product decomposition is known. For instance, for some representations of $\mathbf{SL}(2,\mathbb{R})$'s discrete series, one has (notations are Miller's ones except $\uparrow_0$ which, here, means $D(0)$ – see [22]):

$$\uparrow_{-\frac{m}{2}} \otimes \uparrow_{-\frac{n}{2}} \simeq \bigoplus_{k=0}^{\infty} \uparrow_{-\frac{m+n+2k}{2}}$$

with m, n positive integers.

Let us fix some notations: for each topological vector space E, E^* denote the strong dual of E and $\mathcal{L}_b(E)$ its endomorphism space endowed with the bounded convergence topology. Let V_m^∞ ($m \in \mathbb{N}$) be a closed subspace of $\mathcal{E}(\mathbf{SL}(2,\mathbb{R}))$ such that the restriction of the left regular representation to V_m^∞ is $\uparrow_{-\frac{m}{2}}$-like. The subspace V_m of $\mathbf{K}$-finite vectors, endowed with the finest locally convex topology, is a $\mathfrak{U}$-module. Now, using the notations

$$\mathcal{A}_{\uparrow m} = \mathcal{L}_b(V_m), \quad \mathcal{H}_{\uparrow m} = \mathcal{L}_b(V_m^*),$$

$$\mathcal{A}_{\uparrow} = \prod_{m=0}^{\infty} \mathcal{A}_{\uparrow m}, \quad \mathcal{H}_{\uparrow} = \bigoplus_{m=0}^{\infty} \mathcal{H}_{\uparrow m}$$

(the coefficients of the representation V_m generate the space $\mathcal{H}_{\uparrow m}$) with the following isomorphisms

$$\mathcal{A}_{\uparrow m} \simeq \mathcal{H}^*_{\uparrow m}, \qquad \mathcal{H}_{\uparrow m} \simeq \mathcal{A}^*_{\uparrow m},$$

$$\mathcal{A}_{\uparrow} \simeq \mathcal{H}^*_{\uparrow}, \qquad \mathcal{H}_{\uparrow} \simeq \mathcal{A}^*_{\uparrow},$$

$$(\mathcal{A}_{\uparrow}\widehat{\otimes}\mathcal{A}_{\uparrow})^* \simeq \mathcal{H}_{\uparrow}\overline{\otimes}\mathcal{H}_{\uparrow}, \quad (\mathcal{H}_{\uparrow}\widehat{\otimes}\mathcal{H}_{\uparrow})^* \simeq \mathcal{A}_{\uparrow}\overline{\otimes}\mathcal{A}_{\uparrow},$$

$$(\mathcal{A}_{\uparrow}\overline{\otimes}\mathcal{A}_{\uparrow})^* \simeq \mathcal{H}_{\uparrow}\widehat{\otimes}\mathcal{H}_{\uparrow}, \quad (\mathcal{H}_{\uparrow}\overline{\otimes}\mathcal{H}_{\uparrow})^* \simeq \mathcal{A}_{\uparrow}\widehat{\otimes}\mathcal{A}_{\uparrow}$$

($\overline{\otimes}$ and $\widehat{\otimes}$ are respectively the inductive and the projective topological tensor product – see [17]). Then:

Proposition 3 : *The enveloping algebra $\mathfrak{U}$ is a dense subspace of $\mathcal{A}_{\uparrow}$ and its bialgebra structure extends into a single topological bialgebra structure on $\mathcal{A}_{\uparrow}$ (the separately continuous product is the composition of endomorphisms). By transposition, $\mathcal{H}_{\uparrow}$ is a topological bialgebra: this algebraic structure is identical to the one induced by $\mathcal{E}(\mathbf{SL}(2,\mathbb{R}))$.*

The contragredient of a $\uparrow_{-\frac{m}{2}}$-like representation is not $\uparrow_{-\frac{n}{2}}$-like: this explains the lack of antipode. There exists some elements of $\mathcal{H}_{\uparrow}$ which are not elements of $\mathcal{E}(\mathbf{SL}(2,\mathbb{R}))$: actually, the algebraic structure of $\mathcal{E}(\mathbf{SL}(2,\mathbb{R}))$ is induced only on the (non topological) space $\widetilde{\mathcal{H}}$ spanned by the coefficients, then this structure extends to $\mathcal{H}_{\uparrow}$ by density of $\widetilde{\mathcal{H}}$. In fact, the coproduct of a discrete-series coefficient is not in $\widetilde{\mathcal{H}} \otimes \widetilde{\mathcal{H}}$, then a non trivial topology has to be introduced: with the topology induced either by $\mathcal{H}_{\uparrow}$ or by $\mathcal{E}(\mathbf{SL}(2,\mathbb{R}))$, the coproduct maps a coefficient in $\widetilde{\mathcal{H}}\widehat{\otimes}\widetilde{\mathcal{H}}$.

From deformations like Curtright Zachos ones ([8] – the deformation used in [2] and in [6] is a particular case of those ones), a family of topological bialgebras is obtained:

Proposition 3 (deformed version) : *If q is not a root of unity, the q-deformed enveloping algebra $\mathfrak{U}_q$ is a dense subspace of $\mathcal{A}_{\uparrow}$ and its bialgebra structure extends to $\mathcal{A}_{\uparrow}$ on a single topological bialgebra structure (the separately continuous product is always the endomorphisms composition).*

There exists a formal version of this proposition: the e^{h}-deformed enveloping algebra is dense in $\mathcal{A}_{\uparrow}[[\mathrm{h}]]$ and its bialgebra structure extends to $\mathcal{A}_{\uparrow}[[\mathrm{h}]]$ on a single topological bialgebra structure (all the structures are defined on $\mathbb{C}[[\mathrm{h}]]$).

The structure of the coefficients algebra is very interesting: if we denote $\mathcal{H}^0_{\uparrow} = \mathcal{H}_{\uparrow 0}$, $\mathcal{H}^1_{\uparrow} = \mathcal{H}_{\uparrow 1}$, $\mathcal{H}^2_{\uparrow} = \bigoplus_{k\geqslant 0} \mathcal{H}_{\uparrow 2k+2}$ and $\mathcal{H}^3_{\uparrow} = \bigoplus_{k\geqslant 0} \mathcal{H}_{\uparrow 2k+3}$, we have:

Proposition 4 : *Let $\{c^i_1\}_i$ be a basis of $\mathcal{H}^1_{\uparrow}$. The family*

$$\{\mathbb{1}, c^i_1, c^i_1 \times_{\mathrm{h}} c^j_1, c^i_1 \times_{\mathrm{h}} c^j_1 \times_{\mathrm{h}} c^k_1\}_{i\leqslant j\leqslant k}$$

is a basis of $\mathcal{H}_\uparrow$. *Moreover,*

$$\begin{cases} \mathcal{H}_\uparrow^1 \times_{\mathrm{h}} \mathcal{H}_\uparrow^1 = \mathcal{H}_\uparrow^2, \\ \mathcal{H}_\uparrow^i \times_{\mathrm{h}} \mathcal{H}_\uparrow^i \subset \mathcal{H}_\uparrow^2, \ i = 2, 3, \\ \mathcal{H}_\uparrow^0 \times_{\mathrm{h}} \mathcal{H}_\uparrow^i = \mathcal{H}_\uparrow^i \times_{\mathrm{h}} \mathcal{H}_\uparrow^0 = \mathcal{H}_\uparrow^i, \ i = 0, 1, 2, 3, \\ \mathcal{H}_\uparrow^1 \times_{\mathrm{h}} \mathcal{H}_\uparrow^2 = \mathcal{H}_\uparrow^2 \times_{\mathrm{h}} \mathcal{H}_\uparrow^1 = \mathcal{H}_\uparrow^3, \\ \mathcal{H}_\uparrow^1 \times_{\mathrm{h}} \mathcal{H}_\uparrow^3 = \mathcal{H}_\uparrow^3 \times_{\mathrm{h}} \mathcal{H}_\uparrow^1 \subset \mathcal{H}_\uparrow^2, \\ \mathcal{H}_\uparrow^2 \times_{\mathrm{h}} \mathcal{H}_\uparrow^3 = \mathcal{H}_\uparrow^3 \times_{\mathrm{h}} \mathcal{H}_\uparrow^2 \subset \mathcal{H}_\uparrow^3. \end{cases}$$

In other terms, the (deformed) algebra $\mathcal{H}_\uparrow$ is isomorphic to a polynomial space with an infinity of indeterminates $\{c_1^i\}$ quotiented by an infinity of relations such that each element can be written as a polynomial with degree 3.

In fact, it is possible to write *explicitly* all those relations since the basis changes are given. To have an example of this explicit formulas, see [2] where the chosen representations are the **SU**(1,1)'s left regular representation restricted to the space generated by the functions $\varphi_m^k : x \longmapsto \dfrac{\beta(x)^k}{\alpha(x)^{m+k}}$, α and β being the coefficients of the classical two-dimensional representation of **SU**(1,1):

$$x = \begin{pmatrix} \alpha(x) & \overline{\beta(x)} \\ \beta(x) & \overline{\alpha(x)} \end{pmatrix}, \ |\alpha(x)|^2 - |\beta(x)|^2 = 1.$$

The above construction allows to write explicitly the (deformed or not) product of the functions $x \longmapsto \dfrac{\beta(x)^k \overline{\beta(x)}^l}{\overline{\alpha(x)}^{m+k+l}}$ $(k, l, m$ positive integers).

References

1. Bayen (F.), Flato (M.), Fronsdal (C.), Lichnerowicz (A.) and Sternheimer (D.) : *Quantum mechanics as a deformation of classical mechanics*, Letters in mathematical Physics, vol. 1 (1977) pp. 521-530; *Deformation theory and quantization* I & II, Annals of Physics, vol. 111 (1978) pp. 61-110, 111-151.
2. Bidegain (F.) : *A candidate to noncompact topological quantum groups*, to appear in Letters in Mathematical Physics.
3. Bidegain (F.) and Pinczon (G.) : *A star-product approach to noncompact quantum groups*, Letters in Mathematical Physics, vol. 33 (1995) pp. 231-240.

4. Bonneau (P.) : *Cohomology and associated deformations for not necessarily co-associative bialgebras*, Letters in Mathematical Physics, vol. 26 (1992) pp. 277-283; *Groupes quantiques: déformations et cohomologies*, thesis, Dijon, 1993.

5. Bonneau (P.), Flato (M.), Gerstenhaber (M.) and Pinczon (G.) : *The hidden group structure of quantum groups: strong duality, rigidity and preferred deformations*, Communications in Mathematical Physics, vol. 161 (1994) pp. 125-156.

6. Bonneau (P.), Flato (M.) and Pinczon (G.) : *A natural and rigid model of quantum groups*, Letters in Mathematical Physics, vol. 25 (1992) pp. 75-84.

7. Cartier (P.) : *Développements récents sur les groupes de tresses. Applications à la topologie et à l'algèbre*, Séminaire Bourbaki (1989-90) n°716.

8. Curtright (T. L.) and Zachos (C. K.) :*Deforming maps for quantum algebras*, Physics Letters B, vol. 243 n°3 (1990) pp. 237-244; Curtright (T. L.), Ghandour (G. I.) and Zachos (C. K.) :*Quantum algebras deforming maps, Clebsch-Gordan coefficients, coproducts, R and U matrices*, Journal of Mathematical Physics, vol. 32 n°3 (1991) pp. 676-688.

9. Drinfel'd (V. G.) : *Hamiltonian structures on Lie groups, Lie bialgebras and the geometric meaning of the classical Yang-Baxter equation*, Soviet Mathematics - Doklady, vol. 27 n°1 (1983) pp. 68-71; *A new realization of Yangians and quantized affine algebras*, *id.* vol. 36 n°2 (1988) pp.212-216.

10. Drinfel'd (V. G.) : *Quantum groups*, Proceedings of the International Congress of Mathematicians, Berkeley 1986 (1987) pp. 798-820.

11. Drinfel'd (V. G.) : *On almost cocommutative Hopf algebras*, Leningrad Mathematics vol. 1 n°2 (1990) pp. 321-342; *Quasi-Hopf algebras*, *id.* n°6 (1990) pp. 1419-1457.

12. Faddeev (L. D.), Reshetikhin (N. Y.) and Takhtajan(L. A.) : *Quantization of Lie groups and Lie algebras*, Leningrad Mathematical Journal vol. 1 (1990) pp.193.

13. Flato (M.) and Sternheimer (D.) : *On a possible origin of quantum groups*, Letters in Mathematical Physics, vol. 22 (1992) pp. 155-160; see also Flato (M.) and Lu (Z.) : *Remarks on quantum groups*, *id.*, vol. 21 (1991) pp. 85-88.

14. Flato (M.) and Sternheimer (D.) : *Quantum groups, star products and cyclic cohomologies*; *Topological quantum groups, star-products and their relations*, conférences, 1994.

15. Gerstenhaber (M.) : *The cohomology structure of an associative ring*, Annals of Mathematics vol. 78 (1963) pp. 267-288; *On the deformations of rings and algebras*, *id.* vol. 79 (1964) pp. 59-103.

16. Gerstenhaber (M.) and Schack (S. D) : *Algebras, bialgebras, quantum groups and algebraic deformations*, Contemporary Mathematics vol. 134 (1992) pp. 51-.

17. Grothendieck (A.) : *Produits tensoriels topologiques et espaces nucléaires*, Memoirs of the American Mathematical Society, n°16, American Mathematical Society, 1955.

18. Harish-Chandra : *Lie algebras and the Tannaka duality theorem*, Annals of Mathematics, vol. 51 (1950) pp. 299-.

19. Jimbo (M.) : *A q-difference analogue of* $\mathrm{U}(\mathfrak{g})$ *and the Yang-Baxter equation*, Letters in Mathematical Physics, vol. 10 (1985) pp. 63-69; *A q-analogue of* $\mathrm{U}(\mathfrak{gl}(N+1))$, *Hecke algebra, and the Yang-Baxter equation*, *id.*, vol. 11 (1986) pp. 247-252; *Quantum R-matrix for the generalized Toda system*, Communications in Mathematical Physics, vol. 102 (1986) pp. 537-547.

20. Kulish (P. P.) and Reshetikhin (N. Yu) : *Quantum linear problem for the sine-Gordon equation and higher representation* (trad. du russe), Journal of Soviet Mathematics, vol. 23 (1983) pp. 2435.

21. Kulish (P. P.) and Sklyanin (E. K.) : *Quantum version of the method of inverse scattering problem* (trad. du russe), Journal of Soviet Mathematics, vol. 19 n°5 (1982) pp.1596-1620.

22. Miller, jr (W.) : *Lie theory and special functions*, Academic Press, 1968.

23. Milnor (J. W.) and Moore (J. C.) : *On the structure of Hopf algebra*, Annals of Mathematics, vol. 85 (1965) pp. 211-264.

24. Sklyanin (E. K.) : *Quantum version of the method of inverse scattering problem* (trad. du russe), Journal of Soviet Mathematics, vol. 19 n°5 (1982) pp.1546-1596.

25. Taylor (J. L.) : *Homology and cohomology for topological algebras*, Advances in Mathematics, vol. 9 (1972) pp.137-182;*A general framework for a multi-operator functional calculus*, *id.* pp.183-252.

26. Warner (G.) : *Harmonic analysis on semi-simple Lie groups I*, Springer-Verlag, 1972.

27. Woronowicz (S. L.) : *Compact matrix pseudogroups*, Communications in Mathematical Physics vol. 111 (1987) pp. 613-665; *Differential calculus on compact matrix pseudogroups*, *id.* vol. 122 (1989) pp. 125-170.

SYMMETRIC SYMPLECTIC MANIFOLDS AND DEFORMATION QUANTIZATION

P.BIELIAVSKY AND M.CAHEN
Département de Mathématiques,
Université Libre de Bruxelles, Campus Plaine, CP 218, 1050 Brussels, Belgium

AND

S.GUTT
Département de Mathématiques,
Université Libre de Bruxelles, Campus Plaine, CP 218, 1050 Brussels, Belgium
et Département de Mathématiques,
Université de Metz, Ile du Saulcy, F-57045 Metz Cedex 01, France

Abstract. In this paper, which we are happy to dedicate to our friend Guy Rideau, we introduce the notion of symplectic symmetric spaces.

1. Introduction

Symmetric symplectic manifolds (briefly denoted hereafter S.S.M.) are symplectic manifolds which admit a large family of "symmetries". This basic assumption implies the existence of a preferred symplectic connection for which the "symmetries" are affine transformations. Endowed with this connection they become affine symmetric manifolds and hence can be studied by purely algebraic methods.

The aim of this paper is threefold:

(i) To derive a formula for the preferred symplectic connection on a S.S.M. and to infer from there that they are affine symmetric manifolds.

J. Bertrand et al. (eds.), Modern Group Theoretical Methods in Physics, 63–73.

(ii) To indicate how the study of these manifolds can be reduced to an algebraic problem. In particular we indicate the structure of the S.S.M.'s the holonomy of which is completely reducible.

(iii) To recall the construction of a $\star$ product (deformation quantization) on any S.S.M. This $\star$ product is build recursively in a totally explicit way and the cochains defining this $\star$ product are universal polynomials in the curvature of the preferred connection.

We hope that this will convince the reader that S.S.M.'s form a nice class of symplectic manifolds where $\star$ quantization may be studied in detail concerning convergence properties and spectrum of preferred observables.

2. Definitions; preferred connection

Definition 1. A connected symplectic manifold (M,ω) is said to be a S.S.M. if and only if:

(i) For any x in M, there exists a symplectic diffeomorphism s_x of (M,ω), which is involutive ($s_x^2 = id_{|M}$) and which admits x as isolated fixed point. The symplectic diffeomorphism s_x is called the symmetry at x.

(ii) For any x, y in M one has:

$$s_x s_y s_x = s_{s_x y}$$

(iii) The map $M \times M \to M : (x,y) \to s_x y$ is a smooth map. We shall denote such a space by (M,ω,s).

Definition 2. Let (M,ω,s) be a S.S.M. The automorphism group of (M,ω,s), denoted $Aut(M,\omega,s)$, is the subgroup of the group of symplectic diffeomorphisms composed of those symplectic diffeomorphisms φ such that

$$\varphi s_x \varphi^{-1} = s_{\varphi(x)}$$

for any x in M.

Definition 3. Let (M,ω,s) be a S.S.M. The transvection group of (M,ω,s), denoted $G(M,\omega,s)$, is the group generated by products of an even number of symmetries.
Definition 2 and 3 imply that $G(M,\omega,s)$ is a normal subgroup of $Aut(M,\omega,s)$.

Definition 4. An affine connection ∇ on a S.S.M. (M,ω,s) is said to be preferred if:

(i) The symplectic form ω is covariantly constant (i.e. $\nabla\omega = 0$).

(ii) For any x in M, the symmetry at x, s_x, is contained in the affine group of (M,∇).

Theorem 1. *Let (M, ω, s) be a S.S.M. Then there exists at most one preferred affine connection ∇. Furthermore if X, Y, Z are smooth vector fields on M and if x is in M, the formula:*

$$\omega_x(\nabla_X Y, Z) = \frac{1}{2} X_x \omega(Y + s_{x*}Y, Z) \tag{1}$$

defines a preferred affine connection. This connection has the following properties :

(i) It is torsion free (i.e. $T^\nabla = 0$) and hence is a symplectic connection.

(ii) (M, ∇) is an affine symmetric space.

(iii) The automorphism group of (M, ω, s) is the intersection of the affine group of (M, ∇) with the group of symplectic diffeomorphisms of (M, ω).

This implies in particular that the automorphism group $Aut(M, \omega, s)$ is a Lie transformation group of M acting transitively on M.

It is worthwhile observing that formula (1) is valid, "mutatis mutandi" in pseudo riemannian symmetric spaces.

Let us, from now on, choose a base point o, of the S.S.M. (M, ω, s) and consider the involutive automorphism $\tilde{\sigma}$ of $Aut(M, \omega, s)$ obtained by conjugation by s_o (i.e. $\tilde{\sigma}(\varphi) = s_o \varphi s_o$ for any φ in $Aut(M, \omega, s)$.)

Proposition 1. *The transvection group $G(M, \omega, s)$ is the smallest subgroup of $Aut(M, \omega, s)$ acting transitively on M and stable by $\tilde{\sigma}$; it is a connected Lie subgroup of $Aut(M, \omega, s)$.*

Definition 5. Let (M_i, ω_i, s^i) $(i = 1, 2)$ be S.S.M. They are said to be isomorphic if there exists a symplectic diffeomorphism $\varphi : M_1 \to M_2$ such that for any x_1 in M_1,

$$\varphi \circ s^1_{x_1} = s^2_{\varphi(x_1)} \circ \varphi$$

Definition 6. A triple $(\mathcal{G}, \sigma, \boldsymbol{\Omega})$ where $\mathcal{G}$ is a finite dimensional real Lie algebra, where σ is an involutive automorphism of $\mathcal{G}$, and where $\boldsymbol{\Omega}$ is an element of $\wedge^2 \mathcal{G}$, is called a symmetric triple (briefly denoted hereafter S.T.) if the following properties are satisfied:

(i) Let $\mathcal{G} = \mathcal{K} \oplus \mathcal{P}$ when $\mathcal{K}$ (resp. $\mathcal{P}$) is the $+1$ (resp. -1) eigenspace of σ. Then $[\mathcal{P}, \mathcal{P}] = \mathcal{K}$ and the representation of $\mathcal{K}$ on $\mathcal{P}$, given by the adjoint action, is faithful.

(ii) For any X in $\mathcal{K}$, $i(X)\boldsymbol{\Omega} = 0$. Furthermore $\boldsymbol{\Omega}$ is invariant by the adjoint action of $\mathcal{K}$ and $\boldsymbol{\Omega}$ restricted to $\mathcal{P} \times \mathcal{P}$ has maximal rank.

Definition 7. Let $(\mathcal{G}_i, \sigma_i, \boldsymbol{\Omega}_i)$ $(i = 1, 2)$ be S.T. They are said to be isomorphic if there exists a Lie algebra isomorphism $\psi : \mathcal{G}_1 \to \mathcal{G}_2$ such that (i) $\psi \circ \sigma_1 = \sigma_2 \circ \psi$ and (ii) $\psi^* \boldsymbol{\Omega}_2 = \boldsymbol{\Omega}_1$.

Proposition 2. *There is a bijective correspondence between the set of isomorphism classes of simply connected S.S.M. and the set of isomorphism classes of S.T.*

Let us briefly indicate, in one direction, how this correspondence is made. If (M, ω, s) is a S.S.M., let us associate to it a S.T. $(\mathcal{G}, \sigma, \mathbf{\Omega})$ as follows. The Lie algebra $\mathcal{G}$ is the Lie algebra of the transvection group $G(M, \omega, s)$. Choose a base point o in M and let $\tilde{\sigma}$ be the involutive automorphism of $G(M, \omega, s)$ obtained by "conjugation" by s_o. Then σ is the involutive automorphism of $\mathcal{G}$ which is the differential of $\tilde{\sigma}$ at the neutral element e of G. If $\pi : G(M, \omega, s) \rightarrow M : g \rightarrow g.o$ denotes the projection associated to the choice of base point o, then $\mathbf{\Omega}$ is the 2 form on $\mathcal{G}$ (identified with G_e), $\mathbf{\Omega} = \pi^* \omega_o$.

As a S.S.M. (M, ω, s) is a homegeneous symplectic manifold, under the action of $G(M, \omega, s)$, it is natural to ask under what condition, M is a covering space of a certain coadjoint orbit of $G(M, \omega, s)$ in $\mathcal{G}^*$. The answer is given by:

Proposition 3. *Let (M, ω, s) be a S.S.M. and let $(\mathcal{G}, \sigma, \mathbf{\Omega})$ be an associated triple. Then:*

(i) The 2 form $\mathbf{\Omega}$ is a Chevalley 2-cocycle of $\mathcal{G}$ with values in **R**, *relative to the trivial representation of $\mathcal{G}$ on* **R**.

(ii) If $G(M, \omega, s)$ has a strongly hamiltonian action on (M, ω), $\mathbf{\Omega}$ is a coboundary; i.e. there exists $\alpha \in \mathcal{G}^*$ such that $\mathbf{\Omega} = \partial \alpha$.

(iii) If M is simply connected and if $\mathbf{\Omega} = \partial \alpha$ the action of $G(M, \omega, s)$ is strongly hamiltonian and the moment map $J : M \rightarrow \mathcal{G}^*$ is a symplectic covering map of the orbit of α.

3. Symplectic symmetric manifolds with completely reducible holonomy

Let (M, ω, s) be a S.S.M., with base point o; let $G(M, \omega, s)$ be its transvection group and let $\mathcal{K}$ be the stabilizer of o in $G(M, \omega, s)$. Denote by ∇ the unique preferred symplectic connection described in (2. Theorem 1) and let $\psi(o)$ denote the linear holonomy group of (M, ∇) at o.

Proposition 1.[1] **Let (M, ω, s) be a connected S.S.M.** *with base point o. Then the isotropy group K of o in $G(M, \omega, s)$ is isomorphic to the linear holonomy group $\psi(o)$.*

In this paragraph we assume that $\psi(o)$ acts completely reducibly on M_o. If $(\mathcal{G}, \sigma, \omega)$ is a S.T. associated to (M, ω, s) and if $\mathcal{G} = \mathcal{K} \oplus \mathcal{P}$ is the already mentioned decomposition of $\mathcal{G}$ in eigenspaces relative to σ, then $\mathcal{K}$ is the Lie algebra of the isotropy subgroup K of o. The assumption on the holonomy is equivalent to the following assumption on the S.T.: the adjoint action of $\mathcal{K}$ on $\mathcal{P}$ is completely reducible. If M is simply connected, it is a classical result that K is connected [2] and the assumption on the holonomy is equivalent to requiring that the representation of $\mathcal{K}$ on $\mathcal{P}$ given by the adjoint action is completely reducible.

Proposition 2. *Let (M,ω,s) be a S.S.M. with base point o. If the linear holonomy group $\psi(o)$ acts completely reducibly on M, the transvection group $G(M,\omega,s)$ is a reductive group. Furthermore if M is simply connected, there exists a simply connected S.S.M. (M_1,ω_1,s_1) with semi-simple transvection group $G(M_1,\omega_1,s_1)$ and an isomorphism of S.S.M. between (M,ω,s) and $(\mathbf{R}^{2n},\omega_o)\times(M_1,\omega_1)$.*

Here $(\mathbf{R}^{2n},\omega_o)$ denotes the standard symplectic vector space of dimension $2n$ with standard euclidean symmetries (which is clearly a S.S.M.) and the symplectic structure (resp. the symmetries) on the product is (resp. are) defined in the obvious way.

Definition 1. A S.S.M. (M,ω,s) will be called semi simple (resp. simple) if its transvection group $G(M,\omega,s)$ is semi simple (resp. simple). Similarly a S.T. $(\mathcal{G},\sigma,\mathbf{\Omega})$ will be called semi simple (resp. simple) if the algebra $\mathcal{G}$ is semi simple (resp. simple).

Consider $(\mathcal{G},\sigma,\mathbf{\Omega})$ a semi simple S.T. and let β denote the Killing form of $\mathcal{G}$. Then $\mathcal{G}=\mathcal{K}\oplus\mathcal{P}$ and $\mathcal{K}$ and $\mathcal{P}$ are β-orthogonal; in particular the restriction of β to $\mathcal{P}\times\mathcal{P}$ (resp. $\mathcal{K}\times\mathcal{K}$) is non degenerate. If X (resp. ξ) is in $\mathcal{G}$ (resp. $\mathcal{G}^*$) we shall denote by $X^\flat$ (resp. $\xi^\#$) the element of $\mathcal{G}^*$ (resp. $\mathcal{G}$) defined by $X^\flat=\beta(X,.)$ (resp. $\beta(\xi^\#,.)=\xi$).

Lemma 1. [3] *Let $\mathcal{G}$ be a semi simple algebra and let τ be any 2-cocycle of $\mathcal{G}$ with values in $\mathbf{R}$. Then there exists one and only one element X of $\mathcal{G}$ such that $\tau=\partial X^\flat$, i.e. for any Y,Z in $\mathcal{G}$, $\tau(Y,Z)=-\beta(X,[Y,Z])$.*

In particular if $(\mathcal{G},\sigma,\mathbf{\Omega})$ is a semi simple S.T., there exists a unique element u in $\mathcal{G}$ such that $\mathbf{\Omega}=\partial u^\flat$.

Lemma 2. *Let $(\mathcal{G},\sigma,\mathbf{\Omega})$ be a semi simple S.T. Then the condition $i(X)\mathbf{\Omega}=0$ for all X in $\mathcal{K}$ is equivalent to requiring that the element u of $\mathcal{G}$, such that $\mathbf{\Omega}=\partial u^\flat$, belongs to the center of $\mathcal{K}$.*

Proposition 3. *Let $(\mathcal{G},\sigma,\mathbf{\Omega})$ be a semi simple S.T. Then the center of the isotropy algebra $\mathcal{K}$ is different from 0. In particular if $\mathcal{G}=\oplus_{i=1}^{r}\mathcal{G}_i$ is the decomposition of $\mathcal{G}$ in simple ideals, each of these ideals is stable by σ and hence the S.T. $(\mathcal{G},\sigma,\mathbf{\Omega})$ is isomorphic to the direct sum of S.T. $\oplus_{i=1}^{r}(\mathcal{G}_i,\sigma_i,\mathbf{\Omega}_i)$ where $\sigma_i=\sigma_{|\mathcal{G}_i}$ and $\mathbf{\Omega}_i=\mathbf{\Omega}_{|\mathcal{G}_i\times\mathcal{G}_i}$.*

Proposition 3 implies that any simply connected S.S.M. (M, ω, s) with semi simple transvection group $G(M, \omega, s)$ is isomorphic to a direct product of simply connected S.S.M. (M_i, ω_i, s^i) with simple transvection group.

Lemma 3. *Let $(\mathcal{G}, \sigma, \mathbf{\Omega})$ be a simple S.T. and let V be a proper subspace of $\mathcal{P}$ stable by ad $\mathcal{K}$. Then*

(i) V is a maximal isotropic subspace of $\mathcal{P}$ relative to the restriction to $\mathcal{P} \times \mathcal{P}$ of the Killing form β of $\mathcal{G}$.

(ii) V is an abelian subalgebra of $\mathcal{G}$.

(iii) The action of $\mathcal{K}$ on V is irreducible.

Lemma 4. *Let $(\mathcal{G}, \sigma, \mathbf{\Omega})$ be a simple S.T. and let $u \neq 0$ be an element of the center of $\mathcal{K}$. Then $\partial u^\flat_{|\mathcal{P}\times\mathcal{P}}$ is a 2-form whose rank equals the dimension of $\mathcal{P}$.*

Lemma 5. *Let $(\mathcal{G}, \sigma, \mathbf{\Omega})$ be a simple S.T. Assume $\mathcal{G}$ admits an almost complex structure J commuting with the adjoint action of $\mathcal{G}$. Then J commutes with σ.*

If $\mathcal{G}$ is a simple real Lie algebra, its complexified algebra $\mathcal{G}^{\mathbf{C}}$ is either simple or the sum of two isomorphic simple complex ideals; in the first case $\mathcal{G}$ is said to be absolutely simple; in the second case $\mathcal{G}$ may be considered as a simple complex algebra, viewed as a real algebra.

If $(\mathcal{G}, \sigma, \mathbf{\Omega})$ is a simple S.T. and $\mathcal{G}$ is absolutely simple, $(\mathcal{G}^{\mathbf{C}}, \sigma^{\mathbf{C}}, \mathbf{\Omega}^{\mathbf{C}})$ may be viewed as a simple complex S.T..On the other hand if $\mathcal{G}$ is not absolutely simple, it is a simple complex algebra and $(\mathcal{G}, \sigma, \mathbf{\Omega})$ can be viewed as a simple complex S.T..

Hence, we shall from now on consider simple complex S.T. $(\mathcal{G}, \sigma, \mathbf{\Omega})$.

Lemma 6. *Let $(\mathcal{G}, \sigma, \mathbf{\Omega})$ be a simple complex S.T. and let u be a non zero element of the center of $\mathcal{K}$. Then:*

(i) The element u is semi simple.

(ii) The centralizer of u in $\mathcal{G}$ is $\mathcal{K}$.

(iii) The dimension of the center of $\mathcal{K}$, $\mathcal{Z}(\mathcal{K})$ is one.

It is worthwhile observing that (ii) implies that the rank of $\mathcal{G}$ is equal to the rank of $\mathcal{K}$. Note also that (iii) implies that when $\mathcal{G}$ is absolutely simple, $\dim_{\mathbf{R}} \mathcal{Z}(\mathcal{K}) = 1$ and when $\mathcal{G}$ is complex simple, $\dim_{\mathbf{R}} \mathcal{Z}(\mathcal{K}) = 2$.

As u is semi simple, let $\mathcal{H}$ be a Cartan subalgebra of $\mathcal{G}$ containing u. By (ii) $\mathcal{H}$ is contained in $\mathcal{K}$. So if we write:

$$\mathcal{G} = \mathcal{H} \oplus_{\alpha \in \phi} \mathcal{G}^\alpha$$

where ϕ denotes the set of roots of $\mathcal{H}$ in $\mathcal{G}$, then:

$$\mathcal{K} = \mathcal{H} \oplus_{\alpha \in \phi_1} \mathcal{G}^\alpha$$

where $\phi_1 = \{\alpha \in \phi | \alpha(u) = 0\}$ and:

$$\mathcal{P} = \oplus_{\alpha \in \phi \setminus \phi_1} \mathcal{G}^\alpha$$

In particular the root spaces in $\mathcal{P}$ correspond to roots which do not vanish on u.

From now on u will designate the element of $\mathcal{Z}(\mathcal{K})$ such that $\partial u^\flat = \mathbf{\Omega}$. One can find a non zero complex number λ such that $\alpha(\lambda u \stackrel{def}{=} \tilde{u})$ is a real number for any $\alpha \in \phi$. Choose a basis $\tilde{u} \stackrel{def}{=} \{e_1, e_2, \ldots, e_\ell\}$ of $\mathcal{H}_{\mathbf{R}}$ (see [4] for a precise definition) and define the set ϕ^+ of positive roots as follows:

$$\alpha \in \phi^+ \Leftrightarrow \alpha(e_i) = 0, \quad \forall i < j \text{ and } \alpha(e_j) > 0.$$

Let $\Delta = \{\alpha_1, \ldots, \alpha_\ell\}$ be the basis of ϕ corresponding to this choice of positive roots. The basis Δ contains a system of simple roots of $\mathcal{K}$ and as $\dim \mathcal{Z}(\mathcal{K}) = 1$ we may assume that $\{\alpha_1, \ldots, \alpha_{\ell-1}\}$ is the set of simple roots of $\mathcal{K}$. In particular if $\mathcal{G}^\alpha$ is in $\mathcal{P}$,

$$\alpha = \sum_{i \leq \ell-1} n_i \alpha_i + n_\ell \alpha_\ell \text{ and } n_\ell \neq 0.$$

As $[\mathcal{P}, \mathcal{P}] = \mathcal{K}$ the coefficient n_ℓ can only have values ± 1.

Theorem 1. *Let $(\mathcal{G}, \sigma, \mathbf{\Omega})$ be a complex simple S.T..Then $(\mathcal{G}, \sigma, \mathbf{\Omega})$ is isomorphic to the simple complex triple $(\mathcal{G}, \sigma_1, \mathbf{\Omega_1})$ constructed as follows. Let $\mathcal{H}$ be a Cartan subalgebra of $\mathcal{G}$ and let ϕ be the corresponding set of roots of $\mathcal{G}$.Denote by ϕ^+ a choice of positive roots and by Δ the associated system of simple roots. Let α be an element of Δ which appears with coefficient one in the maximal root μ of ϕ (with respect to the chosen order).Denote by ξ the element of $\mathcal{H}_{\mathbf{R}}$ such that $\alpha(\xi) = 1$ and $\alpha'(\xi) = 0$ for all α' in Δ ($\alpha' \neq \alpha$).Then $\sigma_1 = exp\ \pi i\ ad\ \xi$ and there exists a non zero complex number λ such that $\mathbf{\Omega}_1 = \partial u^\flat$ with $u = \lambda\xi$.*

Theorem 1 is the basis of a complete classification of simple S.T. One may in particular determine easily which simple S.S.M. admits a Kähler structure .Using results of [2] one may also describe all compact S.S.M. For these developments we refer to [5, 6].

4. Deformation quantization of symplectic symmetric spaces

Deformation quantization was introduced in [7] as a autonomous formulation of quantum mechanics on classical phase space. Let us recall the basic definitions.

Definition 1. A deformation quantization (also called a $\star$ *product* on a symplectic manifold (M,ω) is an associative algebra structure on the space $E(N,\nu)$ of formal power series in a complex parameter ν,with coefficients in N (which is either the algebra of smooth functions on M , $\mathcal{C}^\infty(M,\mathbf{R})$, or a subalgebra of $\mathcal{C}^\infty(M,\mathbf{R})$)satisfying the following properties.

If u, v are elements of N and if we write :

$$u \star v = \sum_{r=0}^{\infty} \nu^r C_r(u,v) \tag{2}$$

then:

(i) $C_0(u,v) = uv$ and $C_1(u,v) = \{u,v\}$ where $\{,\}$ denotes the Poisson bracket of functions induced by the symplectic structure ω.

(ii) $(u \star v) \star w = u \star (v \star w)$ i.e formal associativity is satisfied.

(iii) The C_r' s are bidifferential operators vanishing on the constants; hence $1 \star u = u \star 1 = u$, for any u in N.

When $(M,\omega) = (\mathbf{R}^{2n},\omega_0)$ one may choose for C_r the r-th power of the Poisson bracket divided by $r!$; this gives rise to the so-called Moyal $\star$ product.When furthermore $\nu = i\hbar/2$ ($\hbar = \frac{1}{2\pi}\times$ Plank's constant) and u and v are smooth rapidly decreasing functions (i.e elements of $S(\mathbf{R}^{2n})$) formula (2) is the expression ,at the level of functions, of the composition law of the quantum operators Q_u and Q_v which are associated to u and v by the Weyl correspondence (see for instance [8] and references therein).

Observe that property (ii), written at order t in ν, reads :

$$\begin{aligned} E_t(u,v,w) &\stackrel{def}{=} \sum_{\substack{r+s=t \\ r,s>0}} C_r(C_s(u,v),w) - C_r(u,C_s(v,w)) \\ &= \delta C_t(u,v,w) \end{aligned} \tag{3}$$

where δ is the coboundary operator defining the Hochschild cohomology of N; i.e if $C : N \times N \to N$ is a 2-cochain :
$\delta C(u,v,w) = uC(v,w) - C(uv,w) + C(u,vw) - C(u,v)w$.

If associativity is satisfied up to order $(t-1)$, E_t is necessarily a 3-cocycle ; if it is a 3-coboundary and one choose a C_t satisfying (3), then associativity of the $\star$ product obtained with this C_t , satisfies associativity up to order t.

What we describe here under is a recurrence procedure allowing to construct C_t from E_t in such a way that at each order t' , $E_{t'}$ is indeed a coboundary. The explicit formulas depend on the fact that (M, ω) is an S.S.M.

On a S.S.M (M, ω) one has a unique preferred connection ∇.Thus any bidifferential operator C_t has a unique expression of the form :

$$C_t(u,v) = \sum_{\substack{p \leq P(t) \\ q \leq Q(t)}} C_{(t)\ p,q}^{i_1 \ldots i_p, j_1 \ldots j_q} \nabla_{(i_1 \ldots i_p)} u \nabla_{(j_1 \ldots j_q)} v$$

We have chosen a coordinate system x^k $(k \leq m = dim\ M)$ in the neighborhood of the point x where we evaluate the function $C_t(u,v)$; we write briefly $\nabla_{i_1 \ldots i_p} u$ for the $i_1 \ldots i_p$ component of the form $\nabla \ldots \nabla u$, i.e. :

$$\nabla_{i_1 \ldots i_p} u = (\nabla \ldots \nabla u)(\frac{\partial}{\partial x^{i_1}}, \ldots, \frac{\partial}{\partial x^{i_p}})$$

The symbol ( ) means complete symmetrization, i.e. :

$$\nabla_{(i_1 \ldots i_p)} u = \frac{1}{p!} \sum_{\sigma \in S_p} \nabla_{i_{\sigma(1)} \ldots i_{\sigma(p)}} u$$

where S_p is the group of permutations of p elements. In this expression the coefficients $C_{(t)\ p,q}^{i_1 \ldots i_p, j_1 \ldots j_q}$ are the components of $(p+q)$ contravariant tensors, which are completely symmetric in the p first indices and in the q last indices.
Similarly the tridifferential operator E_t has a unique expression of the form

$$E_t(u,v,w) = \sum_{\substack{p \leq P(t) \\ q \leq Q(t) \\ r \leq R(t)}} E_{(t)\ p,q,r}^{i_1 \ldots i_p, j_1 \ldots j_q, k_1 \ldots k_r} \nabla_{(i_1 \ldots i_p)} u \nabla_{(j_1 \ldots j_q)} v \nabla_{(k_1 \ldots k_r)} w \tag{4}$$

and the coefficients $E_{(t)\ p,q,r}$ are tensors which are completely symmetric in each block of indices.
A Hochschild 3-cocycle is exact if and only if its totally antisymmetric part, which is necessarily a tridifferential operator of order one in each of

its arguments, vanishes identically [9]. If E_t is exact one may define C_t such that $\delta C_t = E_t$ by :

$$
\begin{aligned}
C^{i,j}_{(t)\ 1,1} &= 0 \\
C^{i_1\dots i_p,j}_{(t)\ p,1} &= \frac{1}{4(p+1)}\sum_{r=1}^{p}\left(E^{i_1\dots\hat{i}_r\dots i_p,i_r,j}_{(t)\ p-1,1,1} - E^{i_1\dots\hat{i}_r\dots i_p,j,i_r}_{(t)\ p-1,1,1}\right)\ (p>1)
\end{aligned}
$$

$$
\begin{aligned}
C^{i_1\dots i_p,j_1 j_2}_{(t)\ p,2} &= \frac{1}{2}\sum_{\sigma\in S_2}(E^{i_1\dots i_p,j_{\sigma(1)},j_{\sigma(2)}}_{(t)\ p,1,1} \\
&+ \frac{1}{(p+2)}\sum_{r=1}^{p}(E^{i_1\dots\hat{i}_r\dots i_p j_{\sigma(1)},j_{\sigma(2)},i_r}_{(t)\ p,1,1} \\
&- E^{i_1\dots\hat{i}_r\dots i_p j_{\sigma(1)},i_r,j_{\sigma(2)}}_{(t)\ p,1,1}))
\end{aligned}
\tag{5}
$$

$(p \geq 1)$

$$
\begin{aligned}
C^{i_1\dots i_p,j_1\dots j_q}_{(t)\ p,q} &= \frac{1}{q}\sum_{r=1}^{q}(E^{i_1\dots i_p,j_1\dots\hat{j}_r\dots j_q,j_r}_{(t)\ p,q-1,1} \\
&- \frac{1}{4(p+q)}\sum_{s=1}^{p}(E^{i_1\dots\hat{i}_s\dots i_p j_1\dots\hat{j}_r\dots j_q,i_s,j_r}_{(t)\ p+q-2,1,1} \\
&- E^{i_1\dots\hat{i}_s\dots i_p j_1\dots\hat{j}_r\dots j_q,j_r,i_s}_{(t)\ p+q-2,1,1}))
\end{aligned}
$$

$(q > 2, p \geq 1)$

One observes that the terms of order (p,q) in C_t depend only on terms of E_t of order (k,l,m) with $k+l+m = p+q$. If (M,ω,s) is a S.S.M., the curvature tensor R of the preferred symplectic connection is parallel (i.e. $\nabla R = 0$). Hence one may write identities of the form :

$$\nabla_{(i_1\dots i_r)}\nabla_{(j_1\dots j_s)}u = \nabla_{(i_1\dots i_r j_1\dots j_s)}u + \sum_{k\geq 1} P^{a_1\dots a_{r+s-2k}}_{(k)\ i_1\dots i_r,j_1\dots j_s}(R)\nabla_{(a_1\dots a_{r+s-2k})}u$$

where the $P_{(n)}$'s are universal polynomials in the curvature tensor R.

Theorem 1. *[10] Assume that the bidifferential operators are such that (i) $\nabla C_{(r)\ p,q} = 0$ for all r, p, q and (ii) $C_{(r)\ p,q} = 0$ if $p+q$ is odd, for all $r < t$. Then $E_{(t)}$, when written in terms of the $E_{(t)\ k,l,m}$ is such that $\nabla E_{(t)\ k,l,m} = 0$ and $E_{(t)\ k,l,m} = 0$ if $k+l+m$ is odd. In particular $E_{(t)\ 1,1,1} = 0$ and hence $E_{(t)}$ is a coboundary. The formulas (5) define a 2-cochain $C_{(t)}$ such that $\nabla C_{(t)\ p,q} = 0$ and $C_{(t)\ p,q} = 0$ if $p+q$ is odd. So recursively we have a $\star$ product on (M, ω).*

This totally explicit construction of deformation quantization should be compared to the general Fedosov's construction [11].

References

1. Loos, O. (1969) *Symmetric spaces*, Benjamin, p. 96.
2. Koh, S. (1965) *Trans. Am. Math. Soc.*, **119** , pp. 291-301.
3. Jacobson, N. (1962) *Lie algebras*, Interscience, p. 77.
4. Humphreys, J.E. (1970)*Introduction to Lie algebras and representation theory*, Springer, p. 40.
5. Bieliavsky, P. (1995) Thèse de doctorat, U.L.B.
6. Bieliavsky, P., Cahen, M. and Gutt, S. (1995) preprint U.L.B.
7. Bayen, F., Flato, M., Fronsdal, C., Lichnerowicz, A. and Sternheimer,D. (1978) *Ann. Phys.*, **111** , pp. 61-151.
8. Arnal, D (1988) The $\star$ exponential in *Quantum theories and Geometry*, Kluwer Ac. Pub., pp. 23-52.
9. Vey, J. (1975) *Com. Math. Helv.* , **90** , pp. 421-454.
10. Cahen, M. and Gutt, S. (1983) *C.R. Acad. Sc. Paris*, **297, I**, pp. 497-420.
11. Fedosov, B. (1994) *J. of diff. geometry*, **40** , pp.213-238.

THE CONFORMAL INVARIANCE OF THE KLEIN-GORDON EQUATION IN 1+1 DIMENSION

STEPHAN DE BIEVRE AND JACQUES RENAUD
Laboratoire de Physique Théorique et Mathématique
Université Paris VII
2 place Jussieu F-75251 Paris Cedex 05, FRANCE
debievre@mathp7.jussieu.fr
renaud@ccr.jussieu.fr

Abstract. The group $SO_o(2,2)$ is the finite dimensional conformal group of the 1+1 dimensional Minkowski, de Sitter and anti-de Sitter spacetimes M. We identify the indecomposable representation of $SO_o(2,2) \cong SO_0(2,1)\times SO_0(2,1)$ that acts on the one-particle physical space of the massless scalar field on M. This representation is obtained by means of a modified geometric quantization from the nilpotent orbits of $SO_0(2,1)$. Moreover we construct an invariant symbol calculus linking these orbits and this representation.

1. Introduction

On Minkowski spacetime, the scalar massless field has been studied by several authors as a non-trivial example of a non-positive Wightman field exhibiting various complicating features such as gauge invariance and infrared problems (MPS). The conformal invariance of the model is also discussed in these works, but the representation of the conformal group $SO_o(2,2)$ on the physical one-particle space seems to have never been identified clearly, even in that case. The realization of the one particle space as a space of solutions of the Klein-Gordon equation is natural and standard for massive fields. In the case of the massless field it is made difficult by infrared problems, and the usual realization is not on a space of solutions of Klein-Gordon equation, and this introduces "correction terms" for the action of the special conformal transformations on the one particle states such as those that one can find in (SV).

J. Bertrand et al. (eds.), Modern Group Theoretical Methods in Physics, 75–86.

We shall see that one can obtain the one-particle space as a set of solutions of the Klein-Gordon equation on which the representation of the conformal group is well-defined without any "correction terms". This result holds not only for the 1+1 dimensional Minkowski spacetime but also for the 1+1 dimensional de Sitter and anti-de Sitter space time.

More explicitly, we shall construct a subspace $\mathcal{K}$ of the set of distributional solutions of the Klein-Gordon equation. This subspace is selected using an analyticity criterion and is equipped with a positive semi-definite inner product which coincides with the Klein-Gordon product when the latter makes sense. Quotienting out by the one-dimensional kernel of this inner product one obtains a Hilbert space. There is no process of completion here, and hence no Krein space. Moreover the action of the conformal group on $\mathcal{K}$ is the natural one:

$$V_g\psi(x) = \psi\left(g^{-1}\cdot x\right).$$

The representation obtained can be described briefly in the following way. Let us recall that the relativity group for the Minkowski spacetime is the Poincaré group $P^{1,1}$ whereas the relativity group of de Sitter or anti-de Sitter spacetime is $SO_0(1,2)$. In all cases the conformal group is $SO_0(2,2) \simeq SO_0(1,2) \times SO_0(1,2)$. Note that the inclusion of the relativity group in the conformal group is the diagonal one:

$$\begin{aligned} SO_0(2,1) &\rightarrow SO_0(2,2) \simeq SO_0(2,1)\times SO_0(2,1) && (1)\\ g &\rightarrow (g,g). && (2)\end{aligned}$$

It is a remarkable fact that it is the same representation that is involved in the three cases considered, namely a one dimensional indecomposable extension of the representation $U^1 \oplus U^1$ of $SO_0(2,2) \simeq SO_0(1,2) \times SO_0(1,2)$, where U^1 is the first term of the holomorphic discrete series of unitary irreducible representations of $SO_0(1,2)$. More details and complete references can be found in (DBR1)(DBR2)(R).

In section two we construct this representation by means of a modified geometric quantization. In section three and four we show how to implement this representation in the above spacetimes. In section five we define an invariant symbol calculus linking this representation and the corresponding coadjoint orbit or classical phase space.

2. Construction of the representation

The coadjoint conical orbits in $so(1,2)^* \simeq su(1,1)^*$ are the phase spaces of the zero mass particles on the two-dimensional (anti-)de Sitter spacetime.

They contain also, as an open dense subset, the phase spaces for massless particles on two dimensional Minkowski spacetime when one identifies the Poincaré group with a subgroup of the conformal group $SO_0(2,2) \simeq SO_0(1,2) \times SO_0(1,2)$. The quotient of each cone by dilations is diffeomorphic to the set of right (or left) moving lightlike geodesics on those spacetimes. This observation is at the origin of the central role played by the dilations on $so(1,2)^*$ in our construction.

No completely satisfactory orbit correspondence exists for the nilpotent orbits of reductive groups (V). Actually, one main point of our work is that the apparent failure of the orbit method to associate an *irreducible* unitary representation to the nilpotent cone in $so(1,2)^*$ should be expected on physical grounds. Indeed, the appearance of indecomposable representations is typical in the quantum description of massless systems. It is linked to the presence of gauge degrees of freedom and conformal invariance. In the present case, the application of the orbit method in its usual form to the nilpotent cone in $so(1,2)^*$ yields a representation $\mathcal{V}$ including two indecomposable representations. Our geometrical construction shows how to extract the physically correct indecomposable representation V from $\mathcal{V}$. In addition it yields the natural indefinite inner product.

Let us now summarize this method. We choose a basis N_0, N_1, N_2 for the Lie algebra $so(1,2)$ with the commutation relations

$$[N_0, N_1] = N_2,\ [N_0, N_2] = -N_1,\ [N_1, N_2] = -N_0.$$

We write N_i^* for the dual basis in $so(1,2)^*$. The problem is how to quantize the coadjoint orbit of $N_0^* + N_2^*$, which is conical or equivalently nilpotent.

Turning now to the role of dilations, let $\mathfrak{g}^* = so(1,2)^* \setminus \{0\}$ and define the following equivalence relation on $\mathfrak{g}^*$. For $f_1, f_2 \in \mathfrak{g}^*$,

$$f_1 \sim f_2 \qquad \text{iff} \qquad \exists\, \mu > 0 \quad \text{so that} \quad f_2 = \mu f_1.$$

Let $\mathcal{F} \equiv \mathfrak{g}^*/\sim$. We shall write [f] for the elements of $\mathcal{F}$. Since the coadjoint action Ad^* is linear, it passes to the quotient $\mathcal{F}$ and we shall write Ad^* for this action as well.

The main idea is to reformulate the orbit method on $\mathcal{F}$ rather than on $\mathfrak{g}^*$. For the semisimple orbits this yields nothing substantially new. On the other hand, the orbit method does not apply well to the nilpotent orbits such as $\mathcal{O}_{N_0^*+N_2^*}$. The crucial observation that allows us to bypass this difficulty is double. First, $\mathcal{O}_{[N_0^*+N_2^*]}$ is the boundary of $\mathcal{O}_{[N_0^*]}$. Next, $\mathcal{O}_{[N_0^*]}$ carries a Kählerian polarisation which gives it the structure of a complex manifold. Moreover there exists a natural identification I with the unit disc $\mathcal{D} = \{z \in \mathbb{C};\ |z| < 1\}$. One verifies that I is an isomorphism of complex manifolds and intertwines the Ad* action on $\mathcal{O}_{[N_0^*]}$ with the usual homographic action of $SU(1,1)/\mathbb{Z}_2 \simeq SO_o(1,2)$ on $\mathcal{D}$. Moreover, I can be prolonged in an obvious way to $\overline{\mathcal{O}_{[N_0^*]}}$. The quantization of the orbit $\mathcal{O}_{[N_0^*+N_2^*]}$ is then performed in two steps. First, one constructs a representation $\mathcal{V}$ of $SO_o(1,2)$ by induction from a representation of its isotropy group with differential equal to $N_0^*+N_2^*$. Since $N_0^*+N_2^*$ vanishes on the isotropy Lie algebra, this means inducing from the trivial representation and $\mathcal{V}$ is realized on functions on $\mathcal{O}_{[N_0^*+N_2^*]}$ by the formula $\mathcal{V}_g f(t) = f(g^{-1}\cdot t)$ where $g^{-1}\cdot t$ is the natural action of the group on $[N_0^* + N_2^*]$. The second step consists in using the two crucial observations above to select a space of functions on $\mathcal{O}_{[N_0^*+N_2^*]}$ which are boundary values of holomorphic functions on $\mathcal{O}_{[N_0^*]}$.

We proceed in the following manner. First, let us recall what is the first Bargmann space $\mathcal{B}^1$ on which the first term of the discrete series of representation of SU(1,1) can be realized :

$$\mathcal{B}^1 = \{f : \mathcal{D} \to \mathbb{C};\ f \text{ holomorphic and } f \in L^2(\mathcal{D}, \frac{1}{\pi} dz\, d\bar{z})\},$$

where $dz\, d\bar{z}$ is the Lebesgue measure on the disk $\mathcal{D} = \{z \in \mathbb{C};\ |z| < 1\}$. The representation U^1 is defined by

$$(U_g f)(z) = (-\bar{\beta}z + \alpha)^{-2} f\left(\frac{\bar{\alpha}z - \beta}{-\bar{\beta}z + \alpha}\right),$$

for

$$g = \begin{pmatrix} \alpha & \beta \\ \bar{\beta} & \bar{\alpha} \end{pmatrix} \in \mathrm{SU}(1,1).$$

Now we define the set $\mathcal{H}$ of holomorphic functions on $\mathcal{D}$ whose derivative is in $\mathcal{B}^1$. This space is included in the Hardy space of the disk, and hence each element of $\mathcal{H}$ has a radial boundary value on the circle. The restriction of the representation $\mathcal{V}$ on $\mathcal{H}$ named V is defined by

$$(V_g f)(z) = f\left(\frac{\bar{\alpha}z - \beta}{-\bar{\beta}z + \alpha}\right).$$

The representations V and U^1 are linked in the following way. One can prove that the following exact sequence holds.

$$0 \to \mathbb{C} \xrightarrow{i} \mathcal{H} \xrightarrow{D} \mathcal{B}^1 \to 0$$

where $D = d/dz$ is the complex derivative, $\mathcal{H} = D^{-1}\mathcal{B}^1$, and i and D intertwine the trivial representation on $\mathbb{C}$, the representation V on $\mathcal{H}$ and the first term of the discrete series U^1 on $\mathcal{B}^1$. One can then easily verify that V is the unique indecomposable one-dimensional extension of the first term of the discrete series of representations (PS). Moreover we put on $\mathcal{H}$ the inner product pulled back from the scalar product on $\mathcal{B}^1$. *We claim that the space $\mathcal{H}$ with this degenerate inner product and the indecomposable representation V is the massless representation of $SO_0(1,2)$ in the sense that we shall develop below.*

For later references we remark that, using the Cayley transform these representations can be rewritten on the half-plane. The first term of the holomorphic discrete series of representations of SL$(2,\mathbb{R})$ can be realized on the lower half-plane $\mathcal{P}$ in the following way. Let

$$\mathcal{B}_1 = \{f \in L^2(\mathcal{P}, d\lambda);\ \text{f holomorphic on } \mathcal{P}\}$$

where $d\lambda$ is the Lebesgue measure. For $g = \begin{pmatrix} a & b \\ c & d \end{pmatrix} \in \mathrm{SL}(2,\mathbb{R})$, we write

$$U_g^1\psi(z) = (-cz+a)^{-2}\psi\left(\frac{dz-b}{-cz+a}\right).$$

Recall that U^1 is a representation of $\mathrm{SO}_0(2,1) \cong \mathrm{SL}(2,\mathbb{R})/\mathbb{Z}_2$. Let $D = \dfrac{d}{dz}$ be the operator of derivation on the space of holomorphic functions on $\mathcal{P}$. We define $\mathcal{H} = D^{-1}(\mathcal{B}_1)$ to be the set of primitives of elements of $\mathcal{B}_1$. Each $\psi \in \mathcal{H}$ has a non-tangential limit on the boundary $y = 0$ of $\mathcal{P}$. Consider the representation V defined on $\mathcal{H}$ by

$$V_g\psi(z) = \psi\left(\frac{dz-b}{-cz+a}\right) \quad \text{for } g = \begin{pmatrix} a & b \\ c & d \end{pmatrix} \in \mathrm{SL}(2,\mathbb{R}).$$

Since $DV_g = U_g^1 D$, one obtains the following exact sequence of representations:

$$0 \to \mathbb{C} \xrightarrow{i} \mathcal{H} \xrightarrow{D} \mathcal{B}_1 \to 0.$$

Here i is the canonical injection of $\mathbb{C}$ into the set of constant functions, and i and D intertwine the trivial representation, V and U^1.

3. Minkowski Spacetime

The conformally invariant Klein-Gordon equation is, in light-cone coordinates $u = (u^+, u^-)$ with $u^\varepsilon = x^0 - \varepsilon x^1$

$$\Box\phi = \frac{\partial^2}{\partial u^+ \partial u^-}\phi = 0.$$

Let us write

$$\mathcal{L} = \{\phi \in S'(\mathbb{R}^2);\ \Box\phi = 0\}.$$

We shall call a distributional solution $\phi \in \mathcal{L}$ right-moving if $\partial_{u^-}\phi = 0$ and left-moving if $\partial_{u^+}\phi = 0$. Any distributional solution can be written in the form $\phi(u) = g(u^+) + f(u^-)$, but the splitting into a right-moving solution and a left-moving one is not unique because of the presence of the constant solutions which are both right- and left-moving. We have

$$\mathcal{L} = \mathcal{L}^+ + \mathcal{L}^-, \qquad \mathcal{L}^+ \cap \mathcal{L}^- = \{\text{constant functions}\},$$

where

$$\mathcal{L}^\varepsilon = \{\phi \in S'(\mathbb{R}^2);\ \partial_{u^{-\varepsilon}}\phi = 0\}.$$

The conformal group is $SO_0(2,2) \simeq G^+ \times G^-$ where $G^\varepsilon \simeq SO_0(1,2)$. Moreover G^ε acts trivially on $\mathcal{L}^{-\varepsilon}$ and only locally on $\mathcal{L}^\varepsilon$ because the special conformal transformations act by

$$e^{\theta L^\varepsilon} : \quad u^\varepsilon \to \frac{u^\varepsilon}{1+\theta u^\varepsilon}.$$

They are not diffeomorphisms on M, hence they do not act on general distributional solutions of $\Box\phi = 0$. This is exactly the origin of the difficulties arising when dealing with conformal invariance for massless fields. We shall nevertheless show that there is a natural subspace of $\mathcal{L}$, called $\mathcal{K} = \mathcal{K}^+ + \mathcal{K}^-$ below, on which $SO_o(2,2)$ *acts globally by the usual action on spacetime and preserves the usual indefinite Klein-Gordon scalar product.* The space $\mathcal{K}$ will be shown to be the one-particle physical space of the massless free field as defined in the literature.

We will construct $\mathcal{K}$ as a subspace of the space of positive energy solutions of the Klein-Gordon equation, that is to say of those elements in $\mathcal{L}$ the Fourier transform of which is supported on the future light cone. As is well known, those solutions are distributional boundary values of holomorphic functions on the future tube which in light cone coordinates is $\mathcal{P}^+ \times \mathcal{P}^-$ with $\mathcal{P}^\varepsilon = \{z = u^\varepsilon + iy^\varepsilon \quad y^\varepsilon < 0\}$. It turns out that it is much easier to identify the representation of $SO_0(2,2)$ using the analyticity properties of the positive energy solutions than their Fourier transform properties. The principal reason for this is the following remark. The non trivial part of the action of $SO_o(2,2)$ (*i. e.* of $SO_0(2,1)$) on the line $u^{-\varepsilon} = 0$ induces an action on $\mathcal{P}^\varepsilon$ that is easily seen to be the action of $SL(2,\mathbb{R})$ by fractional transformations, and in particular the special conformal transformations act by

$$e^{\theta L^\varepsilon} : \quad z \quad \to \quad \frac{z}{1+\theta z}.$$

which are of course diffeomorphisms on $\mathcal{P}^\varepsilon$, so that the difficulties mentioned above have disappeared. Consider now as in section 2, the Bargman space $\mathcal{B}^1$ defined on the half plane $\mathcal{P}^\varepsilon$ and define $\mathcal{K}^\varepsilon = D^{-1}(\mathcal{B}_1)$ to be the set of primitives of elements of $\mathcal{B}_1$. The define the representation V as above p5. The space $\mathcal{K}^\varepsilon \simeq \mathcal{H}$ can be equipped with the inner product

$$(\phi, \psi)_\varepsilon := (D\phi, D\psi)_{\mathcal{B}_1^\varepsilon},$$

which is preserved by V and makes $\mathcal{K}^\varepsilon$ into an indefinite inner product space. The image of $\mathcal{K}^\varepsilon$ under the Cayley transform is included in $H^2(\mathcal{D})$, the Hardy space on the disk. As a consequence, any element $\psi(z)$ of $\mathcal{K}^\varepsilon$ has almost everywhere a non tangential limit on the boundary $y^\varepsilon = 0$ of $\mathcal{P}^\varepsilon$, denoted by $\psi(u^\varepsilon)$ as well. One can moreover verify that such a ψ belongs to $L^2(\mathbb{R}, du^\varepsilon/(1+(u^\varepsilon)^2) \subset \mathcal{L}^\varepsilon$, so that $\mathcal{K}^\varepsilon$ is realized as a subspace of $\mathcal{L}^\varepsilon$. Now set $\mathcal{K} = \mathcal{K}^+ + \mathcal{K}^- \subset \mathcal{L}$ and equip it with the inner product

$$(\phi, \psi) := (\phi^+, \psi^+)_+ + (\phi^-, \psi^-)_-, \tag{3}$$

with evident notations. Although the decomposition $\psi = \psi^+ + \psi^-$ is not unique, the inner product is well defined because constant functions are in the kernel of $(\,,\,)_\varepsilon$. Note that the representation of $\mathrm{SO}_o(2,2)$ on $\mathcal{K}$ is also well defined and is given explicitly (with some obvious abuse of notation) by

$$(V_g\psi)(u) = (V_{g_+}\psi^+)(u^+) + (V_{g_-}\psi^-)(u^-),$$

where $g = (g_+, g_-) \in \mathrm{SO}_o(2,2) \cong \mathrm{SO}_0(2,1)\times\mathrm{SO}_0(2,1)$ and

$$(V_{g_\varepsilon}\psi^\varepsilon)(u^\varepsilon) = \lim_{y^\varepsilon\to 0} \psi^\varepsilon(g_\varepsilon^{-1}\cdot(u^\varepsilon + iy^\varepsilon)) =: \psi^\varepsilon(g_\varepsilon^{-1}\cdot u^\varepsilon). \tag{4}$$

To summarize, we have obtained a subspace $\mathcal{K}$ of the set of solutions of the Klein-Gordon equation, carrying a well-defined representation V of the conformal group (not only of its one parameter subgroups) verifying $V_g\psi(x) = \psi(g^{-1}\cdot x)$, in the sense of (4).

We now have to show that $\mathcal{K}$ is indeed the physical one particle space of the massless free field and address some related questions. What is the relation between the inner product (3) on $\mathcal{K}$ and the Klein-Gordon inner product? How does U^1 relate to the unitary irreducible massless representations of the Poincaré group, usually realized on $L^2(\mathbb{R}^{+*}, d\xi/\xi)$? Why are there no correction terms in the action of the special conformal transformations on $\mathcal{K}$?

To answer these questions, let us consider the subspace $\mathcal{W}^\varepsilon$ of all the rational functions $P(z)/Q(z)$ in $\mathcal{K}^\varepsilon$ with P and Q polynomials. One can verify that $d^oP \le d^oQ$. We denote by $\mathcal{W}_o^\varepsilon$ the set of elements P/Q of $\mathcal{W}^\varepsilon$ such that $d^oP < d^oQ$. One can easily verify that $\mathcal{W}^\varepsilon$ is invariant under $\mathrm{SO}_o(2,2)$ and that $\mathcal{W}_o^\varepsilon$ is invariant under the action of the Poincaré group. Moreover, $D(\mathcal{W}^\varepsilon)$ is dense in $\mathcal{B}_1^\varepsilon$. We can then give an explicit formula for the inner product (3) on $\mathcal{W} = \mathcal{W}^+ + \mathcal{W}^- \subset \mathcal{K}$ as follows. For $\phi, \psi \in \mathcal{W}^\varepsilon$ we obtain by means of the Stokes formula on $\mathcal{P}^\varepsilon$:

$$(\phi, \psi)_\varepsilon = \int_{\mathbb{R}} \overline{\phi(u^\varepsilon)}\, i \frac{\partial}{\partial u^\varepsilon} \psi(u^\varepsilon)\, du^\varepsilon.$$

Now, treating right and left moving solutions simultaneously, and using again the Stokes formula on M we obtain

$$(\phi, \psi) = \frac{i}{2} \int_{x^0=0} \overline{\phi(x)}\, \overleftrightarrow{\partial}_{x^0}\, \psi(x) dx^1, \tag{5}$$

for $\phi, \psi \in \mathcal{W} = \mathcal{W}^+ + \mathcal{W}^-$. Hence the inner product on $\mathcal{W}$ is exactly the usual Klein-Gordon inner product. Note that the right hand side of (5) does not make sense on all of $\mathcal{K}$, whereas the left hand side does, being defined in (3).

To better understand the representation of the Poincaré group on $\mathcal{K}$, note first that for $\phi \in \mathcal{W}^\varepsilon$ one can explicitly calculate the Fourier transform and verify that:

$$T_\varepsilon(\phi) := \xi \hat{\phi}(\xi) = \frac{\xi}{\sqrt{2\pi\hbar}} \int_{-\infty}^{\infty} e^{\frac{i\xi u^\varepsilon}{\hbar}} \phi(u^\varepsilon)\, du^\varepsilon \text{ belongs to } L^2(\mathbb{R}^{+*}, \frac{d\xi}{\xi}).$$

This confirms in particular that ϕ is a positive energy solution. Note that T_ε vanishes on constant functions. Moreover, one can readily verify that for $\phi, \psi \in \mathcal{W}^\varepsilon$ we have

$$(\phi, \psi)_\varepsilon = \int_{\mathbb{R}} \overline{\phi(u^\varepsilon)}\, i \frac{\partial}{\partial u^\varepsilon} \psi(u^\varepsilon)\, du^\varepsilon \tag{6}$$

$$= \int_{\mathbb{R}^+} \overline{T_\varepsilon(\phi)(\xi)} T_\varepsilon(\psi)(\xi) \frac{d\xi}{\xi}. \tag{7}$$

It is finally well known that T_ε intertwines the representations of the Poincaré group:

$$T_\varepsilon(V_g \psi) = W_g^\varepsilon T_\varepsilon(\psi),$$

where g belongs to the Poincaré group and W^ε is the usual realization of the massless irreducible unitary representation of the Poincaré group corresponding to right-movers ($\varepsilon = +$) or left-movers ($\varepsilon = -$) on $L^2(\mathbb{R}^{+*}, d\xi/\xi)$. Moreover, one can extend T_ε to $\mathcal{K}^\varepsilon$ and, treating again left and right moving solutions simultaneously and setting $\eta = \varepsilon\xi$, we obtain the operator

$$T: \quad \mathcal{K} \to L^2(\mathbb{R}^*, \frac{d\eta}{|\eta|}),$$

which is surjective, preserves the inner products, intertwines the representations of the Poincaré group and the kernel of which is exactly the set of constant functions. Summarizing, we have the exact sequence of representations of the Poincaré group:

$$0 \to \mathbb{C} \xrightarrow{i} \mathcal{K} \xrightarrow{T} L^2(\mathbb{R}^*, \frac{d\eta}{|\eta|}) \to 0. \tag{8}$$

To show that $\mathcal{K}$ can be identified with the usual physical one particle space, we proceed as follows. Choosing a section *i.e.* a right inverse for T one obtains an isomorphism

$$\mathcal{K} \simeq \mathbb{C} \oplus L^2(\mathbb{R}^*, \frac{d\eta}{|\eta|}).$$

As usual this isomorphism is not canonical, it depends on the section. As a result, the representation of $\mathrm{SO}_0(2,2)$ on $\mathbb{C} \oplus L^2(\mathbb{R}^*, d\eta/|\eta|)$ is not easily described, whereas it is on $\mathcal{K}$. Nevertheless this identifies $\mathcal{K}$ with the subspace of one particle states with physical interpretation.

Turning again to the indecomposable representation of the conformal group acting on $\mathcal{K}$, note that this representation does not involve any "correction term" for the special conformal transformation. Correction terms to this expression arise when choosing a section for T. For example, if we consider the restriction of T_ε to $\mathcal{W}^\varepsilon$ it has a "natural" section S defined by $T_\varepsilon S = \mathrm{Id}$ and $ST_\varepsilon(\phi) \in \mathcal{W}_o^\varepsilon$ for any $\phi \in \mathcal{W}^\varepsilon$. More precisely, we have

$$ST_\varepsilon(\phi)(u^\varepsilon) = \phi(u^\varepsilon) - \lim_{u^\varepsilon \to \infty} \phi(u^\varepsilon).$$

Using S, we can define by $V'_g = ST_\varepsilon V_g$ a unitarizable representation of the conformal group on the space of solutions $\mathcal{W}_o^\varepsilon$. The representation V' differs from the natural representation V of the conformal group on $\mathcal{W}_o^\varepsilon$ by a correction term that is readily computed. For ψ in $\mathcal{W}_o^\varepsilon$, the representation of the special conformal transformations by V' is:

$$V'_{e^{\theta L^\varepsilon}}\psi(u^\varepsilon) = \psi\left(\frac{u^\varepsilon}{1+\theta u^\varepsilon}\right) - \psi\left(\frac{1}{\theta}\right) = V_{e^{\theta L^\varepsilon}}\psi(u^\varepsilon) - \psi\left(\frac{1}{\theta}\right).$$

This is precisely the formula containing the so-called correction term obtained in (SV). Note that for g in the Poincaré group we do recover V, without any correction term. This is so because $\mathcal{W}_o^\varepsilon$ is a Poincaré invariant subspace of $\mathcal{K}$.

4. De Sitter and anti-de Sitter spacetime

The construction of the above section can be repeated for the de Sitter and anti-de Sitter spacetimes in a very similar way. In the following we only summarize the results.

Both the de Sitter and anti-de Sitter spacetimes M_κ can be represented as the hyperboloid in $\mathbb{R}^3$ with equation

$$(y^1)^2 - (y^2)^2 - (y^3)^2 = -\kappa^{-2}$$

for some $\kappa > 0$.

The (Lorentzian) metric on M_κ is obtained by restricting the metric with signature $(+,-,-)$ on $\mathbb{R}^3$ to M_κ. The isometry group of M_κ is $O(1,2)$; the component connected to the identity is denoted by $SO_o(1,2)$. In the AdS interpretation, timelike vectors have negative norm and time is compact. In the dS interpretation, timelike vectors have positive norm and space is compact.

As a result, the timelike geodesics of the AdS version of M_κ are all closed and wrap around the "waist" of the hyperboloid. Each intersects the spacelike hypersurface $y^2 = 0$, $y^3 > 0$ exactly in one point. They are interpreted as the worldlines of massive particles. On dS spacetime the timelike geodesics are unbounded, and intersect the dS spacelike hypersurface $y^1 = 0$ in exactly one point.

The lightlike geodesics are the same in both interpretations of M_κ. They are the straight lines which are the generators of the hyperboloid. We interpret them as the worldlines of massless particles, to which we shall refer with some abuse of language as photons. Note that half of them do not intersect the spacelike hypersurface $y^2 = 0$, $y^3 > 0$ of AdS-M_κ. This is a result of the fact that AdS-M_κ is not globally hyperbolic, a remark that will be important in the following as well. On the other hand, each of them intersects $y^1 = 0$ exactly once, in accordance with the global hyperbolicity of dS-M_κ.

The conformally invariant Klein-Gordon equation on a two dimensional spacetime M is given by

$$\Box\phi = 0. \tag{9}$$

In order to identify the space $\mathcal{K}$ and the representation V in this context, let us begin with right-moving solutions of (9). They can be identified to functions on the Cauchy surface $y^1 = 0$ which is a circle. As before we select among these functions those which are boundary values of holomorphic functions on the disk whose derivative are in $\mathcal{B}^1$: $\mathcal{K}^+ = D^{-1}(\mathcal{B}^1)$. *This furnishes a subspace of solutions of Klein-Gordon carrying the natural representation of the relativity group, which is isomorphic to the representation V of section 2* . Note that unlike the plane case, the indecomposability appears for the relativity group itself, whereas V extends easily to the conformal group using (2).

Moreover, in the context of de Sitter space, $y^1 = 0$ is space-like and the inner product on $\mathcal{K}^+$ is the Klein-Gordon inner product. Whereas in the context of anti-de Sitter space, we can prove using Stokes formula that this product is equal to the Klein-Gordon inner product on the curve $y^2 = 0$ which has two components with $y^3 > 0$ and $y^3 < 0$.

The constructions of this section can now readily be repeated for left-moving fields, leading to an identification of $\mathcal{H}$ with a set of left-moving solutions to the Klein-Gordon equation, equipped again with the Klein-Gordon inner product.

In conclusion we have obtained as in the minkowski case a subspace of solutions of the Klein-Gordon equation which is invariant under the natural action of the relativity and conformal groups and naturally equipped with the Klein-Gordon inner product.

5. Symbol calculus

As we have seen before the massless systems on these spacetimes yield on the quantum side the indecomposable representation V and on the classical side the nilpotent coadjoint orbits of so(1,2). These are the cones which we parametrize with $(\lambda, t) \in \mathbb{R}^{+*} \times S^1$. At this step arises a natural question: Does there exist a symbol calculus for operators on this representation space with symbols living on those orbits? Using in an essential manner the construction explained in section 2, we have obtained a correspondence Φ between the representation $\mathfrak{A}$ of the enveloping algebra of so(1,2) induced by V and functions on the nilpotent orbit C. More precisely, let $K^+ = V(N_2 - iN_1)$ and let V_n be the su(1,1)-submodule of $\mathfrak{A}$ generated by $(K^+)^n$, then

$$\mathfrak{A} = \bigoplus_{n=0}^{\infty} V_n.$$

For any operator B on $\mathcal{B}^1$ let $B(z, z)$ be the Berezin symbol of B on the disk (B). Now, we can define the symbol of one operator in the following manner:

$$\text{for} \quad \mathfrak{A} \ni A = \sum_n A_n \qquad A_n \in V_n,$$

we define

$$\Phi(A)(\lambda, t) = \sum_n \left(\frac{\lambda}{\hbar}\right)^n \frac{1}{(n+1)!} \lim_{\substack{z = re^{-it} \\ r \to 1}} (1 - z\bar{z})^n (DA_n D^{-1})(z, \bar{z}).$$

This calculus has some nice properties.

Theorem 5.1 *The transformation*

$$\begin{array}{rcl} \Phi : \mathfrak{A} & \to & \mathcal{C}^\infty(C) \\ A & \mapsto & \Phi(A), \end{array}$$

has the following properties:

1. $\mathrm{Id} = 1$ *and* $\Phi(K_j) = k_j$. *Here* $k_j : C \to \mathbb{R}$ *are the Hamiltonian of* N_j *and* K_j *the corresponding generators of* V.
2. *The calculus is invariant :*

$$\Phi\left(U_g^{-1} A U_g\right)(x) = \Phi(A)(g \cdot x) \quad \forall g \in SU(1,1)\ \forall A \in \mathfrak{A}\ \forall x \in C.$$

3. *The symbol map* $A \mapsto \Phi(A)$ *is injective.*
4. $\Phi(A^*) = \overline{\Phi(A)}\ \forall A \in \mathfrak{A}$.
5. *For all* A, B *in* $\mathfrak{A}$, *the classical limit has the right properties:*

$$\begin{aligned} \lim_{\hbar \to 0} \Phi(AB) &= (\lim_{\hbar \to 0} \Phi(A))(\lim_{\hbar \to 0} \Phi(B)), \\ \lim_{\hbar \to 0} \frac{1}{i\hbar} \Phi([A, B]) &= \left\{ \lim_{\hbar \to 0} \Phi(A), \lim_{\hbar \to 0} \Phi(B) \right\}, \end{aligned}$$

where the limits are pointwise ones and $\{,\}$ *is the Poisson bracket on* C.

Remark: We can construct a star-product (BF²LS) indexed by $\hbar$ as the pullback on $\mathcal{C}^\infty(C)$ of the operator product. The fifth property shows that this product is a deformation of the Poisson product on C restricted to some subspace of $\mathcal{C}^\infty(C)$. The second property proves that this star-product is invariant. Note that the quantizable observables (*i.e.* $\Phi(\mathfrak{A})$) are those polynomials in λ, e^{it} for which the degree in λ is greater or equal than the degree in e^{it}.

References

(B) F. A. Berezin, General concept of quantization, *Commun. math. Phys.* **40** *(1975) 153-174.*

(BF²LS) F. Bayen, M. Flato, C. Fronsdal, A. Lichnerowicz and D. Sternheimer, Deformation theory and Quantization. I. Deformations of symplectic structures, *Ann. Phys.* **111** *(1978) p61-110.*

(DBR1) S. De Bievre and J. Renaud, Quantization of the nilpotent orbits in $so(1,2)^*$ and massless particles on (anti-)de Sitter spacetime, *J. Math. Phys.* **35** *8 (1994) 3775-3793.*

(DBR2) S. De Bievre and J. Renaud, The indecomposable representation of $SO_0(2,2)$ on the one-particle space of the massless field in 1+1 dimension, *To appear in Lett. Math. Phys. (1995).*

(MPS) G. Morchio, D. Pierotti and F. Strocchi, Infrared and vacuum structure in two-dimensional local quantum field theory models. The massless scalar field, *J. Math. Phys.* **31** *6 (1990) 1467-1477.*

(PS) G. Pinczon and J. Simon, Extensions of representations and cohomology, *Rep. Math. Phys.* **16** *(1979) 1 49-75.*

(R) J. Renaud, Symbolic calculus on the nilpotent orbits of $SO_0(1,2)$, *Submitted.*

(SV) J. A. Swieca and A. H. Völkel, Remarks on conformal invariance, *Commun. math. Phys.* **29** *(1973) 319-342.*

(V) D. A. Vogan, "Unitary representation of reductive Lie groups", *Princeton Univ. Press, Princeton, 1987.*

COMPACT TOPOLOGICAL QUANTUM GROUPS

PHILIPPE BONNEAU
Université de Bourgogne
Laboratoire Algèbre et Analyse: Théorie des Représentations
B.P. 138, F-21004 DIJON Cédex, FRANCE
email: physmath@satie.u-bourgogne.fr

To Guy Rideau, with all my friendship.

Abstract.
Using vector spaces topologies we unify the different models of quantum groups. Duality and reflexivity are built in. The Drinfeld deformation can be extended to the distributions on a simple compact Lie group and dually to the infinitely differentiable functions. The topological quantum double is similarly defined and a uniqueness result is obtained.

Introduction

This paper is a review of the main results of the four following papers [B1] , [B2] , [BFP] , [BFGP] . These articles introduced a *topological* model for quantum groups. I would like to stress that this model is not another different definition of quantum groups. On the contrary it presents a *unifying point of view* which has many advantages:

- The models of Drinfeld [D1] and Faddeev-Reshetikhin-Takhtajan [FRT] are explicitly described as a realization of the same deformation on two *reflexive, topological* Hopf algebras which are *strongly dual* one to the other.
- The Hopf algebras involved give very powerful and relatively easy techniques of demonstration which were not available before (for example, harmonic analysis and the use of finite dimensional matrix algebras).
- We deform only the product (Drinfeld) or the coproduct (FRT) but not both.

J. Bertrand et al. (eds.), Modern Group Theoretical Methods in Physics, 87–96.

- We can extend the "quantum group" deformation to the two dual topological (and well known) Hopf algebras $\mathcal{C}^\infty(G)$ and $\mathcal{D}'(G)$ ($\mathcal{C}^\infty$-functions and distributions on a simple compact Lie group G) which sets quantum groups into *the theory of star-products* ([BFFLS] and the contribution of D. Sternheimer to these proceedings).
- Finally these results can be extended to noncompact simple Lie groups (see the contribution of F. Bidegain to these proceedings and [BP]) and to the Lorentz group [MZ] .

Obviously this formalism is susceptible of generalizations and I hope that this review will motivate future works in this direction.

I. Deformation theory and first applications ([B1] and [BFP])

First, we need a rigorous deformation theory for the category of interest. Second we will consider the different definitions in the light of this theory. The category of quasi-Hopf algebras [D3] seems, for us, the natural one. Since we are mainly concerned with representation theory, we just want to preserve equivalence of representations: the coproducts $\triangle$ and $F.\triangle.F^{-1}$(F in the algebra tensorized by itself) define the same tensor products of representations thus we would like them to be equivalent.

To make cohomology calculations simpler, we will say that a bialgebra B is an associative algebra equipped with an algebra homomorphism $\triangle : B \to B \otimes B$ and we will study deformations of bialgebras in this sense. Of course, rigidity for bialgebras implies rigidity for quasi-Hopf algebras.

Following Gerstenhaber [GGS] we write, for a bialgebra $(B, \mu_o, \triangle_o)$:

$$\mu_t = \mu_o + t\mu_1 + t^2\mu_2 + ..., \qquad \triangle_t = \triangle_o + t\triangle_1 + t^2\triangle_2 + ...,$$

and we find the complex

$$\{\mathrm{Hom}_{\mathbb{C}}(\oplus^n B, B) \oplus \mathrm{Hom}_{\mathbb{C}}(\oplus^{n-1} B, B \otimes B), D^n\},$$

where $D^n = (d^{n1} + (-1)^n d_*^{n-1}) \oplus d^{(n-1)2}$, $D^0 = d^{01} + d_*^{01}$ and d^{np}, d_*^{np} are Hochschild and co-Hochschild differentials acting on $\mathrm{Hom}_{\mathbb{C}}(\otimes^n B, \otimes^p B)$. We denote by $H_{bi}^\bullet(B,B)$ the associated cohomology. Then $H_{bi}^2(B,B)$ classifies the inequivalent infinitesimal deformations (the $(\mu_1, \triangle_1)$ part) and $H_{bi}^3(B,B)$ provides us the obstructions to "continue" a step further. We have:

Theorem 1.

$$... \to H_h^{i-1}(B, B\otimes B) \to H_{bi}^i(B,B) \to H_h^i(B,B) \to H_h^i(B, B\otimes B) \to ...$$

is a long exact sequence. ($H_h^\bullet$= *Hochschild's cohomology*)

Corollary 1. *If* $H_h^1(B, B \otimes B) = H_h^2(B, B) = \{0\}$ *then* $H_{bi}^2(B, B) = \{0\}$.

This give a very simple sufficient condition to show the *rigidity* in the category of bialgebras and then also in the one of *quasi-Hopf algebras.*

Now, what precise meaning do these results give to the different definitions of QGs? Let us review these models (G denote a simple complex Lie algebra):

- *Drinfeld model* [D1] (denoted by $\mathcal{U}_t\mathcal{G}$):
  Drinfeld showed that $\mathcal{U}_t\mathcal{G}$ is a trivial deformation in our sense [D2] (without showing $H_{bi}^2(\mathcal{U}_t\mathcal{G}, \mathcal{U}_t\mathcal{G}) = \{0\}$).
- *Jimbo model* [Ji] (denoted by $J_q\mathcal{G}$, the one with generators K_i and K_i^{-1}).

We obtained the following results :

1. $J_q\mathcal{G}$ is not a deformation of $\mathcal{U}\mathcal{G}$ but one of $\mathcal{U}\mathcal{G}\otimes\mathbf{C}[C_1,\ldots,C_n] / <C_i^2-1>_{i=1,\ldots,n}$ (n = rank of $\mathcal{G}$) i.e. $\mathcal{U}\mathcal{G}$ tensorized by parities (origin of the double of representations).
2. $J_q\mathcal{G}$ is not a trivial deformation of $J_1\mathcal{G}$.
3. For $q = e^{it}$, not root of unity, $J_{t_o+t}\mathcal{G}$ is a non trivial deformation of $J_{t_o}\mathcal{G}$. Hence $J_{t_o}\mathcal{G}$ is not rigid.

To sum up, contrary to what is usually written, $J_q\mathcal{G}$ and $\mathcal{U}_t\mathcal{G}$ are not so similar.

- *FRT - model* (Faddeev - Reshetikhin - Takhtajan [FRT]) (denoted by $\mathcal{O}_q(G)$, where $\mathcal{O}(G)$ is the algebra of polynomial functions on a simple compact Lie group G).
  Let us recall a result of Gerstenhaber : there is a deformation of $\mathcal{O}(G)$, equivalent to $\mathcal{O}_q(G)$, which is *preferred*, that is, for which the coproduct stays unchanged.
  These deformations are explicitly given in [GGS] for $SU(n)$. With a very different method, we obtain these deformations for any simple compact Lie group.

II. The $SU(2)$ example [BFP]

One of the purposes of this work is to obtain some rigidities, more precisely some trivial cohomology spaces. We thought that a "topologization" (in a very heuristic way) could make more cocycles exact. From some explicit formulas, found by C. Fronsdal, giving the "deformed" generators from the classical ones, we had the idea to define

$\mathcal{A}(SU(2)) = \prod_{n\in 1/2\mathrm{N}} \mathrm{End}\,(V^{2n+1})$ (denoted by $\mathcal{A}$) equipped with its natural product topology (Fréchet), giving a setting where these formulas converge.

We found that :

- $\mathbb{C}[SU(2)], \mathcal{U}sl(2), \mathcal{U}_t sl(2), J_{e^{it}}sl(2), \forall t \in \mathbb{C}\backslash 2\pi\mathbb{Q}$ are dense sub-Hopf algebras of $\mathcal{A}$.
- $\forall t \in \mathbb{C}\backslash 2\pi\mathbf{Q}$, every deformed coproduct $\triangle_t$ extends itself to $\mathcal{A}$ and $\exists P_t \in \mathcal{A}\hat{\otimes}\mathcal{A}$ (projective tensor product) such that $\triangle_t(a) = P_t \triangle(a) P_t^{-1}$. Then the quasi-Hopf algebras $(\mathcal{A}, \triangle_0)$ and $(\mathcal{A}, \triangle_t)$ are isomorphic for every $t \in \mathbb{C}\backslash 2\pi\mathbb{Q}$.
- $H^2_{bi}(\mathcal{A}, \mathcal{A}) = \{0\}$ i.e. $\mathcal{A}$ is rigid.
- *Strong topological dual* : $\mathcal{A}(SU(2))^* \simeq \mathcal{H}(SU(2))$ (denoted by $\mathcal{H}$ which is the algebra of coefficients of $SU(2)$). It is known that $\mathcal{H} \simeq^{\mathrm{C}[a,b,c,d]} /_{\langle ad-bc-1\rangle}$ and by transposition of the structures of $(\mathcal{A}, \mu, \triangle_t)$ we exactly obtain FRT model $\mathcal{O}_q(SU(2))$ $\quad (q = e^{it})$.

III. The general case [BFGP]

Remark : we just saw that the (topological) dual of the Hopf algebra $\mathcal{A}$ is a Hopf algebra ($\mathcal{H}$). This is not the general case : usually, in infinite dimension, we only have $B^* \otimes B^* \subset (B \otimes B)^*$ (strict inclusion) and the product cannot be dualized. But the relations of type $[\mathcal{A}^* = \mathcal{H}, \mathcal{A}^{**} = \mathcal{A}]$ can be generalized : if V and W are, or both Fréchet and nuclear (A), or both Fréchet's dual and nuclear (B), then

$$V^* \hat{\otimes} W^* \simeq (V \hat{\otimes} W)^* \ , V^{**} \simeq V \ , V^* \hat{\otimes} W \simeq \mathcal{L}(V, W).$$

Therefore if a topological Hopf algebra (continuous composition laws) verifies (A) or (B), we can dualize it and it is reflexive (for topological arguments see [Gr]). We call such an algebra a *well behaved algebra.* An important example in the following is, for a compact group G, $\mathcal{D}'(G)$ and its dual $\mathcal{C}^\infty(G)$ (distributions and $\mathcal{C}^\infty$-functions).

The general case does not go exactly as $SU(2)$. We take t as a formal parameter and the triviality result of Drinfeld in [D2] replaces explicit formulas of the $SU(2)$ example. Let G be a simple compact Lie group and let $\hat{G}$ be its unitary dual, we define

$$\mathcal{A}(G) = \prod_{\pi \in \hat{G}} End(V_\pi)$$

endowed with its natural product topology. We have :

$$\mathcal{A}(G)^* \simeq \mathcal{H}(G) = \bigoplus_{\pi \in \hat{G}} End(V_\pi)$$

where $\mathcal{H}(G)$ is the algebra of coefficients of G, and

Proposition 1: $\mathcal{U}\mathcal{G} \subset \mathcal{A}(G)$ and $\mathbb{C}G \subset \mathcal{A}(G)$ as sub-Hopf algebras, and

$$\overline{\mathcal{U}\mathcal{G}} = \mathcal{A}(G), \qquad \overline{\mathbb{C}G} = \mathcal{A}(G).$$

($\mathcal{G}$ is the complexified Lie algebra of G).

Remarks:

1. *The elements of the Lie group* G (and also, as usually, the ones of its Lie algebra) *are into the algebra we are going to deform.*
2. $\mathcal{A}$ verifies:

   - the finite dimensional representations of $\mathcal{A}(G)$ are exactly these of G extend to $\mathcal{A}$.
   - center: $Z(\mathcal{A}(G)) = \prod_{\pi\in\hat{G}} \mathbb{C}1_\pi$.

sketch of proof: as $\hat{G}$ is a *complete* set of representations for G, $i : G \longrightarrow \mathcal{A}(G)$ such that $i(g) = (\pi(g))_{\pi\in\hat{G}}$ is continuous and one to one . Then $G \subset \mathcal{A}(G)$. We use the same kind of argument for $\mathcal{UG}$ taking care to the correspondence between $\mathcal{UG}$-modules and G-modules. ∎

In a classical way, we define a theory of topological deformations of bialgebras "classified" by a continuous H^2_{bi}. This need some technical results, but nothing really surprising. We have :

Theorem 2. *$\mathcal{A}(G)$ and $\mathcal{D}'(G)$ are rigid.*

(The interest of $\mathcal{D}'(G)$ will soon appear).

Sketch of proof:

We show $H^2_{bi,cont}(\mathcal{A}(G),\mathcal{A}(G)) = H^2_{bi,cont}(\mathcal{D}'(G),\mathcal{D}'(G)) = \{0\}$. To prove it we used the long exact sequence and for the Hochschild cohomology, classical arguments about finite dimensional algebras and cohomology of compact groups.

∎

Drinfeld shows in [D2] that :

(1) $\mathcal{U}_t\mathcal{G} \simeq \mathcal{UG}[[t]]$ (algebra homomorphism)

(2) $\exists P_t \in (\mathcal{UG}\otimes\mathcal{UG})[[t]]$ s.t. $\Delta_t = P_t\Delta_o P_t^{-1}$

As $\mathcal{UG}[[t]] \subset \mathcal{D}'(G)[[t]] \subset \mathcal{A}(G)[[t]]$ we can extend Δ_t (with (2)) to $\mathcal{A}(G)$ and (by(1)) we keep the undeformed product. Therefore we have, formally this time, the preceding $SU(2)$ results.

We have a new and very interesting result : we have constructed a QG-model on $\mathcal{D}'(G)$ and dually a deformation of $\mathcal{C}^\infty(G)$. This gives us the explicit link between QG and star-products. Indeed, if we take G a Poisson-Lie group, we have a deformation of the Poisson bracket.

This leads to the first result of the following theorem :

Theorem 3. *Let G be a simply connected, connected, simple compact Lie group, $\mathcal{G}$ its complexified Lie algebra.*

1. *We can extend the Drinfeld deformation $\mathcal{U}_t\mathcal{G}$ to $\mathcal{A}(G)[[t]]$ and to $\mathcal{D}'(G)[[t]]$, while keeping the product unchanged.*
2. *Dually, we obtain a preferred deformation of $\mathcal{H}(G)$ and $\mathcal{C}^\infty(G)$ which is, in the case of $\mathcal{H}(G)$, exactly the FRT-model for G (when it exists).*

We also show that $(\mathcal{H}(G/\Gamma), \times_t) \simeq (\mathcal{H}(G)^\Gamma, \times_t)$ (invariant elements under a normal subgroup Γ) which gives the same theorem when G is *not simply connected* (there are some possibilities to take a non normal Γ to have a notion of *quantum homogeneous space*)

Ideas for the proof of Th. 3: the first point is done just before the theorem. The second needs some results. Let $\mathcal{H} = \mathcal{H}(G)$ and C_{ij} be the coefficients of the "generating" representation π_G of $\mathcal{H}$.

Proposition 2. *Let $\mathcal{H}_t$ be a preferred deformation of $\mathcal{H}$, then $\{C_{ij}\}$ is a set of generators of $\mathcal{H}_t$ and if T is $[C_{ij}]$matrix, $\exists\, \mathcal{R} \in \mathcal{L}(V_{\pi_G} \otimes V_{\pi_G})$ invertible s.t. $\mathcal{R}\, T_1\, T_2 = T_2\, T_1\, \mathcal{R}$ (with $T_1 = T \otimes Id$ et $T_2 = Id \otimes T$).*

Sketch of proof: The main argument is that a deformation of a continuous representation π of G (compact) is trivial (i.e. equivalent to π on $\mathbb{C}[[t]]$) (see [BFGP]). ∎

The isomorphism $\varphi : \mathcal{U}_t\mathcal{G} \simeq \mathcal{U}\mathcal{G}[[t]]$ is not unique. We will call such maps *Drinfeld isomorphisms.* To each Drinfeld isomorphism corresponds a different way to include $\mathcal{U}_t\mathcal{G}$ in $\mathcal{A}(G)[[t]]$ and therefore a different deformation. We show :

Proposition 3. *Two Drinfeld isomorphisms define equivalent deformations of the associative algebra $\mathcal{H}(G)$.*

This result allows us to take a particular, "good" isomorphism (the one of the next lemma) in the further proofs.

Sketch of proof: We show an equivalence of coproducts on each component of $\mathcal{A}$ with the help of the rigidity of representations. ∎

Lemma 1. *Let ρ_t be a representation of $\mathcal{U}_t\mathcal{G}$ s.t. $\rho_{t=0} = \pi \in \hat{G}$. Then there exists a Drinfeld isomorphism φ s.t. $\rho_t = \pi \circ \varphi$.*

For the proof we use the rigidity of π, and the density theorem of Jacobson. ∎

End of the proof of Theorem 3 :

we will do it first for $G = SU(n), \mathcal{G} = sl(n)$ following step by step Drinfeld's way in ([D1] , section 7).

Let ρ_t be the representation of $\mathcal{U}_t\mathcal{G}$ considered in [D1] , $\rho_{t=0} = \pi_s$ be the natural standard representation of G, φ be the associated Drinfeld isomorphism (Lemma 1) i.e. $\rho_t = \pi_s \circ \varphi \quad (*)$. Let C_{ij} be the coefficients of π_s, ρ_{ij} be those of ρ_t. We have $\rho_{ij} \in (\mathcal{U}_t\mathcal{G})^{*t} \overset{T_\varphi}{\simeq} (\mathcal{U}\mathcal{G}[[t]])^{*t}(= \mathcal{U}\mathcal{G}^*[[t]])$ and by $(*)$ we obtain ${}^T\varphi(C_{ij}) = \rho_{ij}$. The relations given by Drinfeld on the ρ_{ij} are ,therefore, exactly the ones verified by the C_{ij} i.e. the ones of the FRT model for $SU(n)$.

For the other cases the proof is the same, the difficulty being the choice of the good representation ρ_t. For this choice a method is given in [FRT] . Then we exactly refind the relations of [FRT] for $SO(n)$ or $Sp(n)$.

■(Th 3)

Consequences:

- These models give a simple link with the harmonic analysis on G and allow the use of some techniques of this theory.
- We have a very simple (topological) duality between two of the main models, $\mathcal{A}$ (analogous to Drinfeld's) and FRT. Everything is reflexive (this answers a question asked in [FRT]).
- All the deformations are preferred. Then QGs just deform (and trivially!) the way to take tensor product of representations and leave the product unchanged (for "$\mathcal{A}$-side" and dually for $\mathcal{H}$).
- We obtained the existence and the way to calculate (difficult) new FRT-models e.g. for Spin(n) or the exceptional groups.
- The subalgebra of central functions of $\mathcal{H}(G)$ is not deformed, the usual QG product is here the classical commutative product on $\mathcal{H}(G)$.

IV. The topological quantum double [B2]

This work will also give us a more "coherent" definition of the Drinfeld's quantum double [D1] using a well defined and simple notion of duality: the strong topological duality. Then we will have *reflexivity* and even a kind of *self-duality*

IV. 1.*Construction*

We consider a Hopf algebra A, a Hopf algebra A' "in duality" (to be defined) with A, A^o being A' equipped with the opposite coproduct. We denote S' the opposite antipode of A.

The quantum double of A is the vector space $A^o \otimes A$ equipped with the canonical tensor product, and with the following product ("double cross-product" [Ma1])

$$\begin{aligned}\mu_{D(A)}((f \otimes a) \otimes (g \otimes b)) &= \sum_{(a)} f < g,\, S'(a_{(3)}) ? \, a_{(1)} > \otimes a_{(2)} b \\ &= \sum_{(a)(g)} f\, \alpha(a_{(1)} \otimes g_{(1)}) \otimes \beta(a_{(2)} \otimes g_{(2)}) b\end{aligned}$$

where $\alpha : A \otimes A^o \to A^o, \quad c \otimes h \mapsto \sum_{(c)} < h, S'(c_{(2)})\ ?\ c_{(1)} >$ defines a A-module structure on A^o.

and $\beta : A \otimes A^o \to A, \quad c \otimes h \mapsto \sum_{(c)} < h, S'(c_{(3)})\ c_{(1)} >$ defines a A^o-module structure on A (? stands for the variable).

Considering now a well behaved topological Hopf algebra A and its strong topological dual A^*, in place of A' we have α continuous: $\alpha : A \bar{\otimes} A^o \to A^o$, and the same for β. Then we define a continuous product on the inductive topological tensor product [Gr]

$$D(A) = A^o \bar{\otimes} A$$

thanks to the "double cross-product". This is our *definition of double.*

For $\mathcal{A}_t(G)$ (resp. $\mathcal{D}'_t(G)$) (t generic parameter or $t = 0$) one can dualize the structure and one obtains the coproduct ${}^t\mu_D$ on the vector space

$$\mathcal{A}_t(G)\hat{\otimes}\mathcal{H}_t(G) = D(\mathcal{A}_t(G))^* \ (resp.\ \mathcal{D}'_t(G)\hat{\otimes}\mathcal{C}^\infty_t(G) = D(\mathcal{D}'_t(G))^* \ ).$$

Moreover $D(A)^{**} = D(A)$ in these two cases.

IV. 2. *Uniqueness*

When A is cocommutative we easily show that the product on $A^o \bar{\otimes} A$ is the "smash-product" ([Sw1])

$$(f \otimes a).(g \otimes b) = \sum_{(a)} f(a_{(1)} \bullet g) \otimes a_{(2)} b$$

with $x \bullet h(y) = \sum_{(x)} < h, S(x_{(1)}) y\, x_{(2)} >$, the coadjoint action of A on A^o.

We know by [Sw2] that the "smash-product" is the trivial class of an extension theory.Then a natural question is: is there other non trivial extension of "double" type?

As in the preceding results we want these extensions in the category of quasi-Hopf algebras. Adapting [Sw2] in the way of [B1] we obtain a complex(see [B2]). Its cohomology is denoted $H^\bullet_{bisw}(A, \Lambda)$ with A a cocommutative Hopf algebra and a trivial $\Lambda - comodule$, Λ being a commutative Hopf algebra with some compatibility conditions. Extensions are classified by the H^2.
As always the extensions as quasi-Hopf algebras are a subset of $H^2_{bisw}(A, \Lambda)$.

Now we consider the case of $D(\mathcal{D}'(G))$.

Everything can be written with continuous cochains.Then we show:

Theorem 4. $H^2_{bis\omega,c}(\mathcal{D}'(G),\mathcal{C}^\infty(G)) = \{0\}$

Sketch of proof: We show

$$H^2_{s\omega,c}(\mathcal{D}'(G),\mathcal{C}^\infty(G)) \simeq H^2_{gr,c}(G,\mathcal{C}^\infty(G)^{inv}) = \{0\}$$

And the same for $H^1_{sw,c}(\mathcal{D}'(G),\mathcal{C}^\infty(G)^{\bar{\otimes}2})$
($H^\bullet_{gr,c}$ is the continuous group cohomology).
We conclude with a result of theorem1-type.

∎

This theorem gives a *uniqueness result*: the double of $\mathcal{D}'(G)$ is the unique -trivial- extension of $\mathcal{D}'(G)$ with $\mathcal{C}^\infty(G)$. I conjecture that it must be the same for $\mathcal{A}(G)$.

This result have another interest: this is a first example suggesting techniques and concepts for a possible more general theory of extensions for quantum groups (and even for general Hopf algebras). The aim is to find triviality results for extensions as in the classical semi-simple case but such a theory would be also useful to consider particular examples as those given in [Ma2] or [TV] or to try to construct a "Poincaré quantum group".

References

[BFFLS] F. BAYEN, M. FLATO, C. FRONSDAL, A. LICHNEROWICZ and D. STERNHEIMER: "Deformation theory and quantization I and II" *Ann. Phys.* **111** (78), 61-151.

[B1] P. BONNEAU: "Cohomology and associated deformations for non necessarily coassociative bialgebras". *Lett. Math. Phys.* **26** (1992), 277-283.

[B2] P. BONNEAU: "Topological quantum double". *Rev. Math. Phys.* **vol 16, n°2** (1994), 305-358

[BFP] P. BONNEAU, M. FLATO, G. PINCZON: "A natural and rigid model of quantum groups". *Lett. Math. Phys.* **25** (1992), 75-84.

[BFGP] P. BONNEAU, M. FLATO, M . GERSTENHABER, G. PINCZON: "Hidden group structure of quantum groups: strong duality, rigidity and preferred deformations". *Comm. Math. Phys.* **16** (1994), 125-156.

[BP] F. BIDEGAIN, G. PINCZON: "A star-product approach to noncompact quantum groups", *Lett. Math. Phys.* **33**, (1995), 231-240.

[D1] V.G. DRINFELD: "Quantum groups" in *Proc. ICM 1986*, AMS, Providence (1987), Vol. 1, 798-820 (ed. A. M. Gleason).

[D2] V.G. DRINFELD: "On almost cocommutative Hopf algebras". *Leningrad Math. J.* **1** (1990), 321-342.

[D3] V.G. DRINFELD: "Quasi-Hopf algebras". *Leningrad Math. J.*, **1** (1990), 1419-1457.

[FRT] L.D. FADDEEV, N.Y. RESHETIKHIN, L.A. TAKHTAJAN: "Quantization of Lie groups and Lie algebras". *Leningrad Math. J.* **1** (1990), 193-226.

[GGS] M. GERSTENHABER, A. GIAQUINTO, S.D. SCHACK: "Quantum symmetry" in *Lect. Notes in Math.* **1510** (1991), 9-46; Springer Verlag.

[Gr] A. GROTHENDIECK: "Produits tensoriels topologiques et espaces nucléaires". *Mem. Amer. Math. Soc.* **16** (1955).

[Ji] M. JIMBO: "A q-difference analog of $\mathcal{U}(g)$ and the Yang-Baxter equation". *Lett. Math. Phys.* **10** (1985), 63-69.

[Ma1] S. MAJID: "Physics for algebraists: non cocommutative and non commutative Hopf algebras by a bicross product construction" *J. of Alg.* **30** (1990), 17-64.

[Ma2] S. MAJID: "More examples of bicrossproduct and double crossproduct Hopf algebras". *Israël J. Math.* **72** (1990), 133-148.

[MZ] C. MARTIN, M. ZOUAGUI: private communication on work in progress.

[Sw1] M. E. SWEEDLER: "Hopf algebras" Benjamin, N. Y. (1969).

[Sw2] M.E. SWEEDLER: "Cohomology of algebras over Hopf algebras". *Trans.* AMS **133** (1968), 205-233.

[TV] P. TRUINI, V.S. VARADARAJAN: "The concept of a quantum semi-simple group". *Lett. Math. Phys.* **21** (1991), 287-292.

GROUPES DE TRANSFORMATIONS DES VARIETES STATISTIQUES

A Guy Rideau , en remerciement

G. BURDET AND M. PERRIN
Centre de Physique Théorique
CNRS-Luminy, case 907, F13288 Marseille CEDEX 9, France.

Abstract. A conformal implementation of projectively related statistical manifolds ($c \cup p$–geometry) is used to introduce the notion of $c \cup p$–transformations, associated to the 1-rigid action of a Lie subgroup of $Conf \cap Proj.$ groups.

1. Variétés statistiques

Le nom de variété statistique a été donné par Amari [1], en 1985, au triplet (S, g, t) constitué d'une variété de Riemann (S, g) et d'un champ t de tenseurs 3-covariants complètement symétriques. Cette notion a pour origine:

- d'une part les travaux de Rao [2] qui montrent que, sous des hypothèses de régularité élémentaires, tout espace probabilisé (ou système statistique) possède une structure de variété différentiable munie d'une métrique Riemannienne définie par la matrice de Fisher[3],

- d'autre part l'introduction par Amari [1, 4] et Chentsov [5] d'une famille à un paramètre α réel, de connexions linéaires, symétriques mais non métriques i.e. telles que :

$$\overset{\alpha}{\nabla} g = \alpha t. \tag{1}$$

Pour $\alpha = 0$ on retrouve ainsi la connexion métrique canonique (Riemannienne). Désignons par p la densité, par rapport à la mesure dominante, d'un élément P de la famille de probabilités et par $l = -\log p$ la fonction information correspondante, les champs de tenseurs g et t sont alors réalisés de la façon suivante

J. Bertrand et al. (eds.), Modern Group Theoretical Methods in Physics, 97–104.

$$g = E_P[dl \otimes dl]\,, \qquad t = E_P[dl \otimes dl \otimes dl]\,, \tag{2}$$

où E_P désigne l'espérance relative à P.

Introduisons le fibré $Gl(S)$ des repères linéaires de la variété statistique, ayant pour groupe structurel $Gl(n, \mathbb{R})$ où n désigne la dimension réelle de S. Notons θ la forme de soudure $\mathbb{R}^n$-valuée de $Gl(S)$, φ^α et Φ^α désignant respectivement la 1-forme et la 2-forme de courbure de $\overset{\alpha}{\nabla}$.

On peut alors écrire que

$$\varphi^\alpha = \varphi^0 + \alpha\tau \tag{3}$$

où τ est une matrice de 1-formes réelles, à valeurs dans le complément de l'algèbre de Lie $\mathcal{L}(O(n))$ dans $\mathcal{L}(Gl(n, \mathbb{R}))$, donnée par

$$\tau = -\frac{1}{2} g^{-1} \cdot t \cdot \theta \tag{4}$$

les points signalant des contractions évidentes. De la même manière la 2-forme de courbure se décompose en

$$\Phi^\alpha = \Phi^0 + \alpha\, D^0\tau + \alpha^2\, \tau \wedge \tau \tag{5}$$

où la dérivée covariante est donnée par

$$D^0\tau = d\,\tau + \varphi^o \wedge \tau + \tau \wedge \varphi^o\,. \tag{6}$$

2. Groupes de transformations

Une variété statistique met donc en présence deux champs de tenseurs g et t complètement symétriques. A travers les relations (3-4) il est clair que la donnée du triplet $(S, g, \overset{\alpha}{\nabla})$ est équivalente à celle du triplet (S, g, t). Le problème posé se traite alors en deux temps. Premièrement, introduisons $\mathcal{I}(S, g)$, le groupe des isométries de la variété Riemannienne, i.e. le sous-groupe des difféomorphismes de S qui laissent invariante la métrique, se remontent en des automorphismes du fibré des repères orthonormés $O(S, g)$, et préservent la connexion Riemannienne φ^o. Deuxièmement, pour α fixé, considérons le groupe $\mathcal{A}(S, \overset{\alpha}{\nabla})$ des transformations affines de S qui se remontent en des automorphismes de $Gl(S)$ laissant invariantes la forme de soudure θ et la forme connexion φ^α. En tenant compte de l'expression de φ^α écrite en (3), φ^o étant maintenue invariante , il suffit de conserver τ pour engendrer une transformations affine et ceci quelque soit $\alpha \neq 0$. Il vient donc assez trivialement la

Proposition 1. *Le groupe $Inv(S,g,t)$ des transformations laissant invariante une variété statistique est le sous-groupe(de Lie) des difféomorphismes de S défini par l'intersection $\mathcal{I}(S,g) \cap Inv(\tau)$.De plus $Inv(S,g,t)$ est un sous-groupe de $\mathcal{A}(S, \overset{\alpha}{\nabla})$, indépendamment de la valeur de $\alpha \neq 0$.*

De la même façon, au niveau infinitésimal, toujours en tenant compte des relations (3-4), on peut énoncer la

Proposition 2. *Tout champ de vecteurs X, engendrant un sous-groupe à un paramètre de $Inv(S,g,t)$, est tel que*

$$L_X g = 0, \qquad L_X t = 0. \tag{7}$$

Le relevé naturel $\tilde{X}$ de X dans $Gl(S)$ est tel que

$$L_{\tilde{X}}\theta = 0, \qquad L_{\tilde{X}}\varphi^o = 0, \qquad L_{\tilde{X}}\tau = 0. \tag{8}$$

Etudier le groupe d'invariance d'une structure géométrique donnée est le premier pas d'une démarche qui se poursuit naturellement par la recherche de groupes de transformations répondant à des problèmes d'équivalence au sens de Cartan. C'est dans ce cadre que s'inscrivent par exemple les géométries affines, projectives et conformes, et nous décrivons ci-après une structure géométrique adaptée à l'étude des systèmes statistiques.

3. La c ∪ p - géométrie

Le symbole c ∪ p signifie "conforme union projectif ", et par c ∪ p-géométrie nous désignons l' implémentation conforme de la géométrie projective que nous allons décrire.

Définition 1. *Etant données une variété statistique (S,g,t) et une 1-forme exacte ψ sur S, une variété statistique $(S,\tilde{g},\tilde{t})$ est dite c ∪ p- reliée à (S,g,t) si*

$$\tilde{g} = \eta g \qquad \text{et} \qquad \tilde{t} = \eta(t + sym(g \otimes \psi)), \tag{9}$$

où η est une fonction sur S, suffisamment dérivable,telle que: $d(Log\eta) = -\alpha\psi$ pour tout α non nul.

Un calcul usuel montre alors que la 1-forme connexion, canoniquement associée à la métrique $\tilde{g}$ conformément reliée à g, est donnée par

$$\tilde{\varphi}^o = \varphi^o - \frac{\alpha}{2}(\theta \otimes \psi + (\psi \cdot \theta)\mathbb{I}_n -^t \psi \otimes^t \theta). \tag{10}$$

D'autre part on détermine la matrice de 1-formes associée à $\tilde{t}$ comme dans la relation (4), ce qui conduit à :

$$\tilde{\tau} = \tau - \frac{1}{2}(\theta \otimes \psi + (\psi \cdot \theta)\mathbb{I}_n -^t \psi \otimes^t \theta). \tag{11}$$

Puis on en déduit, par une relation identique à (3), que

$$\tilde{\varphi}^\alpha = \varphi^\alpha - \alpha(\theta \otimes \psi + (\psi \cdot \theta)\mathbb{I}_n). \tag{12}$$

Ceci démontre la

Proposition 3. *Pour tout $\alpha \neq 0$, les α-familles de connexions de deux variétés statistiques c ∪ p - reliées sont reliées projectivement.*

Par un calcul simple on établit alors la

Proposition 4. *Les 2-formes de courbure de deux variétés statistiques c ∪ p- reliées vérifient*

$$\tilde{\Phi}^\alpha = \Phi^\alpha + \alpha \quad \theta \wedge \overset{\alpha}{\nabla} \psi + \alpha^2(\theta \otimes \psi) \wedge (\psi \cdot \theta). \tag{13}$$

Pour finir mentionnons que cette c ∪ p - géométrie est caractérisée au choix par deux des trois tenseurs suivants: le tenseur de Weyl projectif, le tenseur de Weyl conforme et leur différence. Nous renvoyons à la Réf.[6] pour plus de détails.

4. Les c ∪ p - transformations

Introduisons maintenant les difféomorphismes qui réalisent la c ∪ p - géométrie.

Définition 2. *Deux variétés statistiques (S, g, t) et $(S, \tilde{g}, \tilde{t})$ seront dites c ∪ p - équivalentes s'il existe f, un difféomorphisme de S, tel que :*

$$\tilde{g} = f^*(\eta g) \qquad \text{et} \qquad \tilde{t} = f^*(\eta(t + sym(g \otimes \psi))). \tag{14}$$

L'ensemble des difféomorphismes ayant cette propriété constitue le groupe des c ∪ p - transformations de la variété statistique noté $C \cup P\ (S, g, t)$.

Comme dans la Sect.2, nous prenons en compte l'aspect conforme du problème et extrayons de la Réf. [7] la propriété selon laquelle il existe une correspondance biunivoque entre les classes d'équivalence conforme de (S, g) et les plongements du fibré des repères conformes $CO(S, g) \hookrightarrow Gl(S)$ i.e. avec les CO-structures. Puis nous tenons compte de l'aspect projectif exprimé par la Prop.3. Ceci nous permet d'énoncer la :

Proposition 5. *Les c ∪ p - transformations sont en bijection avec les automorphismes de $Gl(S)$ qui conservent les plongements $CO(S, g) \hookrightarrow Gl(S)$*

qui leur correspondent, et assurent l'équivalence projective des 1-formes des α - connexions :

$$\tilde{\varphi}^{\alpha} = f^{*}(\varphi^{\alpha} - \alpha(\theta \otimes \psi + (\psi \cdot \theta)\mathbb{I}_n)). \tag{15}$$

Ainsi $C \cup P\ (S, g, t)$ est constitué des transformations simultanément projectives et conformes . Ces deux types de transformations engendrent des groupes de Lie de dimension finie (pour dim $S > 2$), conséquence de l'existence de connexions de Cartan [7] correspondant à des structures 2-rigides [8]. Les dimensions maximales sont :

$\frac{1}{2}(n+1)(n+2) = dim\ O(n+1,1)$ pour le cas conforme ,

$n(n+2) = dim\ PGl(n,\mathbb{R})$ pour le cas projectif.

Le groupe $C \cup P\ (S, g, t)$ est donc aussi de dimension finie, mais sa dimension ne peut être déterminée simplement par la recherche du sous-groupe maximal de $O(n+1,1) \cap PGl(n,\mathbb{R})$ car il faut tenir compte de l'aspect *groupe de transformations*. Procèdons alors au niveau infinitésimal.

Définition 3. *Un champ de vecteurs X est une $c \cup p$ - transformation infinitésimale i.e. appartient à $\mathcal{L}(C \cup P\ (S, g, t))$, si le groupe de transformations locales à 1-paramètre engendré par X appartient à $C \cup P\ (S, g, t)$.*

Ce groupe local prolongé à $Gl(S)$ induit un champ de vecteurs $\tilde{X}$, on démontre alors la

Proposition 6. *Il y a équivalence des conditions suivantes :*

i) X est une $c \cup p$ - transformation infinitésimale de (S, g, t),

ii)

$$L_X g = \varepsilon\, g, \qquad L_X t = \varepsilon\, t + sym(g \otimes \xi), \tag{16}$$

où le facteur conforme infinitésimal ε est relié à la forme infinitésimale ξ de ψ par $d\varepsilon = -\alpha\, \xi$,

iii)

$$\begin{cases} L_{\tilde{X}}\theta &= \frac{1}{2}\varepsilon\, \theta \\ L_{\tilde{X}}\varphi^{o} &= \frac{1}{4}(\theta \otimes d\varepsilon + (d\varepsilon \cdot \theta)\mathbb{I}_n -^{t} d\varepsilon \otimes^{t} \theta) \\ L_{\tilde{X}}\varphi^{\alpha} &= -\alpha(\theta \otimes \xi + (\xi \cdot \theta)\mathbb{I}_n) \end{cases} \tag{17}$$

où ε et ξ désignent ici les tirés en arrière sur $Gl(S)$ des objets correspondants définis sur S.

Cette proposition signifie simplement que les champs de vecteurs doivent être simultanément conformes et projectifs.

Sur l'espace Euclidien $\mathbb{R}^n$, où les groupes des transformations conformes et projectives atteignent leur dimension maximale, l'expression des champs de vecteurs correspondants est bien connue, à savoir :

$$\begin{cases} X_{proj} &= M \cdot x + A + (B \cdot x)x \\ X_{conf} &= L \cdot x + A + ax + Cx^2 - 2(C \cdot x)x \end{cases} \tag{18}$$

où : A $\in \mathbb{R}^n$, B and C $\in \mathbb{R}^{n*}$, M $\in \mathcal{L}(Gl(n,\mathbb{R}))$, $L \in \mathcal{L}(O(n))$, $a \in \mathbb{R} \setminus \{0\}$. On constate ainsi que la partie commune à X_{proj} et X_{conf} se limite à :

$$X = L \cdot x + A + a\,x \tag{19}$$

et que l'algèbre $\mathbb{R}^{n*}$ a disparu. Donc $X \in \mathcal{L}(\mathbb{R}^n \ltimes CO(n))$ dont la dimension est $\frac{1}{2}\,n\,(n+1)+1$, d'où la

Proposition 7. $\quad Dim\,\mathcal{L}(C \cup P\,(S,g,t)) \leq \dfrac{1}{2}\,n\,(n+1)+1.$

Nous constatons donc que $Dim\,\mathcal{L}(C \cup P\,(S,g,t)) \leq Dim\,Gl(S)$ ce qui laisse supposer le rôle prépondérant d'une structure 1-rigide dans ce problème. Effectivement, la notion proposée d'équivalence entre variétés statistiques est construite sur la classe conforme de la métrique donnée qui induit une CO-structure unique dont la présence est essentielle, et effectivement

$$Dim\,CO(S) = \frac{1}{2}\,n\,(n+1)+1.$$

5. Un exemple significatif : la loi gaussienne

La loi gaussienne $N(\mu,\sigma^2)$ pour $\mu \in \mathbb{R}$ et $\sigma \in \,]0,+\infty[$ est une famille exponentielle dont la densité s'écrit

$$p(x) = \frac{1}{\sigma\sqrt{2\pi}}\exp[-\frac{(x-\mu)^2}{2\sigma^2}]$$

par rapport à la mesure de Lebesgue sur $\mathbb{R}$. Il lui correspond une variété statistique bidimensionnelle et, dans la carte définie par $\{\mu,\sigma\}$, on utilise les réalisations (2) qui conduisent aux expressions suivantes pour la métrique g et le tenseur t :

$$g \;=\; \tfrac{1}{\sigma^2}\;[\; d\mu \otimes d\mu \;+2\;d\sigma \otimes d\sigma]$$

$$t \;=\; \tfrac{2}{\sigma^3}\;[\; sym(d\mu \otimes d\mu \otimes d\sigma) + 4\;d\sigma \otimes d\sigma \otimes d\sigma\;]$$

On en déduit en particulier que la variété statistique a une courbure scalaire constante

$$R = \alpha^2 - 1 = 2\,K$$

où K désigne la courbure sectionnelle. Ceci montre que, pour $\alpha = 0$, on obtient un espace hyperbolique familier : le demi-plan de Poincaré [4] . Le système différentiel (16) décrivant les c ∪ p - transformations infinitésimales s'écrit et se résoud alors facilement. Comme on est en dimension 2, les champs de vecteurs conformes engendrent une algèbre de Lie de dimension infinie, mais la seconde partie du système la limite fortement puisque le champ de vecteurs résultant est donné par

$$X_{c\cup p} = (a\mu + b)\;\partial_\mu + a\sigma\;\partial_\sigma$$

On vérifie alors qu'il engendre l'algèbre de Lie de $CO(2)$ réalisée par les matrices

$$\begin{pmatrix} \frac{b}{2} & -a \\ a & -\frac{b}{2} \end{pmatrix}$$

le facteur conforme se réduisant à la fonction constante $\epsilon = 2a$.

6. Conclusion

Nous terminerons sur les trois remarques suivantes :

- en théorie asymptotique de l'inférence statistique on utilise une règle de temps d'arrêt, portant sur le nombre d'observations à prendre en compte dans une suite d'expériences, laquelle est liée à une fonction sur la variété statistique: le facteur d'expansion η. Dans un papier récent [9] il a été montré que la prise en compte de cette règle de temps d'arrêt revient à modifier la géométrie de la variéte statistique par l'introduction du "facteur conforme" η , ce qui a motivé le travail présenté ci-dessus,

- les variétés statistiques "spéciales et exactes", introduites dans la référence [10], sont subdivisées en classes qui sont en correspondance biunivoque avec les classes conformes des variétés Riemanniennes sous-jacentes à travers la notion de c ∪ p - équivalence.

- pour les variétés statistiques α-plates existe un analogue (non commutatif) du théorème de Pythagore basé sur la notion d'entropie relative, ainsi qu'une définition de la projection orthogonale d'un point sur une sous-variété minimisant l'entropie relative. Ces deux propriétés font intervenir l'intersection orthogonale de géodésiques correspondant à une paire de connexions conjuguées (α et $-\alpha$-connexions). Il est clair que cette "figure" est conservée entre deux variétés statistiques c ∪ p - reliées .

References

1. Amari, S-i.(1985), *Differential Geometrical Methods in Statistics*, Lecture Notes in Statistics, **28**, Springer-Verlag New York .
2. Rao, C.R. (1945),*Bull.Calcutta Math.Soc.* **37** pp 81-91
3. Fisher R.A. (1925), *Proc. Cambridge Phil. Soc.* **122** pp. 700-725
4. Amari, S-i.(1987) Differential Geometrical Theory of Statistics - Towards New Developments in *Differential Geometry in Statistical Inference*, S.S. Gupta Ed., IMS Lecture Notes-Monographs Series,**10**, pp. 19-94
5. Chentsov, N.N. (1972) *Statistical Decision and Optimal Inference*, Nauka,Moscow (in Russian); translation in English, (1982) American MathematicalSociety, **53**, Providence, R.I.
6. Burdet, G., Perrin, M. Prétirage CPT 94/ P.3001.
7. Kobayashi,S. (1972) *Transformation Groups in Differential Geometry*, Springer Verlag Ed.
8. Gromov, M. (1988). Rigid transformation groups in *Géométrie Différentielle*, Travaux en cours **33**,Hermann, Paris pp 65-139
9. Okamoto, I. Amari,S-i. Takeuchi, K. (1991) *The Annals of Statistics*, **19** pp. 961-981
10. Noguchi, M. (1992)*Differential Geometry and its Applications*, **2** pp. 197-222

NON-FRUSTRATED SIGNED GRAPHS

PH. COMBE
Centre de Physique Théorique
CNRS-Luminy, case 907, F 13288 Marseille Cedex 9, France.

AND

H. NENCKA
Universidade da Madeira, Colégio dos Jesuítas,
Largo do Colegio, P 9000 Funchal, Madeira - Portugal.

Abstract. The problem of the frustration on graphs is investigated in terms of Cramerlike systems on a hypercube. This approach provides a linear algorithmic method to characterise the non-frustrated edge configurations. The case of complete graphs is given as example.

1. Introduction

The notion of frustration has been introduced in physics to analyse the disorder in terms of local invariance [31] and has been extensively used in the investigation of the spin glass models (for a review of the original papers see [25]) and more generally in the case of ill condensed matter [2]). However frustration may be associated to an old mathematical problem, the edge coloring of graphs [12]. The notion of frustration is useful in many physical problems in particular in the case of modelisation based on gauge theory on a lattice. An example the study of the ground-state and the elementary excitations of a glass (vitreous silica) [29], another is given by the modelisation of the collective properties of neural networks.

To understand and simulate cognitive processes it was proposed a modelisation by networks of automata (formal neurons[24]) with a high degree of feedback; the attractor neural networks (see e.g. [3, 19, 20, 22, 30]). The topology of such networks is given by a graph, the formal neurons being at the vertices of the graph and the edges being associated to pairs of connected formal neurons. The dynamics of a formal neural networks, with high degree of feedback, can be identified, in a natural way, to a dynamical

J. Bertrand et al. (eds.), Modern Group Theoretical Methods in Physics, 105–113.

system which is time evolution. An usual model of attractor network with continuous time neurons is the so-called *additive model*[1, 14] defined by the equations:

$$\frac{dx_i}{dt} = -\frac{x_i}{\tau_i} + \sum_{j=1}^{n} W_{ij} f_j(x_j), \qquad W_{ii} = 0, \tag{1}$$

where τ_i is a time constant and f_i is usually a sigmoidal function. This model is invariant under the simultaneous change of the state $x_i \to -x_i$ of the vertex x_i and $W_{ij} \to -W_{ij}$ of the weights of edges ending at vertex x_i.

In this approach recognition abilities of the network are associated to the long-term behaviour of the solutions of the dynamical system (1). In the state space $\mathcal{X}$ of the system, the long-term behaviour of a solution $x(t)$ starting at $x \in \mathcal{X}$, can be characterized by the *ω-limit set* $\omega(x) \subset \mathcal{X}$ [16, 17, 27]. A point $y \in \mathcal{X}$ is an element of the ω-limit set $\omega(x)$ if there exists an increasing sequence $t_k \to +\infty$ such that $x(t_k) \to y$. The ω-limit set can contain not only punctual and cyclic attractors but also chaotic attractors.

The behaviour of the dynamical system (1) depends strongly of the topology of the network as well as of the matrix $[W_{ij}]$ of synaptic weights which models the excitatory or inhibitory nature (sign) and the strength (modulus) of the connections between the formal neurons of the network.

For a large class of dynamics of activation the convergence can be proved under the constraint that the synaptic connections are symmetric [8, 14]. Indeed in this case we can construct a Lyapounov function, which insures that the neural network converges to some equilibrium state.

Such symmetry constraint is very far from the biological reality. To be more close to biological networks we have to relax, at least partially, the symmetry condition. The study of *asymmetric synaptic connections* [21, 23, 28] is more difficult because, in general, we lost the possibility to construct a Lyapunov function. However for the wider class of *symmetrically signed networks* [4, 9, 18]: networks with symmetrically signed synaptic weights,

$$J_{ij} = \mathrm{sgn} W_{ij} = \mathrm{sgn} W_{ji} = \pm 1\,, \tag{2}$$

techniques similar to those of statistical mechanics are still useful.

To prove the convergence of a symmetrically signed network of additive type toward a stable state is to show that up to a gauge transformation the associated signed graph is non-frustrated [4, 9, 18].

In section 2 we recall some fundamental notion in graph theory, section 3 is devoted to the definition of the frustration on graphs. An application to complete graphs K_n are given in section 4.

2. Some basic notions and notations on graph theory

A (finite) *graph* $\mathcal{G} = (\mathcal{V}, \mathcal{L})$ consists of a (finite) collection of points $\mathcal{V} = \{x_1, x_2, \ldots, x_n\}$ called *vertices*, and a collection of *edges* $\mathcal{L} = \{\ell_1, \ell_2, \ldots, \ell_m\}$. Each edge is either homeomorphic to an interval of the real line and joins two distinct vertices, or is homeomorphic to a circle (*loop*) and joins a given vertex to itself [5, 6, 7, 11, 15, 26]. It is assumed that any two distinct edges are either disjoint, or else intersect in a common end point. An edge ℓ has an endpoint set $\partial\ell$ which contains either one or two elements of the vertex set $\mathcal{V}$. In the following we are interested by *simplicial graphs* that is graphs which have no loop and only one edge between two given vertices. For a simplicial graph an edge is defined by a couple of vertices $\ell = (x, y)$, if the pair is directed we speak of *directed graph* and between vertices x and y there is, at most, two edges, and if the pair is non-directed we speak of *(non-directed) graph* in this case between vertices x and y there is, at most, one edge. In the following we limit ourself to non-directed graphs. The number $d(x)$ of edges meeting at $x \in \mathcal{V}$ is called the *degree* of the vertex x.

If $\mathcal{G} = (\mathcal{V}, \mathcal{L})$ and $\mathcal{G}' = (\mathcal{V}', \mathcal{L}')$ are graphs, $\mathcal{G}'$ is a *subgraph* of $\mathcal{G}$ if and only if $\mathcal{V}' \subset \mathcal{V}$, $\mathcal{L}' \subset \mathcal{L}$ and $\{\partial(\ell'), \ell' \in \mathcal{V}'\} \subset \{\partial(\ell), \ell \in \mathcal{L}\}$.

A *walk* η on a graph is an alternating sequence of vertices and edges in which each edge is flanked by its own vertices, no edge is used more than once but at least one edge is used. A *path* is a walk such that no vertex is used more than once. A graph is said *connected* if there is at least one path between any two vertices. A *cycle* is a path which begins and ends at the same vertex. If the cycle contains n edges we speak of *n-cycle* Σ_n. The set of paths Γ can be equipped with a non commutative, non everywhere defined product. Indeed let $\gamma = \{\ell_1, \ell_2, \ldots, \ell_p\}$ and $\gamma' = \{\ell'_1, \ell'_2, \ldots, \ell'_q\}$ being two paths in $\mathcal{G}$ having no common edges and such that $\partial\gamma \cap \partial\gamma' \neq \emptyset$, we can define a new path $\{\ell_1, \ell_2, \ldots, \ell_p, \ell'_1, \ell'_2, \ldots, \ell'_q\}$ denoted by $\gamma\triangle\gamma'$, iff $\partial\gamma \cap \partial\gamma' = \partial\ell_p \cap \partial\ell'_1 \neq \emptyset$. In particular any cycle C can be decomposed as the product of two paths $C = \gamma\triangle\gamma'$. This decomposition is not unique, because any vertex of the cycle can be chosen as origin of the path γ. The product of two cycles C' and C'' having at least one common edge is the cycle $C = C'\triangle C''$ obtain by concatenation of the two cycles and suppression of the common edges. Concatenation is just the symmetric-difference of the two set C' and C''.

The edges of a graph fall into two classes, the *cyclic edges* laying on some cycle, and the *acyclic edges* laying on no cycle. For frustration problems we can limit ourself to graph having only cyclic edges, the contribution of acyclic edges being trivial. A graph $\mathcal{G} = (\mathcal{V}, \mathcal{L})$, with no acyclic edge, have a generating family of cycles $\mathcal{P} = \{P_1, P_2, .., P_r\}$ which is a minimal set of cycles such that any cycle of the graph $\mathcal{G}$ can be written as a finite product of elements of $\mathcal{P}$. The cardinality of the family of fundamental cycles is given by the *cyclomatic number*[5] :

$$r(\mathcal{G}) = \mid \mathcal{P} \mid = \mid \mathcal{L} \mid - \mid \mathcal{V} \mid + 1, \tag{3}$$

where $\mid A \mid$ is the cardinality of the set A. Notice that, in general, this family is not unique even if the fundamental cycles are minimal cycles. A *minimal cycle* is a cycle which does not contain any sub-cycle.

In the following we are interested by *signed graphs* $\mathcal{G} = (\mathcal{V}, \mathcal{L}, J)$, that is graphs where each edge $\ell = \{x_i, x_j\}$ are labeled by a sign. The edge configuration of the graph $\mathcal{G} = (\mathcal{V}, \mathcal{L})$ is given by an application J from $\mathcal{L}$ into $\{-1, +1\}$ $J(\ell) = J_{ij} = \pm 1$.

3. Frustration on graphs

To a signed graph $\mathcal{G} = (\mathcal{V}, \mathcal{L}, J)$ we can associate the formal Hamiltonian H defined on $\{-1, 1\}^n$ by,

$$H(\sigma) = -\frac{1}{2} \sum_{(i,j)\in\mathcal{L}} J_{ij}\sigma_i\sigma_j. \tag{4}$$

This hamiltonian is invariant under the gauge transformation [31]:

$$\begin{cases} \sigma_i & \to -\sigma_i, \\ J_{ij} & \to -J_{ij} \text{ if } (i,j) \in \mathcal{L}. \end{cases} \tag{5}$$

A configuration $\sigma = \{\sigma_1, \ldots, \sigma_n\}$ of the signed graph $\mathcal{G} = (\mathcal{V}, \mathcal{L}, J)$ with formal Hamiltonian 4 is said *non-frustrated* if its energy $H(\sigma)$ is just equal to $- \mid \mathcal{L} \mid$, with $\mid \mathcal{L} \mid$ the number of edges, and *frustrated* otherwise.

By extension a Hamiltonian is non-frustrated if there exists, at least, one non-frustrated configuration and it is frustrated if all the configurations are frustrated.

In the following we assume that the simplicial signed graph $\mathcal{G} = (\mathcal{V}, \mathcal{L}, J)$ has only cyclic edges, so that the graph $\mathcal{G}$ is completely defined if we know a generating family of fundamental cycles $\mathcal{P} = \{P_1, P_2, \ldots, P_r\}$.

The graph $\mathcal{G}$ is non-frustrated if the signs of edges are such that the formal Hamiltonian $H(\sigma)$ is non-frustrated. Then the signed graph $\mathcal{G}$ is non-frustrated if all the cycles, as subgraphs of $\mathcal{G}$, are non-frustrated.

To each cycle C of the graph we can associate a function $\varphi(C)$ which take the value $+1$ if the cycle is non-frustrated and -1 if it is frustrated.

Let $C_k = \{\ell_1, \ell_2, \ldots, \ell_k\}$, $\ell_i = \{x_i, x_{i+1}\}$ with $x_1 = x_{k+1}$ be a k-cycle, then the frustration function [30] φ is defined by

$$\varphi(C_k) = \prod_{i=1}^{k} J(\ell_i) = \prod_{n=1}^{k} J_{i_n i_{n+1}}, \tag{6}$$

The simplicial signed graph $\mathcal{G}$ with no acyclic edges being fully defined by a generating family $\mathcal{P}$, the knowledge of the frustration function for each cycle in $\mathcal{P}$ is sufficient to determine the frustration function of any cycle. Indeed if P_1 and P_2 are two elements of $\mathcal{P}$ having common edges, their product is a cycle $C = P_1 \triangle P_2$ such that:

$$\varphi(C) = \varphi(P_1 \triangle P_2) = \varphi(P_1)\varphi(P_2). \tag{7}$$

It follows that a signed graph $\mathcal{G} = (\mathcal{V}, \mathcal{L}, J)$ generated by the family of fundamental cycles $\mathcal{P} = \{P_1, \ldots, P_r\}$ is non-frustrated if each cycle P_i is non-frustrated, then the condition of non-frustration is given by the k equations,

$$\varphi(P_1) = \varphi(P_2) = \ldots = \varphi(P_r) = 1. \tag{8}$$

Let us remark that the space of configuration of weights for a signed graph $\mathcal{G} = (\mathcal{V}, \mathcal{L}, J)$ is isomorphic to $\{-1,1\}^{|\mathcal{L}|}$ the vertices of the $|\mathcal{L}|$-cube with edges of length 2. In this space each equation in relation (8) defines a family of hyperplanes. Let $P_i = \{\ell_1, \ell_2, \ldots, \ell_k\}$ be a k-cycle and denotes by $J(\ell_j)$ the sign of edge ℓ_j, the non-frustration condition of the k-cycle P_i rewrites:

$$\varphi(P_i) = J(\ell_1)J(\ell_2)\ldots J(\ell_k) = 1, \tag{9}$$

On the k-cube $\{-1,1\}^k \subset \{-1,1\}^{|\mathcal{L}|}$ the equation (9) is equivalent to the family of linear equations:

$$J(\ell_1) + J(\ell_2) + \ldots + J(\ell_k) = k - 4m, \quad 0 \leq m \leq \frac{k}{2}, \quad m \in \mathbb{N} \cup \{0\}. \tag{10}$$

To each solution of (9) is associated a solution in the set of equations (10) which belongs to the k-cube $\{-1,1\}^k$. Conversely to each possible values of m the solution of the corresponding equation in (10) which belong to the k-cube $\{-1,1\}^k$ is associated a solution of (9).

If $k = 2p$ is an even integer, the equality $m = k/2$ can be realized, then to solve equation (9) is equivalent to solve $k/2 + 1 = p + 1$ equations (10).

If $k = 2q + 1$ is an odd integer the highest value reached by m is now: $(k-1)/2 = q$. The number of equations (10) is in this case $(k-1)/2 + 1 = q + 1$.

The condition of non-frustration of a graph $\mathcal{G}$ generated by the family of fundamental cycles $\mathcal{P} = \{P_1, \ldots, P_r\}$ given by the system of r equations 8 is now replaced by the family of N, r-dimensional "Cramer's systems",

$$\left\{ \begin{array}{lcl} \sum\limits_{\ell \in P_1} J(\ell) & = & k_1 - 4m_1, \\ \sum\limits_{\ell \in P_2} J(\ell) & = & k_2 - 4m_2, \\ ---- & - & ----- \\ \sum\limits_{\ell \in P_r} J(\ell) & = & k_r - 4m_r, \end{array} \right. \tag{11}$$

with m_i an integer such that $0 \leq m_i \leq k_i/2$, where k_i is the number of edges in the cycle P_i, $i = 1, 2, \ldots, r$. The number of Cramer's systems being $N = a_1 a_2 \ldots a_r$ where $a_i = p_i + 1$ if $k_i = 2p_i$ and $a_i = q_i + 1$ if $k_i = 2q_i + 1$.

The non-frustration conditions for the graph $\mathcal{G}$ are given by the Cramer's systems which have solutions in $\{-1, 1\}^{|\mathcal{L}|}$.

These family of Cramer's systems (11) provide an algorithmic procedure to determine all the non-frustrated configurations J for an arbitrary signed graph $\mathcal{G}$. The number of Cramer's system increase very quickly with the cardinality of $|\mathcal{V}|$ nevertheless the constraint that the solutions have to be vertices of the hypercube $\{-1, 1\}^{|\mathcal{L}|}$ eliminates the most part of linear systems. In the case of a graphlike structure corresponding to the case where $|\mathcal{V}|$ is not finite, as for tiling, additional spatial symmetries are necessary to solve the problem. Finally if the generating family $\mathcal{P}$ is invariant under a sub-group of the group of permutation of vertices, the solutions might be gathered in a few number of equivalence classes. Let us now illustrate the method on the complete graphs K_n, $n \geq 3$.

4. Non-frustrated configuration on K_n

In the case of the complete graphs K_n with n positive integer larger than 3 the classification of non-frustrated configuration is rather simple because of the symmetry of such structure. Indeed the complete graph K_n has $|\mathcal{V}| = n$ vertices and $|\mathcal{L}| = \frac{1}{2}n(n-1)$ edges, and the minimal cycles are just 3-cycles. A natural generating family is furnished by a choice of $|\mathcal{P}| = \frac{1}{2}(n-1)(n-2)$ triangular cycles, then $k_i = 3$ and $m_i = 0$ or 1 for any cycle in $\mathcal{P}$. The number of Cramer's systems (11) is $N_n = 2^{(n-1)(n-2)/2}$, but few systems have solutions in $\{-1, 1\}^{|\mathcal{L}|}$. The value $m_i = 0$, $\forall i$ corresponds to the trivial case where all the $J_{ij} = +1$. Finally it is convenient to write this system in a way which exhibits a recurrence in equations (12) we have chosen as subgraph $K_{n-1} \subset K_n$ the complete graph with vertices $\{x_2, x_3, \ldots, x_n\}$.

$$
K_n \left\{ \begin{array}{l}
\left\{ \begin{array}{lclclcl}
J_{12} & + & J_{13} & + & J_{23} & = & b_{12} \\
J_{13} & + & J_{14} & + & J_{34} & = & b_{13} \\
------- & - & ---- & - & ---- & - & ----- \\
J_{1(n-1)} & + & J_{1n} & + & J_{(n-1)n} & = & b_{1(n-1)}
\end{array} \right. \\
\\
K_{n-1} \left\{ \begin{array}{lclclcl}
J_{23} & + & J_{24} & + & J_{34} & = & b_{23} \\
------ & - & ---- & - & ---- & - & ----- \\
J_{2(n-1)} & + & J_{2n} & + & J_{(n-1)n} & = & b_{2(n-1)} \\
------ & - & ---- & - & ---- & - & ----- \\
------ & - & --- & - & ---- & - & ----- \\
J_{(n-2)(n-1)} & + & J_{(n-2)n} & + & J_{(n-1)n} & = & b_{(n-2)(n-1)}
\end{array} \right.
\end{array} \right. \tag{12}
$$

where the $b_{ji} = 3$ or -1.

The simplest case is furnished by the graph K_3 which is just a triangle, up to a permutation of the vertices, we obtain the two non-frustrated configurations:

J_{12}	J_{13}	J_{23}
+	+	+
+	−	−

Table 1: Non-frustrated configuration of the complete graph K_3

By recurrence for an arbitrary complete graph K_n with $n = 2p$ or $2p+1$, we have $p+1$ non-frustrated edge-patterns, up to permutations of vertices.

The p non-frustrated configuration of K_{2p} different of $J_{ij} = +1$, are given by:

$$J_{ij}^{(k)} = \begin{cases} +1 & \text{if } j < k \\ -1 & \text{if } j \geq k \end{cases} \quad 1 \leq i < j \leq 2p, \quad p-1 \leq k \leq 2p. \tag{13}$$

The $p+1$ corresponding edge-patterns are characterized by the values of Hamming distances δ to the pattern $J_{ij} = +1$ for all i and j.

$$\delta = 0, 2p-1, 2(2p-2), \ldots, (k-1)(2p-k+1), \ldots, (p-1)(p+1), p^2 \tag{14}$$

The p non-frustrated configuration of K_{2p+1} different from $J_{ij} = +1$ are given by:

$$J_{ij}^{(k)} = \begin{cases} +1 & \text{if } j < k \\ -1 & \text{if } j \geq k \end{cases} \quad 1 \leq i < j \leq 2p+1, \quad p \leq k \leq 2p+1, \tag{15}$$

and the $p+1$ corresponding edge-patterns are characterized by the values

$$\delta = 0, 2p, 2(2p-1), \ldots, (k-1)(2p-k+2), \ldots, (p-1)(p+2), p(p+1) \tag{16}$$

of Hamming distances to the pattern $J_{ij} = +1$ for all i and j.

In Table 2 the non-frustrated configurations of the first complete signed graphs K_n are given, up to permutations of the vertices, with a convenient choice of representatives to show the recurrence.

	J_{12}	J_{13}	J_{14}	J_{15}	J_{16}	J_{17}	J_{23}	J_{24}	J_{25}	J_{26}	J_{27}	J_{34}	J_{35}	J_{36}	J_{37}	J_{45}	J_{46}	J_{47}	J_{56}	J_{57}	J_{67}	δ
K_3	+	+	.	.	.	.	+	.	.	.	.	.	.	.	.	.	.	.	.	.	.	0
	+	−	.	.	.	.	−	.	.	.	.	.	.	.	.	.	.	.	.	.	.	2
K_4	+	+	+	.	.	.	+	+	.	.	.	+	.	.	.	.	.	.	.	.	.	0
	+	+	−	.	.	.	+	−	.	.	.	−	.	.	.	.	.	.	.	.	.	3
	+	−	−	.	.	.	−	−	.	.	.	+	.	.	.	.	.	.	.	.	.	4
K_5	+	+	+	+	.	.	+	+	+	.	.	+	+	.	.	+	.	.	.	.	.	0
	+	+	+	−	.	.	+	+	−	.	.	+	−	.	.	−	.	.	.	.	.	4
	+	+	−	−	.	.	+	−	−	.	.	−	−	.	.	+	.	.	.	.	.	6
K_6	+	+	+	+	+	.	+	+	+	+	.	+	+	+	.	+	+	.	+	.	.	0
	+	+	+	+	−	.	+	+	+	−	.	+	+	−	.	+	−	.	−	.	.	5
	+	+	+	−	−	.	+	+	−	−	.	+	−	−	.	−	−	.	+	.	.	8
	+	+	−	−	−	.	+	−	−	−	.	−	−	−	.	+	+	.	+	.	.	9
K_7	+	+	+	+	+	+	+	+	+	+	+	+	+	+	+	+	+	+	+	+	+	0
	+	+	+	+	+	−	+	+	+	+	−	+	+	+	−	+	+	−	+	−	−	6
	+	+	+	+	−	−	+	+	+	−	−	+	+	−	−	+	−	−	−	−	+	10
	+	+	+	−	−	−	+	+	−	−	−	+	−	−	−	−	−	−	+	+	+	12

Table 2: Non-frustrated configuration of complete graph K_n for $3 \leq n \leq 7$. The last column gives the number δ of negative signs or equivalently for each n the Hamming distance of an edge-pattern to the pattern with only positive edges.

References

1. Amari,S. I. (1982) Competitive and cooperative aspect in dynamic of neural excitation and self-organisation . In S. I. Amari and Arbib (Eds) *Competition and cooperation in neural nets.* Springer verlag New York, 1–28
2. Balian, R., Maynard, R., Toulouse, G. (1979) *Ill-Condensed Matter*, North-Holland, Amsterdam
3. Amit, D. (1989) *Modeling brain function.* Cambridge University Press
4. Benaïm, M.,(1992) *Dynamical Properties of Neural Nets Using Competitive Attraction Mechanisms* . Thèse, Ecole Nationale Supérieure de l'Aéronautique et de l'Espace, Toulouse
5. Berge, C. *Graphes et hypergraphes*, Dunod, Paris 1990
6. Bollobas,B. (1972) *Extremal Graph Theory*, Academic Press, London New York
7. Busaker, R. B., Saaty, T.L. (1965) *Finite Graph and Networks. An Introduction with Applications.* McGraw-Hill Book Company. New York Toronto, London, Sydney
8. Cohen, M. A., Grossberg, S. (1983) Absolute stability of global pattern information and parallel memory storage by competitive neural networks. *IEEE Transactions on Systems, Man and cybernetics SMC* **13**, 815-826
9. Combe, Ph., Nencka, H. (1995) Cooperative Neural Networks and Frustration on Graphs. To appear
10. Coxeter, H. S. M. (1964) Regular compound tesselations of the hyperbolic plane, *Proc. Roy. Soc.* **278**, pp 147–169

11. Coxeter, H.S.M., Moser, W.O.J. (1980) *Generators and Relations for Discrete Groups*, Springer Verlag, Berlin Heidelberg, New York
12. Fiorini, S., Wilson, R. J. (1977) *Edge Coloring of Graphs*, Pitman, London:
13. Gross,J.L., Tucker, T.W. (1987) *Topological Graph Theory*, A. Wiley–Interscience Publications, New York
14. Grossberg, S.(1969) On learning and energy entropy dependence in recurrent and non recurrent signed networks. *Journal of Statistical Physics* **48**, 105–132
15. Harrary, F. (1969) *Graph Theory* Addison-Wesley
16. Hirsch, M. W. (1982) Systems of differential equations which are competitive or cooperative. I: Limits sets. *SIAM J. Math. Anal.* **13**, 167–179
17. Hirsch, M. W. (1985) Systems of differential equations which are competitive or cooperative. II: Convergence almost every where. *SIAM J. Math. Anal.* **16**, 423–439
18. Hirsch, M. W. (1989) Convergent Activation Dynamics in Continuous Time Networks. *Neural Networks* **2**, 331–349
19. Hopfield, J.J. (1982) Neural networks and physical systems with emergent selective computational abilities. *Proc. Nat. Acad. Sci. USA* **79**, 2554–2558
20. Ingber, L. (1982) Statistical mechanics of neocortical interactions. *Physica* **5D**, 83–107
21. Kohonen, T., (1984) *Self organization and associative memory.* Springer Verlag, New York
22. Little, W. (1982) The existence of persistent state in the brain. *Math. Biosciences* **19**, 101–119
23. Maldbourg (von der), C., (1973) Self organization of orientation sensitive cells in the striate cortex. *Kybernetik* **14**, 85–100
24. Mc Culloch, W. S., Pitts, W. (1943) A logical calculus of ideas immanent in nervous activity. *Bull. Math. Biophys.* **5**,115–133.
25. Mezard, M., Parisi, G., Virasoro, A.-A. (1987) *Spin glass theory and beyond.* World Scientific
26. Ore, O. (1970) *Les graphes et leurs applications*, Dunod Paris
27. Palis, J. , De Melo, W. (1982) *Geometric Theory of Dynamical Systems.* Springer Verlag, New york
28. Parisi, G., (1986) Asymmetric neural networks and the process of learning. *J. Phys.* **A19**, L675–L680
29. Rivier, N. (1994) Glass: elementary excitations and geometry;from bounds to tunneling modes, *Philosophical Magazine* . **B 69**, pp 911–924
30. Shaw, G.L., Vasudevant, R. (1974) Persistent states of neural networks and random nature of synaptic transmission. *Math. Biosciences* **21**, 207–218
31. Toulouse, G. (1977) Theory of the frustration effect in spin glasses:I. *Commun. Phys.* **2**, *pp 115–119* . See also reprint in *Spin glass theory and beyond.* M. Mezard, G. Parisi, A.-A. Virasoro Eds. pp 99–103. World Scientific 1987
32. White,A.T. (1973) *Graphs,Groups and Surfaces.* North-Holland Publishing Company, New York

DE L'UBIQUITE DE LA CONDITION KMS

GERARD G. EMCH
Department of Mathematics, University of Florida
Gainesville, FL 32611, USA

Pour Guy, en témoignage d'amitié

Abstract. The roles of the KMS condition in statistical mechanics, quantum field theory, quantum ergodic theory, and geometric quantization, are compared.

1. Mécanique statistique quantique de l'équilibre

La condition KMS doit son nom aux travaux de Kubo [1] et de Martin & Schwinger [2] sur les propriétés analytiques des fonctions de Green en mécanique statistique. L'article de Haag, Hugenholtz & Winnink [3] en donne une formulation intrinsèque en termes de fonctions de corrélation, qui permet de lier cette propriété à la théorie des structures modulaires due à Tomita & Takesaki [4]. Il est remarquable que l'élucidation de telles structures ait déjà joué un rôle [5] dans l'élaboration du programme [6] dont est née la théorie des algèbres de von Neumann [7, 8]. Nous commençons notre revue à ce niveau primordial, ayant tout d'abord en vue la mécanique statistique de systèmes quantiques finis en équilibre thermodynamique canonique.

Soit $\mathcal{H}$ un espace de Hilbert à base dénombrable, de dimension finie ou infinie, $\mathcal{B}(\mathcal{H})$ l'algèbre des opérateurs linéaires bornés sur $\mathcal{H}$, et $\mathcal{S}$ l'*-idéal bilatère de $\mathcal{B}(\mathcal{H})$ formé des opérateurs de Hilbert-Schmidt sur $\mathcal{H}$: $\mathcal{S} = \{\Psi \in \mathcal{B}(\mathcal{H}) \mid Tr(\Psi^*\Psi) < \infty\}$; $\mathcal{S}$ est équipé d'une structure d'espace de Hilbert par le produit scalaire $< \Phi, \Psi >= Tr(\Phi^*\Psi)$. En particulier, notons que les isométries partielles $W_{\xi,\eta}$ définies – pour toute paire (ξ, η) de vecteurs unités dans $\mathcal{H}$ – par

$$W_{\xi,\eta} : \psi \in \mathcal{H} \mapsto (\eta, \psi)\, \xi \in \mathcal{H} \tag{1.1}$$

J. Bertrand et al. (eds.), Modern Group Theoretical Methods in Physics, 115–128.

appartiennent à $\mathcal{S}$ et que nous avons:

$$A\,W_{\xi,\eta} = W_{A\xi,\eta} \quad et \quad W_{\xi,\eta}\,A^* = W_{\xi,A\eta} \qquad \forall \quad A \in \mathcal{B}(\mathcal{H}); \tag{1.2}$$

et, pour toute base $\{\psi_k\}$ orthonormée dans $\mathcal{H}$,

$$\Psi_{kl} \equiv W_{\psi_l,\psi_k} \quad \text{c.-à-d.} \quad \Psi_{kl} : \psi \in \mathcal{H} \mapsto (\psi_l, \psi)\psi_k \in \mathcal{H} \tag{1.3}$$

définit une base $\{\Psi_{kl}\}$ orthonormée dans $\mathcal{S}$.

Les égalités (1.2) suggèrent deux représentations (la première linéaire, la seconde anti-linéaire) de $\mathcal{B}(\mathcal{H})$ agissant sur l'espace de Hilbert $\mathcal{S}$:

$$\pi^L(A)\Psi = A\,\Psi \quad \pi^R(A)\Psi = \Psi\,A^* \quad \forall \quad (A,\Psi) \in \mathcal{B}(\mathcal{H}) \times \mathcal{S} \tag{1.4}$$

Ces représentations jouissent de deux propriétés remarquables:
(a) elles sont le commutant l'une de l'autre, c.-à-d.

$$\pi^L(\mathcal{B}(\mathcal{H}))' = \pi^R(\mathcal{B}(\mathcal{H})) \quad ; \quad \pi^R(\mathcal{B}(\mathcal{H}))' = \pi^L(\mathcal{B}(\mathcal{H})) \tag{1.5}$$

où, pour tout $\mathcal{A} \subseteq \mathcal{B}(\mathcal{S}) : \mathcal{A}' \equiv \{B \in \mathcal{B}(\mathcal{S}) \mid [A,B] = 0 \quad \forall \quad A \in \mathcal{A}\}$;
(b) il existe un opérateur anti-unitaire $J : \mathcal{S} \mapsto \mathcal{S}$ tel que

$$J^2 = I \quad ; \quad J\pi^L(A)J = \pi^R(A) \quad \forall \quad A \in \mathcal{B}(\mathcal{H}); \tag{1.6}$$

cet opérateur est en effet issu de la conjugaison hermitienne dans $\mathcal{B}(\mathcal{H})$:

$$J\Psi = \Psi^* \quad \forall \quad \Psi \in \mathcal{S} \quad . \tag{1.7}$$

Rappelons qu'une algèbre de von Neumann sur un espace hilbertien $\mathcal{H}$ est une sous-*-algèbre $\mathcal{N} \subseteq \mathcal{B}(\mathcal{H})$ telle que $\mathcal{N} = \mathcal{N}''$. Nous serons particulièrement intéressés dans la suite par les algèbres de von Neumann $\mathcal{N}$ pour lesquelles il existe un opérateur J anti-unitaire involutif (c.-à-d. $J^2 = I$), agissant sur $\mathcal{H}$, et tel que $J\mathcal{N}J = \mathcal{N}'$.

Retournons à l'exemple précédent où $\mathcal{N} = \pi^L(\mathcal{B}(\mathcal{H}))$. Pour faire le plus explicitement possible le lien avec la mécanique statistique, introduisons un hamiltonien H agissant dans $\mathcal{H}$ et tel que $Tr\,\exp(-\beta H) < \infty$ pour un $\beta \in (0,\infty)$. Nous avons alors

$$H\psi_k = \epsilon_k\psi_k \tag{1.8}$$

où $\sum_k e^{-\beta\epsilon_k} < \infty$, et $\{\psi_k\}$ est une base orthonormée pour $\mathcal{H}$. La matrice de densité ρ décrivant l'équilibre canonique relatif à H pour la température $T = (k\beta)^{-1}$

$$\rho = \frac{\exp(-\beta H)}{Tr\,\exp(-\beta H)} \tag{1.9}$$

s'écrit alors comme une somme convexe

$$\rho = \sum_k \lambda_k \Psi_{kk} \quad , \quad \lambda_k = \frac{e^{-\beta\epsilon_k}}{\sum_k e^{-\beta\epsilon_k}} \quad ; \tag{1.10}$$

où les opérateurs Ψ_{kl} sont définis par (1.3). Il est alors naturel d'introduire le vecteur $\Phi \in \mathcal{S}$ défini par

$$\Phi \equiv \rho^{\frac{1}{2}} = \sum_k \lambda_k^{\frac{1}{2}} \Psi_{kk} \tag{1.11}$$

qui satisfait aux conditions suivantes:

$$J\,\Phi = \Phi \quad ; \tag{1.12}$$

$$< \Phi, \pi^L(A)\,\Phi >=< \varphi ; A > \tag{1.13}$$

avec

$$\varphi : A \in \mathcal{B}(\mathcal{H}) \mapsto < \varphi ; A >= Tr\,\rho A \in C \quad ; \tag{1.14}$$

de plus Φ est cyclique dans $\mathcal{S}$ pour $\pi^L(\mathcal{B}(\mathcal{H}))$, c.-à-d.

$$\{\pi^L(A)\,\Phi \mid A \in \mathcal{B}(\mathcal{H})\} \quad \text{est dense dans} \quad \mathcal{S}. \tag{1.15}$$

Ces deux dernières propriétés nous disent donc que $\pi^L(\mathcal{B}(\mathcal{H})$ est la représentation canonique de Gelfand pour l'état φ. Enfin, Φ est séparateur pour $\pi^L(\mathcal{B}(\mathcal{H}))$, c.-à-d.

$$\pi^L(A)\Phi = 0 \quad et \quad A \in \mathcal{B}(\mathcal{H}) \quad sii \quad A = 0 \tag{1.16}$$

qui reflète le fait que l'état d'équilibre canonique est fidèle. Il suit de $\pi^R(\mathcal{B}(\mathcal{H})) = \pi^L(\mathcal{B}(\mathcal{H}))'$ que Φ est aussi cyclique et séparateur pour $\pi^R(\mathcal{B}(\mathcal{H})$, ce qui peut d'ailleurs être vérifié directement.

L'hamiltonien (1.8), considéré comme générateur de $U(t) = e^{-iHt}$, définit aussi la dynamique; c.-à-d. dans la forme de Heisenberg:

$$\alpha : (t, A) \in R \times \mathcal{B}(\mathcal{H}) \mapsto \alpha_t[A] \equiv U(t)^* A U(t) \quad , \tag{1.17}$$

ou, dans la forme de Schrödinger:

$$\nu : (t, \Psi) \in R \times \mathcal{S} \mapsto \nu_t[\Psi] \equiv U(t)\,\Psi\,U(-t) \quad . \tag{1.18}$$

Ceci aussi apparait de manière intéressante dans la représentation de Gelfand associée à l'état d'équilibre canonique, pour laquelle le vecteur Φ (cf. 1.11) est cyclique dans $\mathcal{S}$. Considérons en effet l'opérateur défini par

$$S : A\Phi \in \mathcal{S} \mapsto A^*\Phi \quad \forall \quad A \in \mathcal{B}(\mathcal{H}) \tag{1.19}$$

c.-à-d.

$$S\,\Psi_{kl} = (\frac{\lambda_k}{\lambda_l})^{\frac{1}{2}}\,\Psi_{lk} \tag{1.20}$$

ou encore

$$S = J\Delta^{\frac{1}{2}} \tag{1.21}$$

avec

$$\Delta\,\Psi_{kl} = e^{-\beta(\epsilon_k-\epsilon_l)}\,\Psi_{kl} \quad ; \tag{1.22}$$

remarquons en particulier que Δ est un opérateur hermitien positif et inversible. Nous avons alors

$$\Delta^{\frac{it}{\beta}}\,\Psi_{kl} = U(t)\,\Psi_{kl}\,U(-t) \tag{1.23}$$

qui redonne en effet la dynamique dans la forme de Schrödinger (1.18), et

$$J\,\Delta^{it} = \Delta^{it}\,J \quad \forall \quad t \in R \quad . \tag{1.24}$$

L'origine profonde des propriétés de la représentation de Gelfand associée à un état d'équilibre canonique est à chercher dans la condition KMS. Pour les systèmes discutés jusqu'ici – c.-à-d. les systèmes quantiques ayant un nombre fini de degrés de liberté – cette condition est obtenue immédiatement. Avec φ donné par (1.14), considérons – pour toute paire (A,B) d'éléments de $\mathcal{B}(\mathcal{H})$ – la fonction de corrélation

$$f_{A,B}(t) \equiv < \varphi; A\,\alpha_t[B] > \ . \tag{1.25}$$

La cyclicité de la trace implique alors, immédiatement en effet:

$$f_{A,B}(t+i\beta) = < \varphi; \alpha_t[B]A > \quad . \tag{1.26}$$

Il est tout-à-fait remarquable qu'inversément cette propriété d'analyticité caractérise à son tour l'état d'équilibre canonique associé à la dynamique α_t donnée par (1.17). Supposons, en effet, que $\tilde{\rho}$ soit une matrice de densité telle que les fonctions de corrélation correspondantes satisfassent cette propriété, c.-à-d.

$$< \tilde{\varphi}; \alpha_t[B]A > = < \tilde{\varphi}; A\alpha_{t+i\beta}[B] > ; \tag{1.27}$$

pour $A = I$ cette expression se réduit à $f_{I,B}(t) = < \tilde{\varphi}\ ;\ \alpha_t[B] > = f_{I,B}(t+i\beta) \quad \forall \quad t \in R$; la fonction analytique $f_{I,B}$ est donc périodique en $i\beta$, ce qui n'est possible que si elle est constante; l'état $\tilde{\varphi}$ est donc *stationnaire*, et ainsi: $[\tilde{\rho}, H] = 0$. La condition (1.27), à $t = 0$, donne: $Tr\tilde{\rho}BA = Tr\tilde{\rho}Ae^{-\beta H}Be^{\beta H}$; ceci, étant vrai pour tout A et B dans $\mathcal{B}(\mathcal{H})$, implique que $\tilde{\rho}$ est proportionelle à $\exp(-\beta H)$, et donc que $\tilde{\rho} = \rho$ (puisque $< \tilde{\varphi}; I > = 1$ c.-à-d. $Tr\tilde{\rho} = 1$).

Les premières évidences que la condition KMS survit [3] à la limite thermodynamique, entrainant avec elle la structure modulaire, peuvent être trouvées dans l'étude du gaz parfait de Bose [9] et du modèle de Heisenberg à une dimension [10].

Ces travaux sont sous-tendus par la théorie mathématique générale de Tomita-Takesaki: tout état fidèle et normal φ sur une algèbre de von Neumann $\mathcal{N}$ définit une dynamique unique pour laquelle φ satisfait la condition KMS; cette dynamique est obtenue en considérant, dans la représentation canonique de $\mathcal{N}$ correspondant à φ, un opérateur fermable S définit comme dans (1.19) – avec maintenant $\mathcal{N}$ au lieu de $\mathcal{B}(\mathcal{H})$ – et dont la décomposition polaire (1.21) donne, de manière unique:
(a) un opérateur hermitien positif et invertible Δ;
(b) une dynamique (c.-à-d. une représentation de R dans le groupe des automorphismes de $\mathcal{N}$), unitairement implémentée par $\Delta^{\frac{it}{\beta}}$, et pour laquelle φ satisfait la condition KMS;
(c) un opérateur anti-unitaire involutif J tel que, dans cette représentation, $J\mathcal{N}J = \mathcal{N}'$ comme dans (1.5), et $J\Delta = \Delta^{-1}J$, comme dans (1.24).

La réciproque, cependant, n'est plus vraie en général: une dynamique donnée peut avoir plusieurs états satisfaisant la condition KMS pour la même température, ce qui permet de comprendre la coexistence de plusieurs phases thermodynamiques. En effet, les deux résultats [4] suivants conduisent à l'identification de celles-ci comme des états KMS extrémaux; tout d'abord, un état KMS est extrémal – c.-à-d. ne peut être décomposé en somme convexe d'autres états KMS – si et seulement si la représentation de Gelfand relative à cet état conduit à un facteur, c.-à-d. à une algèbre de von Neumann dont le centre est trivial; de plus, l'ensemble des états qui satisfont la condition KMS pour une dynamique et une température données forment un complexe de Choquet, ce qui correspond, en termes physiques, à la condition qu'un état d'équilibre canonique devrait pouvoir être décomposé de manière *unique* comme une combinaison convexe de phases thermodynamiques pures; il est alors possible que les états KMS extrémaux apparaissant dans cette décomposition n'aient qu'une symétrie plus basse que celle de l'état KMS original, conduisant ainsi à un mécanisme de symétries spontanément brisées [11, 12, 13, 14, 15].

La structure mathématique $\{\mathcal{N}, \varphi, \Delta, J\}$ définie par la théorie de Tomita-Takesaki est dite *modulaire*. Physiquement, la condition KMS peut être vue comme une condition de stabilité [16, 17, 18, 19, 20].

2. Théorie quantique des champs

Chaque fois que se présente une algèbre de von Neumann $\mathcal{N}$ admettant un vecteur cyclique et séparateur Φ il est naturel de se poser la question de savoir si les objets Δ et J qui y sont canoniquement associés par la théorie mathématique de Tomita-Takesaki, pourraient avoir une signification physique (géométrique et/ou dynamique). En théorie quantique des champs, l'existence d'une paire $\{\mathcal{N}, \Phi\}$ est assurée par le théorème de Reeh et Schlieder [21, 22] selon lequel, pour la représentation de Gelfand associée à l'état fondamental (vide) φ, le vecteur Φ correspondant est déjà cyclique pour une quelconque sous-algèbre locale $\mathcal{N}(\mathcal{O})$, c.-à-d. une algèbre de von Neumann associée à un ouvert $\mathcal{O}$ dans l'espace de Minkowski; de plus Φ est séparateur pour $\mathcal{N}(\mathcal{O})$ si le complément causal de $\mathcal{O}$ est non-vide ; cette propriété est donc satisfaite pour tout double cône, et pour tout onglet, c.-à-d. une région équivalente, par le groupe inhomogène de Lorentz, à l'onglet typique

$$\mathcal{W} = \{(t, x^1, x^2, x^3) \in M^{1,3} \mid |t| < x^1\} \quad . \tag{2.1}$$

Une interprétation physique de la théorie de Tomita-Takesaki (appliquée à certains champs quantiques particuliers) a été proposée pour les onglets par Bisognano et Wichmann [23], et pour les double-cônes par Hislop et Longo [24]. Il est possible, depuis peu, de donner une présentation simplifiée de l'essentiel de ces travaux, utilisant en fait des hypothèses moins fortes: en particulier, en insistant prioritairement sur le caractère local de la théorie, sans supposer *a priori* que la théorie soit invariante de Lorentz [25] (voir cependant [26, 27]).

Considérons, à cet effet, la situation la plus simple [25], qui pourra servir de modèle pour le cas d'un onglet dans l'espace $M^{1,1}$ de Minkowski à deux dimensions. Soit $\mathcal{N}$ une algèbre de von Neumann, agissant sur un espace de Hilbert $\mathcal{H}$, et admettant un vecteur $\Phi \in \mathcal{H}$ cyclique et séparateur; soit φ l'état fidèle $< \varphi; A >= < \Phi, A\Phi >$, et $\{\Delta, J\}$ la structure de Tomita-Takesaki correspondante. Supposons de plus que nous avons deux représentations unitaires,

$$\{U_k : a \in R \mid U_k(a) \in \mathcal{U}(\mathcal{H}) \quad , \quad k = +, -\} \tag{2.2}$$

de générateurs

$$H_k \quad \text{avec} \quad Sp(H_k) \subseteq R^+ \quad ; \tag{2.3}$$

et telles que :

$$U_k(a)\,\Phi = \Phi \quad \forall \quad a \in R \quad ; \tag{2.4}$$

$$\sigma_+(a)[N] \equiv U_+(a)\,N\,U_+^* \in \mathcal{N} \quad \forall \quad (a, \mathcal{N}) \in R^+ \times N \quad ; \tag{2.5}$$

$$\sigma_-(a)[N] \equiv U_-(a)\,N\,U_-^* \in \mathcal{N} \quad \forall \quad (a, \mathcal{N}) \in R^- \times N \quad . \tag{2.6}$$

Il est alors prouvé [25] par continuation analytique que pour tout $(t, a) \in R^2$:

$$J\,U_\pm(a)\,J^{-1} = U_\pm(-a) \tag{2.7}$$

$$\Delta^{\frac{it}{2\pi}} U_+(a) \Delta^{-\frac{it}{2\pi}} = U_+(e^{-t}a) \quad ; \quad \Delta^{\frac{it}{2\pi}} U_-(a) \Delta^{-\frac{it}{2\pi}} = U_-(e^{+t}a) \quad . \tag{2.8}$$

Supposons enfin que $[U_+(a), U_-(b)] = 0$ pour tout $(a,b) \in R^2$, et définissons

$$U : (t,x) \in R^2 \mapsto U_+(\frac{t+x}{2})\, U_-(\frac{t-x}{2}) \in \mathcal{U}(\mathcal{H}) \quad , \tag{2.9}$$

$$V : \lambda \in R \mapsto \Delta^{\frac{i\lambda}{2\pi}} \in \mathcal{U}(\mathcal{H}) \quad ; \tag{2.10}$$

nous obtenons immédiatement:

$$J\, U(t,x)\, J = U(-t,-x) \quad , \tag{2.11}$$

$$V(\lambda)\, U(t,x)\, V(\lambda)^* = U(t\cosh\lambda - x\sinh\lambda, -t\sinh\lambda + x\cosh\lambda) \quad ; \tag{2.12}$$

c.-à-d. $\{V, U, J\}$ est une représentation du groupe de Poincaré $P^{1,1}$ étendu par CPT, qui donne ainsi l'interprétation cherchée de $\{\Delta, J\}$ pour une théorie des champs, où φ est la manifestation de l'état fondamental (vide) de la théorie définie sur $M^{1,1}$, quand cette théorie est restreinte à un onglet $\mathcal{W}$, et où $\mathcal{N}$ est l'algèbre quasi-locale correspondant à $\mathcal{W}$. Les propriétés du vide se reflètent dans les hypothèses par la condition spectrale (2.3-4), et l'invariance (2.5-6) qui peut être visualisée maintenant comme des translations selon les vecteurs nuls qui envoient $\mathcal{W}$ dans (et non sur) lui-même; ce qui explique *a posteriori* pourquoi les hypothèses ne portent que sur les *semi-groupes d'endomorphismes* (2.5-6). Il est bon d'insister sur le fait que seule cette invariance de translation a été supposée alors que le groupe de Lorentz $\{V(\lambda) \mid \lambda \in R\}$ (voir 2.10) et la réflexion CPT, à savoir J, ont été obtenus comme conséquences. Pour se libérer des hypothèses explicitement faites ci-dessus sur les translations, il faut plonger encore plus profondément dans la théorie et axiomatiser les relations [28, 29] qui doivent exister entre les structures modulaires se rapportant à des onglets différents. Ces axiomes peuvent alors être étendus [30] à des espace-temps non-plats d'intérêt en relativité générale. Physiquement, un intérêt majeur de telles généralisations est qu'elles contribuent à approfondir notre compréhension [31, 32] de l'effet Hawking, selon lequel un champ quantique en présence d'un "trou noir" peut imiter une distribution thermique.

3. Théorie ergodique quantique

La théorie ergodique classique [33], dans sa forme la plus élémentaire, considère un espace de probabilité $\{\Omega, \mathcal{F}, d\mu\}$ (où les sous-ensembles de mesure nulle sont négligés) et un groupe $\{T(t) \mid t \in R\ (ou\ Z)\}$ de transformations mesurables de Ω telles que $\mu \circ T(t) = \mu$, pour tout t. Les questions fondamentales auxquelles cette théorie s'adresse sont de déterminer si ces systèmes dynamiques présentent un comportement asymptotique intéressant, par exemple:

$$\lim_{|t|\to\infty} \frac{\mu(E \cap T(t)[F])}{\mu(E)} = \mu(F) \tag{3.1}$$

pour toute paire (E, F) de sous-ensembles mesurables de Ω. Ce type de questions a suscité la découverte de systèmes dynamiques classiques tels que les flots de Kolmogorov ou ceux de Anosov.

Pour jeter un pont vers la théorie quantique, formulons d'abord la théorie classique en termes algébriques. En considérant les fonctions f appartenant à $\mathcal{L}^\infty(\Omega, \mathcal{F})$ comme des opérateurs $N(f)$ agissant par multiplication sur l'espace de Hilbert $\mathcal{L}^2(\Omega, \mathcal{F}, d\mu)$, l'algèbre $\mathcal{L}^\infty(\Omega, \mathcal{F})$ est représentée comme une sous-algèbre de von Neumann $\mathcal{N}$ dans $\mathcal{B}(\mathcal{L}^2(\Omega, \mathcal{F}, d\mu))$, où elle est de plus maximale abélienne. Un état fidèle et normal (c.-à-d. complètement additif) est alors défini sur $\mathcal{N}$ par l'identification $< \varphi; N(f) >= \int d\mu(\omega) f(\omega)$. Finalement, $\alpha(t)[N(f)] = f \circ T(t) \; \forall \, t \in R$ $(ou\ Z)$ définit une action sur $\mathcal{N}$.

Un système dynamique quantique est alors naturellement défini comme un triplet $\{\mathcal{N}, \varphi, \alpha\}$ constitué par une algèbre de von Neumann non-abélienne $\mathcal{N}$, agissant sur un espace de Hilbert $\mathcal{H}$; un état fidèle normal φ sur $\mathcal{N}$, correspondant à un vecteur cyclique et séparateur $\Phi \in \mathcal{H}$; et une représentation α de R $(ou\ Z)$ dans le groupe $Aut(\mathcal{N})$ des automorphismes de $\mathcal{N}$, telle que $\varphi \circ \alpha(t) = \varphi$ pour tout t. La relation (1.1) s'écrit alors en terme de fonctions de corrélation

$$\lim_{|t|\to\infty} < \varphi; A\, \alpha(t)[B] >=< \varphi; A >< \varphi; B > \quad . \tag{3.2}$$

Un flot de Kolmogorov non-commutatif est défini [34] comme un système dynamique quantique $\{\mathcal{N}, \varphi, \alpha\}$ pour lequel il existe une sous-algèbre de von Neumann $\mathcal{A} \subset \mathcal{N}$ telle que:

$$\mathcal{A} \subset \alpha(t)[\mathcal{A}] \qquad \forall \quad t \geq 0 \quad ; \tag{3.3}$$

$$\bigvee_t \alpha(t)[\mathcal{A}] = \mathcal{N} \quad ; \quad \bigcap_t \alpha(t)[\mathcal{A}] = C\, I \quad ; \tag{3.4}$$

$$\alpha_\varphi(s)[\mathcal{A}] \subseteq \mathcal{A} \qquad \forall \quad s \in R \quad ; \tag{3.5}$$

où $\alpha_\varphi(R)$ est le groupe d'automorphismes de $\mathcal{N}$ associé à φ par la théorie de Tomita-Takesaki.

Dans le cas des systèmes de Kolmogorov classiques — pour lesquels il existe une sous-algèbre raffinante $\mathcal{F}_o \subset \mathcal{F}$ dont le passé est asymptotiquement trivial et dont le futur engendre $\mathcal{F}$ — les fonctions indicatrices des éléments de $\mathcal{F}_o$, regardées comme des projecteurs agissant sur $\mathcal{L}^2(\Omega, \mathcal{F}, d\mu)$, engendrent alors une algèbre de von Neumann $\mathcal{A}$ qui satisfait aux propriétés (3.3-4); (3.5) est trivialement satisfaite ici, puisque $\mathcal{N}$ abélienne entraîne que $\alpha_\varphi(t)$ est réduit à l'automorphisme identité.

Dans le cas quantique où φ n'est pas nécessairement une trace – c.-à- d. où, en général, $< \varphi; AB > \neq < \varphi; BA >$ – l'hypothèse (3.5) avait été introduite [34] pour garantir l'existence d'une espérance conditionnelle – c.-à-d. [35] d'une projection de norme 1 – de $\mathcal{N}$ sur chaque $\alpha(t)\mathcal{A}$; ainsi les sous-espaces fermés $E(t)$ sous-tendus par $\alpha(t)[\mathcal{A}]\,\Phi$ jouissent-ils des propriétés suivantes:

$$E(t_1) \subset E(t_2) \quad \forall \quad t_1 < t_2 \quad ; \tag{3.6}$$

$$\bigvee_{t\in R} E(t) = \mathcal{H} \quad ; \quad \bigcap_{t\in R} E(t) = E_\Phi \quad . \tag{3.7}$$

Ces relations impliquent, par le théorème d'unicité de von Neumann, que le spectre du générateur du groupe unitaire $U(R)$ qui implémente $\alpha(R)$ est l'union d'une partie discrète non-dégénérée, 0, et d'une partie, R, absolument continue par rapport à la mesure de Lebesgue. Le lemme de Lebesgue-Riemann permet alors de prouver (3.2). D'autres propriétés ergodiques [34] s'avèrent aussi satisfaites par ces flots de Kolmogorov quantiques; voir aussi les travaux de Narnhofer et Thirring, en particulier [36]; pour les problèmes qui touchent à la définition d'une entropie dynamique quantique, voir encore [37].

Enfin, une définition de flots de Anosov quantiques [38] a été proposée récemment, inspirée par la conjonction des travaux de Borchers [25] discutés dans la section 2, et par un avatar quantique [39, 40] du système dynamique classique connu sous le nom de 'Arnold CAT' (la traduction de CAT par 'chat' perd malheureusement la saveur de l'acronyme anglais aux dépens du rappel iconographique). Ces systèmes de Anosov quantiques possèdent des propriétés ergodiques très fortes [38], par exemple une propriété d'essaim exponentielle et uniforme.

4. Quantification géométrique

Les structures modulaires sont aussi immanentes en théorie des représentations des groupes. Soient en effet G un groupe avec mesure de Haar $d\mu$ invariante à gauche, et Δ la fonction modulaire correspondante; U^L (resp. U^R) la représentation régulière gauche (resp. droite) de G, agissant sur l'espace de Hilbert $\mathcal{H} = \mathcal{L}^2(G, d\mu)$; c.-à-d., avec $(g_o, g) \in G \times G$ et $\Psi \in \mathcal{H}$:

$$\{U^L(g_o)\Psi\}(g) = \Psi(g_o^{-1}w) \qquad \{U^R(g_o)\Psi\}(g) = \Delta(g_o)^{\frac{1}{2}}\,\Psi(gg_o) \quad . \tag{4.1}$$

Ces deux représentations, qui engendrent le commutant l'une de l'autre, sont remplacées l'une par l'autre par l'action de l'opérateur involutif anti-unitaire

$$\{J\Psi\}(g) = \Delta(g)^{-\frac{1}{2}}\, \Psi(g^{-1})^* \quad , \tag{4.2}$$

c.-à-d. en introduisant les algèbres de von Neumann

$$\mathcal{N}^L = \{U^L(g) \mid g \in G\}'' \quad et \quad \mathcal{N}^R = \{U^R(g) \mid g \in G\}'' \tag{4.3}$$

nous obtenons [7] :

$$\mathcal{N}^L = (\mathcal{N}^R)' = J\mathcal{N}^R J \quad et \quad \mathcal{N}^R = (\mathcal{N}^L)' = J\mathcal{N}^L J \tag{4.4}$$

qui rappellent (1.5-6), quoique $\mathcal{N}^L$ et $\mathcal{N}^R$ ne soient maintenant plus nécessairement des facteurs.

Le lien entre ces structures et celles rencontrées dans le programme de quantification géométrique [41, 42, 43] a fait l'objet de travaux [44, 45, 46] dont nous voulons ici ne rapporter que les aspects essentiels. Alors que le programme habituel construit d'abord une étape intermédiaire (dite de préquantification) en utilisant la méthode des orbites co-adjointes, l'étape intermédiaire ici consiste à considérer la décomposition centrale [7] des algèbres de von Neumann $\mathcal{N}^L$ et $\mathcal{N}^R$, engendrées par les représentations régulières gauche et droite (celles-ci n'étant en général pas des facteurs). Les représentations irréductibles, dont la construction est l'objet de la quantification, doivent ensuite être extraites des représentations intermédiaires obtenues selon l'une ou l'autre de ces deux méthodes. Dans le programme habituel, ces représentations irréductibles sont obtenues par un choix de polarisation; ce qui peut être interprété comme l'attribution d'un rôle privilégié à certaines sous-variétés lagrangiennes des variétés symplectiques que sont les orbites co-adjointes. Ici, ce rôle est joué par des sous-algèbres de von Neumann abéliennes maximales (CSCO) dans la représentation intermédiaire. Nous connaissons des cas où ces deux démarches procèdent en parallèle; mais il est aussi des cas où elles sont complémentaires. Nous donnons ici un exemple de chacune de ces possibilités.

Le premier exemple est celui des règles de commutations canoniques, considérées d'abord pour l'espace de configuration R^n. Le groupe G est le groupe de Weyl W_o^{2n+1}, de dimension $(2n + 1)$, qui correspond aux représentations projectives du groupe R^{2n} dont les générateurs sont les n composantes de l'impulsion et les n composantes de la position. Le groupe W_o^{2n+1}, étant une extension centrale de R^{2n} par S^1, il est particulièrement aisé d'écrire la décomposition centrale des représentations régulières. Pour la représentation régulière gauche, nous obtenons [44] des représentations factorielles unitairement équivalentes à celles que l'on obtient par la méthode de préquantification habituelle, l'équivalence unitaire traduisant l'ambiguité essentielle liée au choix arbitraire d'une jauge de Kostant. Ces représentations sont indexées par le spectre Z du centre de $U^L(G)''$, c.-à-d. par les éléments du dual de S^1, c.-à-d. encore par un multiple entier $k\hbar$ de la constante de Planck $\hbar$ (cette dernière entrant dans la théorie au niveau de la définition de W_o^{2n+1}). Une telle représentation $U_\hbar^L$ (où nous avons choisi $k = 1$ pour alléger l'écriture) peut être décrite comme agissant dans l'espace de Hilbert $L^2(R^{2n}, d^n p\, d^n q)$; ses générateurs sont alors les opérateurs différentiels auto-adjoints:

$$P^k = -i\hbar\partial_{q^k} \quad , \quad Q^k = i\hbar\partial_{p^k} + q^k \quad . \tag{4.5}$$

Streater [47], avait déjà remarqué (tout en attribuant la priorité à Segal) que ce même espace de Hilbert sous-tend aussi une anti-représentation des règles de commutations canoniques, à savoir

$$\tilde{P}^k = i\hbar\partial_{q^k} + p^k \quad , \quad \tilde{Q}^k = -i\hbar\partial_{p^k} \quad ; \tag{4.6}$$

et que celle-ci commute avec la représentation de préquantification; il est facile de vérifier que, de plus, elle en engendre précisément le commutant.

Cette remarque, *a priori* surprenante, est cependant naturelle au vu de (4.3-4): la décomposition centrale de la représentation régulière gauche produit simultanément une décomposition centrale de la représentation régulière droite, et aussi de l'opérateur anti-unitaire J; la représentation $U_\hbar^R$ du groupe de Weyl obtenue ainsi correspond alors à la représentation dont les générateurs sont donnés par (4.6). En appelant $\mathcal{N}_\hbar^L$, $\mathcal{N}_\hbar^R$ et $J_\hbar$ les réductions de $\mathcal{N}^L$, $\mathcal{N}^R$ et de J au sous-espace invariant correspondant, à savoir $\mathcal{L}^2(R^{2n}, d^n p\, d^n q) \subseteq \mathcal{L}^2(G, d\mu)$, nous avons maintenant

$$\mathcal{N}_\hbar^L = (\mathcal{N}^R)'_\hbar = J_\hbar \mathcal{N}_\hbar^R J_\hbar \quad ; \quad \mathcal{N}_\hbar^R = (\mathcal{N}^L)'_\hbar = J_\hbar \mathcal{N}_\hbar^L J_\hbar \quad . \tag{4.7}$$

Extraire une représentation irréductible de $\mathcal{N}_\hbar^L$ correspond précisément au choix, dans son commutant $\mathcal{N}_\hbar^R$, d'une sous-algèbre maximale abélienne $\mathcal{A}_\hbar^R$; l'interprétation de cette algèbre en tant que sous-algèbre d'observables $\mathcal{A}_\hbar^L$ dans $\mathcal{N}_\hbar^L$ s'obtient alors en utilisant la conjugaison $J_\hbar$.

Les résultats présentés jusqu'ici restent valables dans le cas où le groupe de Weyl W_o^{2n+1} (pour l'espace de configuration plat R^n) est généralisé [48] au groupe de Weyl W_λ^{2n+1} (pour l'espace de configuration constitué par la variété homogène $M_\lambda^n = L^{1,n}/O^n$ de courbure négative λ) : le théorème de von Neumann reste vrai (il existe, à une équivalence unitaire près, une seule représentation irréductible du groupe de Weyl généralisé W_λ^{2n+1}) ; et la décomposition centrale de la représentation régulière gauche redonne exactement les représentations de préquantification correspondant aux orbites co-adjointes maximales.

Un exemple pour lequel ces méthodes ne coïncident pas, mais au contraire se complètent, a été récemment discuté [46]. Nous y montrons que toute représentation irréductible, appartenant à la série discrète, d'un groupe de Lie semi-simple, ayant un sous-groupe de Cartan abélien compact H , est la restriction de la représentation régulière gauche de G à un sous-espace qui n'est autre que l'intersection de deux projecteurs (commutant l'un avec l'autre), le premier procédant de la décomposition centrale (c'est donc un projecteur minimal dans le centre de la représentation régulière), le second émanant naturellement de la préquantification habituelle, projetant ainsi sur le sous-espace invariant de $\mathcal{L}^2(G, d\mu)$ qui sous-tend une représentation de G induite par un caractère de H . Si cette seconde décomposition est faite pour commencer [49], alors [46] la décomposition centrale reproduit l'effet de la polarisation complexe conduisant [49] à la représentation irréductible considérée.

Références

1. R. Kubo, "Statistical Mechanics of Irreversible Processes", *Journ. Phys. Soc. Japan* 12:570-586 (1957)
2. P.C. Martin and J. Schwinger, "Theory of Many-Particle Systems", *Phys. Rev.* 115:1342-1373 (1959)
3. R. Haag, N. Hugenholtz and M. Winnink, "On the Equilibrium States in Quantum Statistical Mechanics" *Comm. Math. Phys.* 5:215-236 (1967)
4. M. Takesaki, "Tomita's Theory of Modular Hilbert Algebras and its Applications" *Lectures Notes in Mathematics* 128, Springer, New York (1970); "States and Automorphisms of Operator Algebras – Standard Representations and KMS Boundary Condition", *Lectures Notes in Physics* 20, Springer, New York (1973)
5. F.J. Murray, "The Rings of Operators Papers", *in* The Legacy of John von Neumann, *Proc. Symp. Pure Math.* 50:57-60 (1990)
6. F.J. Murray and J. von Neumann, "On Rings of Operators", *Ann. Math.(2)* 37:116-229 (1936); II, *Trans. Amer. Math. Soc.* 41:208-248 (1937)
7. J. Dixmier, "Les algèbres d'opérateurs dans l'espace hilbertien (Algèbres de von Neumann)", Gauthier-Villars, Paris (1957); 2nd ed. (1969) ; and "Les C*-algèbres et leurs représentations", Gauthier-Villars, Paris (1964)
8. M. Takesaki, "Theory of Operator Algebras. I, II.", Springer, Heidelberg (1979, 1986)
9. H. Araki and E.J. Woods, "Representations of the Canonical Commutation Relations describing a Nonrelativisitic Bose Gas", *Jour. Math. Phys.* 4:637-662 (1963)
10. H. Araki, "Gibbs States of a one-dimensional Quantum Lattice", *Commun. Math. Phys.* 14:120-157 (1969)
11. R. Haag, "The Mathematical Structure of the Bardeen-Cooper-Schrieffer Model", *Nuovo Cim.* 25:287-298 (1962)
12. G.G. Emch and M. Guenin, "Gauge invariant formulation of the BCS Model", *Journ. Math. Phys.* 7: 915-921 (1966)
13. W. Thirring and A. Wehrl, "On the Mathematical Structure of the BCS Model", *Commun. Math. Phys.* 4:303-314 (1967); W. Thirring, "*ibid.*", *Commun. Math. Phys.* 7:181-189 (1968); "The Mathematical Structure of the BCS Model and related Models", *in* The Many-Body Problem, L.M. Garrido, A. Cruz and T.W. Preist, eds., Plenum Press, London (1969)

14. G.G. Emch, H.J.F. Knops and E.J. Verboven,"Breaking of Euclidean Symmetry with an Application to the Theory of Crystallization", *Journ. Math. Phys.*11:1655-1668 (1970); G.G. Emch and H.J.F. Knops, "Pure Thermodynamical Phases and Extremal KMS States", *Journ. Math. Phys.* 11:3008-3018 (1970)
15. G.G. Emch, "Algebraic Methods in Statistical Mechanics and Quantum Field Theory", Wiley-Interscience, New York (1972) ; "Mathematical and Conceptual Foundations of 20th-Century Physics", North-Holland, Amsterdam (1984)
16. R. Haag, D. Kastler and E.B. Trych-Pohlmeyer,"Stability and Equilibrium States",*Commun. Math. Phys.* 38:173-193 (1974); R. Haag and E.B. Trych-Pohlmeyer,"Stability properties of Equilibrium States", *Commun. Math. Phys.* 56:213-224 (1977)
17. H. Araki and G.L. Sewell, "KMS conditions and Local Thermodynamical Stability of Quantum Lattice Systems", *Commun. Math. Phys.* 52:103-109 (1977); G.L. Sewell, "*ibid.* II", *Commun. Math. Phys.* 55:53-61 (1977)
18. M. Fannes and A. Verbeure, "Correlation Inequalities and Equilibrium States", *Commun. Math. Phys.* 57:165-171 (1977)
19. W. Pusz and S.L. Woronowics, "Passive states and KMS states for general Quantum Systems",*Commun. Math. Phys.* 58:273-290 (1978)
20. W. Thirring,"On the Equivalence of Adiabatic Invariance and the KMS Condition", *Foundations of Physics* 13:409-415 (1983)
21. H. Reeh und S. Schlieder, "Bemerkungen zur Unitäräquivalence von Lorentzinvarianten Feldern", *Nuovo Cim.* 22:1051-1068 (1961)
22. H. Araki, "Von Neumann Algebras of Local Observables for the Free Scalar Field", *Journ. Math. Phys.* 5: 1-13 (1964)
23. J. Bisognano and E.H.Wichmann,"On the Duality Condition for a Hermitean Scalar Field", *Journ. Math. Phys.* 16:985-1007 (1975); "On the Duality Condition for Quantum Fields", *Journ. Math. Phys.* 17:303-321 (1976)
24. P.D. Hislop and R. Longo,"Modular Structure of the Local Algebra associated with a Free Massless Scalar Field Theory", *Commun. Math. Phys.* 84:71-85 (1982); R. Longo, "Algebraic and Modular Structure of von Neumann Algebras in Physics", *Proc. Pure Math.* 38(2):551-566 (1982)
25. H.-J. Borchers,"The CPT-Theorem in Two-dimensional Theories of Local Observables", *Commun. Math. Phys.* 143:315-332 (1992)
26. R. Brunetti, D. Guido and R. Longo, "Modular Structure and Duality in Conformal Quantum Field Theory", *Commun. Math. Phys.* 156:201-219 (1993)
27. J. Yngvason, "A Note on Essential Duality", *Lett. Math. Phys.* 31:127-141 (1994)
28. D. Buchholtz and S.J. Summers, "An Algebraic Characterization of Vacuum States in Minkowski Space", *Commun. Math. Phys.* 155:442-458 (1993)
29. H.-J. Borchers, "On Modular Inclusion and Spectrum Condition", *Lett. Math. Phys.* 27:311-324 (1993)
30. S.J. Summers and R. Verch, "An Application of Modular Inclusion to Quantum Field Theory in Curved Space-Time", *DESY preprint* 126 (1993); "Modular Inclusion, the Hawking Temperature and Quantum Field Theory in Curved Space-Time", *preliminary version* (1994)
31. G.L. Sewell, "Quantum Fields on Manifolds; PCT and Gravitationally induced Thermal States", *Ann. Phys.* 141:201-224 (1982)
32. K. Fredenhagen and R. Haag, "On the derivation of Hawking Radiation associated with the formation of Black Holes", *Commun. Math. Phys.* 127: 273-284 (1990)
33. I.P. Cornfeld, S.V. Fomin and Ya.G. Sinai, "Ergodic Theory", Springer, Heidelberg (1982)
34. G.G. Emch, "Generalized K-Flows",*Commun. Math. Phys.* 49:191-215 (1976)
35. M. Takesaki, "Conditional Expectations in von Neumann Algebras", *J. Funct. Anal.* 9:306-321 (1972)
36. H. Narnhofer and W. Thirring, "Quantum K-Systems", *Commun. Math. Phys.* 125:565 (1989); "Algebraic K-Systems", *Lett. Math. Phys.* 112: 691 (1990)

37. A. Connes. H. Narnhofer and W. Thirring, "Dynamical Entropy of C*-algebra and von Neumann Algebra", *Commun. Math. Phys.* 112:691 (1987)
38. G.G. Emch, H. Narnhofer, W. Thirring and G.L. Sewell, "Anosov Actions on Non-Commutative Algebras", *Journ. Math. Phys.* 35:5582-5599 (1994)
39. F. Benatti, H. Narnhofer and G.L. Sewell, "A Non-Commutative Version of the Arnold Cat Map", *Lett. Math. Phys.* 21:157 (1991)
40. H. Narnhofer, "Quantized Arnold Cat Maps can be Entropic K-Systems", *Jour. Math. Phys.* 33:1502-1510 (1992)
41. J. M. Souriau, "Structure des systèmes dynamiques", Dunod, Paris (1966)
42. A. Kirillov, "Eléments de la théorie des représentations", Mir, Moscow (1974)
43. D.J. Simms and N.M.J. Woodhouse, "Lectures in Geometric Quantization", *Lecture Notes in Physics* 53, Springer, New York (1976); N.M.J. Woodhouse, "Geometric Quantization", Clarendon Press, Oxford (1980)
44. G.G. Emch, "Prequantization and KMS Structures", *J. Theor. Phys.* 20:891-904 (1981); "KMS Structures in Geometric Quantization", *Contemporary Mathematics* 62:175-186 (1987); "Geometric Quantization: Regular Representations and Modular Algebras", *in* Group Theoretical Methods in Physics, V.V. Dodonov and V.I. Man'ko, Springer, Heidelberg (1991)
45. S.T. Ali and G.G. Emch, "Geometric Quantization: Modular Reduction Theory and Coherent States", *J. Math. Phys.* 27:2936-2943 (1986)
46. S.T.Ali, A.M. El Gradechi and G.G. Emch, "Modular Algebras in Geometric Quantization", *J. Math. Phys.* 33:6237-6243 (1994)
47. R.F. Streater, "Canonical Quantization", *Commun. Math. Phys.* 2:354-374 (1966)
48. J. Bertrand, G.G. Emch and G. Rideau, "The Cohomology of the Classical and Quantum Weyl CCR in Curved Spaces", *Lett. Math. Phys.* (to appear)
49. A.M. El Gradechi and S. De Bièvre, "Phase Space Quantum Mechanics on the Anti-de Sitter Spacetime and its Poincaré Contraction", *Ann. Phys. (NY)* 235:1-35 (1994)

A MODEL FOR QCD IN THREE DIMENSIONS

CHRISTIAN FRØNSDAL
Department of Physics, University of California
Los Angeles, CA 90024-1547, USA

Dedicated to my friend Guy Rideau.

Abstract. I study the chiral anomaly in $1+1$ and $2+1$ De Sitter space times. New features come to light in dimension $1+1$ that carry over to very interesting things in dimension $2+1$. It turns out that the anomaly equation (bosonization formula) is a relation between two topological gauge theories; dipole fermion fields and Chern-Simons boson fields. Both field theories describe particles that are "confined", in the sense that, in the absence of anomalies, the physical excitations of either kind are only detected at the boundary. In this respect, the model presents some analogies with the quark-gluon section of QCD. Quantum anomalies are known to be present and to imply additional degrees of freedom; one hopes that they may extend the analogy to include hadrons. The theory has affinities to the Seiberg-Witten monopole model and may point towards an alternative approach to Floer's invariants of three-manifolds.

1. Introduction.

1.1. PHYSICS

The challenge posed by QCD is to find a solvable field theory that can serve as a starting point for a new type of perturbation theory for strong interactions. Rather than looking at confinement as a dynamically produced distortion of free particle dynamics of the conventional kind, it is proposed to interpret the existence of hadrons as a result of anomalies of a topological gauge theory of quarks and gluons. It is easy to invent topological field theories that contain fermions, but much harder to envisage the emergence

J. Bertrand et al. (eds.), Modern Group Theoretical Methods in Physics, 129–142.

of hadrons. Nevertheless, I think it worth while to study topological field theories that show some affinity to phenomenological QCD.

By a topological gauge theory we mean one where the physical sector; that is, the BRST cohomology group, is trivial. Some and perhaps most theories of this kind turn out to be topological in another sense; that is, the metric plays a very secondary role, if it is not irrelevant altogether. The first topological field theory to be discovered [1], singleton field theory in four dimensions, was originally formulated on a De Sitter background, but actually one need only to suppose that the metric is asymptotically De Sitter [2]; the Hamiltonians of these theories are surface integrals over the boundary at spatial infinity [1],[3], and Witten's observation [4], in connection with another model, that the energy momentum tensor is BRST-exact, applies.

One of the most fascinating aspects of singletons is that they are natural candidates for constituents of massless particles [5]. If one aims for QCD then it is difficult to see how this leads to hadrons, since hadrons are emphatically not massless, but gluons at least are massless! In this paper it is shown that it is very natural to think of gluons as singleton composites.

1.2. BOSONIZATION AND THE NEUTRINO THEORY OF LIGHT.

A neutrino theory of light was seriously proposed by Jordan [6] in the 1930's. The idea still has some appeal, though it was thoroughly demolished in a famous paper by Pryce [7], in 1938. It had a revival with the discovery, by Coleman [8] and Mandelstam [9], of the phenomenon of bosonization in two dimensions. Bosonization is an important tool in conformal field theory. Most recently, the proposal of Witten [10], to study Donaldson's topological invariants of 4-manifolds by means of monopole equations, invites investigation of the equations

$$F_{ij} = \bar{\psi}\sigma_{ij}\psi, \quad \not{\partial}_A\psi = 0,$$

in which the field strength F is self dual. Floer's topological invariants of three-manifolds [11] are also connected to topological gauge theories and to "bosonization". The paper by Pryce [7] alluded to has some bearing on the problem of bosonization in higher dimensions. The heart of the difficulty is not that of producing two-neutrino states that look like photons, but that there are too many ways to do it. In two dimensions a state of two massive fermions is kinematically the same as a massless boson. In higher dimensions there are some (actually, "almost none") two-fermion states that can be identified with photons but there is no one-to-one correspondence - unless the fermion is a singleton. In three dimensions there is a one-one mapping between states of two singletons (connected by the current to the vacuum) and photons described by a pure Chern-Simons action. In four dimensions there is a one-one mapping between two-singleton states and a family of massless particles with all helicities [2]. Thus it is clear that the peculiar kinematics that characterizes singletons is just what one needs to make bosonization plausible in dimensions higher than 2.

Naturally, mathematicians prefer compact manifolds and Riemannian metrics. While a pseudo-Riemannian metric is preferred in physics, we have no difficulty with compactness. Indeed, space time is often compactified by completion with a boundary at infinity. This leads to manifolds with boundaries, one of the thorniest aspect of Donaldson theory, in which Floer theory plays the role of boundary data.

The work reported here arose from some naive questions about bosonization in De Sitter space, with the expectation that this should lead to singletons. From a general, mathematical point of view there seems at first sight to be little reason to introduce a particular metric, but the discovery of topological invariants is not the only aspect that may turn out to be relevant for physics. The study of a Riemannian manifold with boundary can hardly get far without some discussion of the asymptotic properties of the metric. If the interior metric is really irrelevant then one may perhaps be allowed to choose a metric for greater convenience. A De Sitter metric is very nice to work with, and it seems particularly well adapted to the context of QCD since it eliminates all infrared difficulties [12]. The fact that the most important operators have discrete spectra greatly simplifies the analysis. Both of these considerations are advantages of a De Sitter metric over Minkowski. But even more motivating is the simple fact that the infrared domain is much more interesting with a constant curvature.

1.3. OUTLINE.

Sections 2 and 3 offer a preliminary study of the kinematics of bosonization in $1+1$ dimensions. Not surprisingly, the role of the conformal group is relevant here. The special conformal group in two dimensions is isomorphic to the symmetry group of $2+2$ De Sitter space, and this suggests a natural extrapolation to three dimensions. Section 4 is a short digression on massless, scalar fields in 2+1 dimensions. This already brings us to singletons, and the reader is reminded of the fact that they are left or right movers [2] (in three dimensions!). In Sections 5 and 6 it is shown that massless fermions are singletons, and that their quantization leads to dipole equations, second order field equations for fermi fields (and fourth order equations for scalar bosons). How this comes about is a central point of the whole paper; therefore I devote the next paragraph to it.

In fact, this brings us in close contact with invariant theory, where the deficiencies of the Dirac operator is a central theme, though we are here at first concerned with the covariant Dirac operator in a space with constant, pseudo-Riemannian curvature, in the absence of any Yang-Mills type connection. To discuss the index of the Dirac operator one needs to define it's domain, and boundary conditions inevitably play a role. It turns out that the singletons have a very slow fall-off at infinity. When the domain of $\not\partial$ is taken to include the singleton solutions of the Dirac equation, then it has no self-adjoint extension. A corollary of this is that covariant quantization becomes impossible on the nullspace of the Dirac operator. Another way to see this is to notice that the representation of $SO(2,2)$ induced on $Ker\,\not\partial$ is non-decomposable. The singletons are not in the invariant subspace, so it becomes necessary to enlarge the representation to a full Gupta-Bleuler triplet. This leads to the dipole equation $\Box\psi = 0$ for the fermi field (and $\Box^2\phi = 0$ for the scalar bose field). The Dirac equation becomes the Lorentz condition that defines the "physical subspace" of this gauge theory. With the inclusion of Fadde'v-Popov ghosts and BRST invariance this is the condition that the BRST charge annihilate the physical states. The singletons are the BRST cohomology quotients, the "index" of the Dirac operator.

The dipole fermion leads to two "currents"". The canonical current that we call J is a 2-form and is conserved (closed) by virtue of the field equations; it is of dimension 2 and thus it is natural to couple it to a gauge potential. The "anomaly current" $j_i = \bar{\psi}\gamma_i\psi$ is a 1-form of dimension 1 and it is the extrapolation of the canonical current associated with the Dirac field on the 2-dimensional boundary. Kinematical considerations now tell us that, if j is proportional to a gauge potential, then that potential must be of the pure Chern-Simons type. This theory is studied in Section 7, with the result that the Chern-Simons potential is another manifestation of singletons. This was to be expected, since both theories are topological [13]. In Section 8 we introduce minimal coupling between the Chern-Simons potential a and the singleton-dipole fermion. The field equations are

$$F + e'J = 0, \quad \Box_a\psi = 0,$$

where e' is a real constant, the magnetic charge. The similarity with the Seiberg-Witten equations is the more striking because the field strength F is self dual, in a sense to be made clear. The physicality conditions and the anomaly equation are

$$\partial \cdot a = 0, \quad \not\partial_a\psi = 0, \quad j \propto a.$$

Perhaps these equations point to a simpler reformulation of Floer theory [11].

2. Massless Scalar Fields, $d = 2$.

Notation used throughout this paper is the same as in numerous papers dealing with physics on De Sitter space, for example [1, 2, 3]. Briefly, y_μ are coordinates for the De Sitter hyperboloid, $\xi = y \cdot y'$ is the invariant distance, $K(y, y') = k(\xi)$ the invariant propagator. We study the fields characterized by the wave equation $\mathcal{C}\phi(y) = 0$, where $\mathcal{C}$ is the Casimir operator.

The differential equation for $k(\xi)$, is solved by k=constant and by $k' = (1 - \xi^2)^{-1}$ or $k(\xi) = \tanh^{-1}\xi$. More directly useful is the expansion

$$\begin{aligned} k(\xi) &= \sum_{n=0}^{\infty} \frac{1}{2n+1}\xi^{-2n-1} = \xi^{-1} + \frac{1}{3}\xi^{-3} + \ldots \\ &= 2\frac{1}{y_+}\frac{1}{y'_-} + 4\frac{y_1}{y_+^2}\frac{y'_1}{y'^2_-} + \frac{2}{3}\frac{y_+y_- - 4y_1^2}{y_+^3}\frac{y'_+y'_- - 4y'^2_1}{y'^3_-} + \ldots . \end{aligned} \tag{1}$$

There is just one mode for each energy. The representation of $SO(2,1)$ carried by this propagator is irreducible; it has a non-degenerate energy spectrum with unit spacing and lowest energy 1, so it is $D(1)$ in a standard notation. There is just one other irreducible representation with $\mathcal{C} = 0$ and non-negative energy; it is the identity representation $D(0) = id$, with constant propagator.

Nevertheless, this is far from exhausting the list of interesting propagators for massless fields in 2 dimensions. An important consideration in 2-dimensional field theory is conformal invariance. The massless wave equation is conformally invariant, but the propagator (1) is not. The conformal group acts naturally on the kernel of $\mathcal{C}$; the restriction of this representation of to $SO(2,1)$ is $D(1) \oplus id$. The action is trivial on the zero mode, which therefore defines an invariant subspace, but the complement is not invariant. (There is no invariant complement.) The representation has the form $R \to id$ with R irreducible and not trivial; consequently there is no reproducing kernel. We need to expand the representation to (at least) $id \to R \to id$ (Araki's theorem [14]); the additional 1-dimensional space is provided by the function $\ln y_+ \propto t$.

At first sight this function satisfies the wave equation

$$\mathcal{C}\varphi = (-y^2\partial^2 + N(N+1)))\varphi = 0 . \tag{2}$$

Here $N = y^\mu\partial_\mu$. In fact, $\partial^2 \ln y_+ = 0$ but $N \ln y_+ = 1$, and thus $\mathcal{C}\ln y_+ = 1$. Consequently $\mathcal{C}K(y, y') = 1$ (say). A solution is $k(\xi) = \ln(\xi - 1)$ or

$$K(y, y') = \ln y_+ + \ln y_- + 2\frac{y_1}{y_+}\frac{y'_1}{y'_-} + 2\frac{y}{y_+}\frac{y'}{y'_-} + \ldots . \tag{3}$$

The constant mode is paired with the additional logarithmic mode and, besides, there are now two modes at each positive energy, instead of the non-degenerate spectrum of the propagator (1). On this expanded space $SO(2,1)$ acts by the non-decomposable representation

$$id \to (D(1) \oplus D(1)) \to id \ .$$

The two copies of $D(1)$ are the left and right movers of conformal field theory.

3. Spinor Fields, $d = 2$.

The most interesting spinor fields in 2 dimensions are the chiral ones. The associated irreducible representation is $D(\frac{1}{2})$, with lowest energy $\frac{1}{2}$, and the wave equation is

$$(\kappa + 1)\psi = 0 \ ,$$

where κ is the De Sitter Dirac operator. This is a special case that admits chiral solutions; we may call them "massless". We have two copies of $D(\frac{1}{2})$–left movers and right movers–carried by solutions $\psi_\pm$ that satisfy

$$\not{y}\psi_+ = y\psi_+ \ , \quad \not{y}\psi_- = -y\psi_- \ . \tag{4}$$

Unlike the case of massless scalar fields, there is no known reason to construct more complicated "massless" spinor field theories, for the conformal group acts on the left (right) movers in familiar fashion.

We are interested in two-particle states with very low mass, especially those that contribute to the current

$$j := z^\mu j_\mu := \bar{\psi}(y \cdot z - \not{y}\not{z}) \ \psi,$$

so we calculate the matrix elements for the states of lowest energy:

$$(z_-\partial_- + z_+\partial_+ - z_1\partial_1)\frac{y - iy_1}{y_+} \ .$$

This is just the gradient of the two ground states of $D(1)$, right and left movers, that make up the conformal version of a massless, scalar field. It could not be otherwise.

The conclusion is that, kinematically at least, there is a good basis for identifying the fermionic current j with the gradient of a scalar field,

$$j'_\mu \propto \partial_\mu \phi. \tag{5}$$

We shall take this equation from calculations in flat space, without attempting a direct evaluation of the anomaly in De Sitter space. What is more significant is the interesting fact that the formula definitely favors the conformal scalar field over the much simpler alternative massless field.

4. Massless Scalar Fields, $d = 3$.

The symmetry group is now $SO(2,2)$. The most remarkable representations, the singletons, are left or right movers, in three dimensions. We must explain this concept.

The group $SO(2,2)$ is isomorphic to $SL(2,R) \otimes SL(2,R)$; it is therefore possible to imagine fields that transform trivially under one of the factors but non-trivially with respect to the other. These representation, that may be called chiral, naturally extend the idea of left/right movers from two to three dimensions. But such fields do not have invariant propagators. The only way to realize such representations with local fields is as subquotients of nondecomposable representations. In terms of the factorization of $SO(2,2)$ we have $D(a,b) = D(a) \otimes D(b)$ and in particular

$$D(1,1) = D(1) \otimes id\ , \quad D(1,-1) = id \otimes D(1)\ . \tag{6}$$

The impetus for considering generalized massless fields in 2 dimensions comes from conformal field theory. In three dimensions our interest is aroused by the intriguing possibility of realizing the highly degenerate representations (6) in terms of fields. A connection between the two topics is suggested by the fact that the field modes associated with (6) are left resp. right movers. In fact, it has been shown [15] that the theory of the generalized massless field that is propagated by $\hat{K}$ is a topological field theory with a very small physical sector that is directly interpretable in terms of a conformal field on the 2-dimensional boundary of 3-dimensional space time.

5. Spinor Fields, $d = 3$.

Spinor fields on 2+2 de Sitter space take their values in the fundamental, 4-dimensional, reducible representation

$$D(-\tfrac{1}{2},-\tfrac{1}{2}) \oplus D(-\tfrac{1}{2},\tfrac{1}{2}) = \Big(D(-\tfrac{1}{2}) \otimes id\Big) \oplus \Big(id \otimes D(-\tfrac{1}{2})\Big) \tag{7}$$

of $SO(2,2) = SL(2,R) \otimes SL(2,R)$.

The most interesting spinor fields are those that carry the singletons:

$$D(\tfrac{1}{2},\tfrac{1}{2}) = D(\tfrac{1}{2}) \otimes id\ , \quad D(\tfrac{1}{2},-\tfrac{1}{2}) = id \otimes D(\tfrac{1}{2})\ . \tag{8}$$

Like the representations in (6) these are left and right movers, compare [15].

The wave equation is $(\kappa+1)\psi = 0$, and the representation carried by the solutions is

$$(D(\tfrac{1}{2},\tfrac{1}{2}) \to D(\tfrac{3}{2},-\tfrac{1}{2})) \oplus (D(\tfrac{1}{2},-\tfrac{1}{2}) \to D(\tfrac{3}{2},\tfrac{1}{2})). \tag{9}$$

These non-decomposable representations do not admit reproducing kernels. The smallest representation that includes (9) and that has a propagator is

$$\begin{aligned} D(-\tfrac{1}{2},\tfrac{1}{2}) \otimes D(1,0) &= \Big(D(\tfrac{3}{2},-\tfrac{1}{2}) \to D(\tfrac{1}{2},\tfrac{1}{2}) \to D\Big(\tfrac{3}{2},-\tfrac{1}{2}\Big)\Big) \\ &\quad \oplus \Big(D(\tfrac{3}{2},\tfrac{1}{2}) \to D(\tfrac{1}{2},-\tfrac{1}{2}) \to D(\tfrac{3}{2},\tfrac{1}{2})\Big). \end{aligned} \tag{10}$$

The entire representation (15) is carried by the space of positive energy solutions of the "dipole" wave equation

$$\partial^2\psi = 0\ , \quad N\psi = -\psi\ . \tag{11}$$

Thus we find that fermionic singleton field theory in three dimensions, as in four dimensions, is based on a dipole field equation. This is yet another topological field theory.

6. The Singleton Dipole, $d = 3$.

It is possible to build a classical field theory on the action

$$\int \bar{\psi}(\kappa+1)\psi\ . \tag{12}$$

The current is then

$$j = z^\mu j_\mu = \bar{\psi}(z^\mu y_\mu - \not{y}\not{z})\psi\ . \tag{13}$$

We can analyze the matrix elements between the vacuum and those 2-particle states that have the lowest mass (the irreducible representation with positive energy and the highest of the highest weights). Just as in the 2-dimensional case we find that the result is a gradient,

$$j_\mu \propto \partial_\mu \phi \tag{14}$$

of a scalar field ϕ. The lowest energy modes of φ are $(y_2 \pm iy_1)/y_+$, which shows that ϕ transforms according to the "massless" representations $D(1,\pm1)$. However, it makes no sense to talk about anomalies in classical field theory, and to quantize the fermions we must introduce the dipole.

According to (11) each component of ψ satisfies $\partial^2\psi_a = 0$, $N\psi_a = -\psi_a$. The representation carried by the positive energy solutions is

$$D(-\tfrac{1}{2},\tfrac{1}{2}) \otimes D(1,0). \tag{15}$$

This representation is a Gupta-Bleuler triplet. The Lorentz condition (physicality condition) is (when $N\psi = -\psi$)

$$\partial\!\!\!/\psi = 0 \qquad \text{(Lorentz condition)}; \tag{16}$$

it eliminates the "scalar" modes on the left and leaves a subspace that carries the representation (9), as we have seen.

This "physical" subspace includes a subspace of gauge modes–the representations $D(\frac{3}{2},\pm\frac{1}{2})$ in (9). Unitarity requires that these modes have no interaction; this is the meaning of gauge invariance.

The space of gauge modes is so much larger than the quotient of physical modes that no local measurement can isolate the physical component of the field. The modes associated with $D(\frac{1}{2},\pm\frac{1}{2})$– the only ones that are truly physical–can be distinguished only by their asymptotic behavior at spatial infinity, where gauge modes become ignorable. Interactions must be "topological," in the sense that they are confined to modes that are determined by the boundary values of the field. Simply, though inaccurately: interactions take place only at infinity.

It follows from this that a theory of interacting singletons, in the absence of anomalies, has no local observables. As long as there are no anomalies, singletons will remain unobservable, we shall say that they are "kinematically confined." Our strategy will be to set up a theory with interactions that (ostensibly) only take effect at infinity, then find an anomaly that produces a real 3 dimensional interacting field theory: a type of 3-dimensional QCD.

The action is

$$\int d^3x\sqrt{g}g^{ij}(\partial_i\bar{\psi})(\partial_j\psi), \tag{17}$$

so ψ has dimension one-half. The canonical current has dimension 2 while the current (13) has dimension one, as it must be if we are to follow Witten's bosonization strategy [10].

We begin by introducing an external vector gauge field A_μ. Then we examine the possibility of an identification of the type (anomaly equation)

$$j'_\mu = eA_\mu\ , \tag{18}$$

where j_μ is the current examined above, (13).

There are several inequivalent versions of 3-dimensional gauge theory, even in flat space. The correct choice in the present context is determined by the requirement that the potential field operator carry the same representation of $SO(2,2)$ as the current, and here we come to a central point about singleton kinematics. The free fermionic current carries a very reducible representation, a direct sum of nondecomposable representations, but the physical singleton representation contributes to only two nondecomposable summands. When the Fermi field is restricted to the "physical" subspace $\partial\!\!\!/\psi = 0$ and to positive energies, then these very distinguished summands are

$$(D(1,1) \to D(2,0)) \oplus (D(1,-1) \to D(2,0)) \,. \tag{19}$$

When the restriction to the physical subspace is lifted, then they expand to

$$D(2,0) \to D(1,\pm 1) \to D(2,0) \,, \tag{20}$$

each a complete Gupta-Bleuler triplet. There is a pair of vector field gauge theories in which the free potential field operator carries precisely one or the other of these representations. They are described next.

It is remarkable that the modes in (18), that correspond to one particle states of the gauge potential, are the only two-singleton states that are connected to the vacuum by the current. This is what makes bosonization in three dimensions plausible.

7. The Gauge Field, $d = 3$.

The free field equations are

$$\pm\, \epsilon^{\mu\nu\lambda\rho} y_\nu \partial_\lambda A^\pm_\rho = \partial^\mu (y^\lambda A^\pm_\lambda) \,. \tag{21}$$

We choose the upper sign and drop the sign on A^+_ρ. In intrinsic notation there is a vector field a_i, $i = 0,1,2$ and a scalar field b satisfying

$$\begin{aligned} \epsilon^{ijk}\partial_j a_k - \partial^i b &= s^i \,, \\ 2b + \nabla^i a_i &= 0 \,. \end{aligned} \tag{22}$$

These equations, in which an external source has been included, are the Euler-Lagrange equations of the action

$$\mathcal{L} = \int d^3x \Big\{ \tfrac{1}{2}\epsilon^{ijk} a_i \partial_j a_k - a^i \partial_i b + \sqrt{\rho}\, b^2 - a^i s_i \Big\} \,. \tag{23}$$

The first term in $\mathcal{L}$ is the Chern-Simons form, b is a Nakanishi-Lautrup gauge fixing field. Unitarity demands that b be free, this implies that the source must be conserved,

$$\partial_i s^i = 0 \,. \tag{24}$$

In the flat limit this is familiar as pure Chern-Simons theory. It is then locally trivial in the sense that it has no propagating modes: the physical ("transverse") modes in the center of (19) disappear in the flat limit. In De Sitter space the theory is less trivial since the physical quotient is not empty; locally it is nevertheless relatively trivial because the physical modes are too few to be picked up by a localized antenna: they can be distinguished only by their slow fall-off towards infinity. Another way to see this is by examination of the interaction term $\int d^3x \; a_i s^i$. If the potential satisfies the physicality condition, then it is locally a gradient, $a_i = \partial_i \lambda$ and the interaction becomes $\int d^3x \; \partial_i(\lambda s^i)$, since the source must be conserved. The interaction with the source is thus limited to the boundary of spacetime.

As we have seen, Chern-Simons theory furnishes the type of vector potential that is kinematically favoured to appear in a lifting of the anomaly formula from two to three dimensions; we propose that

$$j'_\mu \propto A_\mu,$$

where j' is the axial partner of the current (13) with ψ the dipole field operator, and A the Chern-Simons gauge potential. To justify this we shall exploit the very intimate relationship that exists between the chiral field theories in two and three dimensions.

8. Boundary Terms in 3-Dimensional Field Theories.

Boundary terms for scalar and spinor dipoles have been treated elsewhere, and it will be enough to quote the results. For Chern-Simons theory we give the details.

The scalar dipole is characterized by the lagrangian

$$\mathcal{L}_3 = \int d^3x \sqrt{g}(g^{ij}\phi_{,i}\, b_{,j} + b^2/2). \tag{25}$$

We have introduced the Nakanishi-Lautrup field b; the Lorentz condition is $b = 0$. Analysis of the field equations reveals that, if the metric is asymptotically De Sitter, then the asymptotic behavior of the fields is as follows,

$$\phi(x) = r^{-\frac{1}{2}}(\hat{\phi} + o(r^{-\frac{5}{2}})), \quad b(x) = r^{-\frac{5}{2}}(\hat{b} + o(r^{-\frac{9}{2}})). \tag{26}$$

Here $\hat{\phi}$ and $\hat{b}$ are fields on the boundary at spatial infinity ($r \to \infty$). The boundary manifold is $S_1 \times S_1$ and it will be parameterized by angles t, θ in terms of which the induced boundary metric is Lorentzian. The fall-off is slow and the variation includes surface term

$$\delta\mathcal{L}_3 = \ldots.. + \int dt\, d\theta\, (\hat{b}\; \delta\hat{\phi})$$

Therefore, to avoid surface terms in the field equations we must add a surface term to the action,

$$\mathcal{L} = \mathcal{L}_3 + \mathcal{L}_2, \quad \mathcal{L}_2 := \tfrac{1}{2}\int dt d\theta (\hat{\phi},_t^2 - \hat{\phi},_\theta^2). \tag{27}$$

All the physics is in the boundary term, the rest of the Lagrangian is just gauge fixing. The four-dimensional version of this theory was the first topological field theory discovered [1, 3].

The fermionic dipole has the gauge-fixed action

$$\mathcal{L}_3 = \int d^3x \sqrt{g}(\bar{\psi}\, \not{\partial}\lambda + \bar{\lambda}\, \not{\partial}\psi + \bar{\lambda}\lambda), \tag{28}$$

and the Lorentz condition is $\lambda = 0$. The asymptotic behaviour is

$$\not{y}\, \psi(x) = \chi + o(r^{-2}), \quad r^2\lambda(x) = \sigma + o(r^{-1}),$$

and the complete Lagrangian is

$$\mathcal{L} = \mathcal{L}_3 + \mathcal{L}_2, \quad \mathcal{L}_2 := \int dt d\theta\, (\bar{\chi}\, \not{\partial}\chi). \tag{29}$$

Now let us turn to the Chern-Simons action

$$\mathcal{L}_3 = \int d^3x \left\{ \tfrac{1}{2}\epsilon^{ijk} a_i \partial_j a_k - a^i \partial_i b + b^2 - a^i s_i \right\}. \tag{30}$$

From the field equation we derive the following information about asymptotic behaviour:

$$\begin{aligned} a_r(x) &= r^{-3}(\hat{a} + o(r^{-2})), \\ b(x) &= r^2(\hat{b}' + o(r^{-2})), \\ a_t(x) &= \hat{a}_t + \hat{a}'_t/r^2 + o(r^{-4}), \\ a_\theta(x) &= \hat{a}_\theta + \hat{a}'_\theta/r^2 + o(r^{-4}). \end{aligned} \tag{31}$$

The implication of the 4 field equations for the leading coefficients are

$$\begin{aligned} 2\hat{b} + \partial \cdot \hat{a} &= 0, \\ -2\hat{b} + \partial_\theta \hat{a}_t - \partial_t \hat{a}_\theta &= 0, \\ \partial_t \hat{b} + \partial_\theta \hat{a}_r + 2\hat{a}'_\theta &= 0, \\ \partial_\theta \hat{b} + \partial_t \hat{a}_r + 2\hat{a}'_t &= 0, \end{aligned} \tag{32}$$

The two equations to the right should be interpreted as conditions on the non-leading coefficients, while those on the left are analogous to the indicial equation that determines the leading term in power series solutions of ordinary differential equations. Note that a complete set of solutions can be constructed by other means, so we know there are no obstructions against solving iteratively for all the non-leading coefficients, for a suitable choice of the leading ones, but we can not be sure that the two equations

$$2\hat{b} + \partial \cdot \hat{a} = 0, \quad -2\hat{b} + \partial_\theta \hat{a}_t - \partial_t \hat{a}_\theta = 0 \tag{33}$$

are sufficient. Indeed one suspects that they are not, since they fail to differentiate between left and right movers.

The two equations on the left of (32) tell us that, on the physical subspace, where $\hat{b} = 0$, there is a field ϕ such that $\hat{a}_i = \partial_i \phi, i = t, \theta$, and that $\Box\phi = 0$.

Now consider the surface terms that arise in carrying out the variation of the above Lagrangian:

$$\delta \mathcal{L}_3 = ... + \tfrac{1}{2} \int dt d\theta \, (\hat{a}_t \, \delta \hat{a}_\theta - \hat{a}_\theta \, \delta \hat{a}_t).$$

There is no way to eliminate this identically (the situation is actually the same as in the other cases), but if we add a boundary term,

$$\mathcal{L} = \mathcal{L}_3 + \mathcal{L}_2, \quad \mathcal{L}_2 := \tfrac{1}{2} \int dt d\theta \, (\hat{a}_t \, \hat{a}_t - \hat{a}_\theta \, \hat{a}_\theta), \tag{34}$$

then the combination that appears in

$$\delta \mathcal{L} = ... + \tfrac{1}{2} \int dt d\theta \, (\hat{a}_t \, \delta \hat{a}_\theta - \hat{a}_\theta \, \delta \hat{a}_t + \hat{a}_t \, \delta \hat{a}_t - \hat{a}_\theta \, \delta \hat{a}_\theta).$$

gives us $\hat{a}_t = a_\theta$, the equation for a left mover.

This result is really quite interesting. Not only have we found a surface term that, when it is included in the Lagrangian, gives us field equations without surface terms. This phenomenon is familiar, as we has seen. What is more amazing is that this surface term contains the data that is required to distinguish between the left and right movers. On reflection, this is seen to be natural, for the distinction between left and right is carried by the singleton sector, itself only detectable at the boundary. We observe a very intimate relationship between left movers in three dimensions and left movers in two dimensions.

The sign of $\mathcal{L}_2$ is determined by positivity of the energy, while that of the Chern-Simons action is not. If we change the sign we obtain the right movers. The handedness of the gauge potential is thus decided by the sign of the main term, once the surface term is included. The field equations are the same for left and for right movers; the boundary conditions differentiate them.

The De Sitter group acts globally on the De Sitter hyperboloid, and this action extends to the compactified space that we get by including the boundary. This is easy to see, since adding the points at which $r = \infty$ is equivalent to compactifying the interior of the De Sitter cone by including the cone itself. The cone (read: the boundary at infinity) is invariant under this action of the group, therefore we get a De Sitter invariant field theory by restriction of the 3-dimensional field theory to the 2-dimensional boundary. There we recover the (abelian) Wess-Zumino model, the Chern-Simons action providing the exotic Wess-Zumino term.

References

1. M. Flato and C. Fronsdal "Quantum Field Theory of Singletons. The Rac.", J.Math.Phys.**22**, 1100 (1981).
2. M. Flato, C. Fronsdal and H. ar-Rashid, "Three-D Singletons and Two-D C.F.T." Int.J.Mod.Phys. **7**, 2193-2206 (1992).
3. M. Flato and C. Fronsdal, "The Singleton Dipole", Commun.Math.Phys. **108**, 469-482 (1987); U. Percocco, "The Spin-1/2 Singleton Dipole", Lett.Math.Phys. **12** 315-322 (1986).
4. E. Witten, "Topological Quantum Field Theories", Commun.Math.Phys. **117**, 353-386 (1988).
5. M. Flato and C. Fronsdal, "One Massless Particle Equals two Dirac Singletons", Lett.Math.Phys. **2**, 421 (1978).
6. P. Jordan, Z.Phys. **93** (1935) 464; **105** 229.
7. M. Pryce, Proc.Roy.Soc. **A165** (1938) 247.
8. S. Coleman, Phys.Rev. **D11** 2088 (1975).
9. S. Mandelstam, Phys.Rev. **D11** 3026 (1975).
10. E. Witten, "Monopoles and Four-manifolds", Preprint, IAASSNS-HEP 94-96 (hep-th 9411102), Princeton, Nov. 94.
11. A. Floer, "An Instanton Invariant for 3-Manifolds", Commun.Math.Phys. **118**, 215-240 (1988).
12. C. Fronsdal, *Essays on Supersymmetry*, D.Reidel 1986.
13. C. Fronsdal, "Variations on QED in Three Dimensions", Ann.Phys. **206**, 27-49 (1991).
14. H. Araki, Commun.Math.Phys. **97** 149 (1985).
15. M. Flato and C. Fronsdal "Three-dimensional Singletons", Lett.Math.Phys. **20**, 65-74 (1990).

POISSON PATH INTEGRAL REPRESENTATION OF GREEN FUNCTIONS FOR CERTAIN FIELD THEORIES

B.GAVEAU
Mathématiques, Université Pierre et Marie Curie (Paris VI), tour 45-46, 4 place Jussieu, 75252 Paris Cedex 05, France.

AND

J.BERTRAND AND G.RIDEAU
LPTM, Université Denis Diderot, 75251 Paris Cedex 05, France.

Abstract. The interaction of a particle with a boson field is considered. The matrix elements of the cut-off Green function are expressed in terms of a discrete time Poisson process and shown to converge to a finite limit, under some stringent conditions on the boson field. A renormalization process is set up to extend the result to more physical situations.

1. Introduction

Poisson stochastic processes can be used to write solutions of partial differential equations in terms of probabilistic expectation values [1]. The method applies in particular to Schrödinger's equation and leads to solutions having the form of a path integral in phase space for a large class of potentials. [2, 3]. In field theory, Poisson processes can be associated directly with the occupation numbers of particles (bosons or fermions) and allow to write down the propagator in various situations. [5, 6]

In the present work, a particle interacting with a boson field is considered. The matrix element of the Green function with cut-off on the energy of the boson field is expressed as an expectation with respect to a Poisson process in space (p, n_k), where p is the momentum of the particle and n_k is the occupation number of the boson field. The expression is first shown to be uniformly bounded, independently of the cut-off (section 3). Then,

J. Bertrand et al. (eds.), Modern Group Theoretical Methods in Physics, 143–156.

a closer look at the paths contributing to the expectation allows to show the existence of a strictly positive lower bound (section 4). The result is that, under certain conditions on the boson field, the matrix elements of the Green function have a finite limit when the cut-off is removed. In more general situations, it is shown how a renormalization of the theory can be performed by redefining the coupling constant. As an application, the radiative correction to the propagator of the free particle can be computed.

2. The Green function in terms of a Poisson path integral

2.1. NOTATIONS CONCERNING THE HAMILTONIANS

We shall work in a cubic box V of size L in dimension $1 \leq d \leq 2$. The possible momenta are vectors k in the lattice $\mathcal{L}$ defined by $k = (2\pi/L)(k_1, ..., k_d)$ with the k_j integers.

The space of states of the free particle in this box is the Hilbert space $\mathcal{H}_{part}$ of square integrable functions on V, or equivalently, the space of functions $f(p)$ on the lattice $\mathcal{L}$ such that

$$\sum_{p\in\mathcal{L}} |f(p)|^2 < +\infty$$

The hamiltonian of the free particle is given by $\epsilon(p) > 0$ in the p-representation. The space of states of the boson field is the usual Fock space $\mathcal{F}_B$ with the orthonormal basis of vectors $|\{n_k\}>$ where, for each k in $\mathcal{L}$, $n_k \geq 0$ is the occupation number of the field mode of momentum k. By definition, all n_k's are 0 except for a finite number of them. The creation and annihilation operators of a mode of momentum k are denoted by a_k^* and a_k respectively

$$a_k^* \, |\{n_l\}> \; = \; \sqrt{n_k+1}\, |\{n_l + \delta_{kl}\}> \tag{1}$$
$$a_k \, |\{n_l\}> \; = \; \sqrt{n_k}\, |\{n_l - \delta_{kl}\}> \tag{2}$$

and the free field hamiltonian is:

$$H_B = \sum_{k\in\mathcal{L}} \phi(k)\; a_k^* a_k \tag{3}$$

where $\phi(k)$ is some positive function of k . In electrodynamics or in the polaron problem, for example, the function ϕ_k is equal to $\sqrt{|k|^2+\alpha^2}$.

The total hamiltonian of the particle interacting with the field is

$$H = \epsilon(p) + H_B + W \equiv H_0 + W \tag{4}$$

where W is the interaction between the particle and the field. In the p-representation, we have

$$W = g \sum_{k\in\mathcal{L}} \frac{1}{2\sqrt{\phi(k)}}(a_k^*\, \tau_{-k} + a_k\, \tau_k) \tag{5}$$

where τ_k acts on a particle wave function as:

$$(\tau_k f)(p) = f(p-k) \tag{6}$$

2.2. THE CUT-OFF GREEN FUNCTION

Let λ be a complex number and $|\psi_0>$ a fixed state in $\mathcal{H}_{part} \otimes \mathcal{F}_B$. We want to solve the equation

$$(H+\lambda)|\psi>=|\psi_0> \tag{7}$$

Formally, we have

$$|\psi>=(H+\lambda)^{-1}|\psi_0>$$

and $(H+\lambda)^{-1}$ can be decomposed as usual:

$$(H+\lambda)^{-1} = \sum_{n\geq 0}(-1)^n \left((H_0+\lambda)^{-1}W\right)^n (H_0+\lambda)^{-1} \tag{8}$$

To define $(H+\lambda)^{-1}$ using a path integral, we begin to study a cut-off Green function corresponding to the cut-off hamiltonian

$$H_K = H_{0,K} + W_K \tag{9}$$

where

$$\begin{aligned} H_{0,K} &= \epsilon(p) + \sum_{k\in\mathcal{L}_K} \phi(k)\, a_k^* a_k \\ W_K &= g \sum_{k\in\mathcal{L}_K} \frac{1}{2\sqrt{\phi(k)}}(a_k^*\ \tau_{-k} + a_k\ \tau_k) \end{aligned} \tag{10}$$

and the summation is on the subset $\mathcal{L}_K$ defined by:

$$\mathcal{L}_K = \{k \in \mathcal{L}, |k| \leq K\} \tag{11}$$

The Hilbert space is now $\mathcal{H}_{part} \otimes \mathcal{F}_K$ where $\mathcal{F}_K$ is the Hilbert space of the boson field with cut off at energy K and, in relation (8), we substitute H_K, W_K and $H_{0,K}$ for H, W and H_0.

2.3. THE PATH INTEGRAL FORMULA FOR $(H_{0,K}+\lambda)^{-1}$.

The wave function associated with a state $|\psi>$ in $\mathcal{H}_{part} \otimes \mathcal{F}_K$ is defined as:

$$F\left(p^{(0)}, \{n_k^{(0)}\}_{|k|\leq K}\right) =< p^{(0)}, \{n_k^{(0)}\}_{|k|\leq K}|\psi> \tag{12}$$

where $|\psi_0> \equiv |p^{(0)}, \{n_k^{(0)}\}_{|k|\leq K}>$ is the state $|\delta(p-p^{(0)})> \otimes \{n_k^{(0)}\}_{|k|\leq K}>$ in $\mathcal{H}_{part} \otimes F_K$.

The action of $(H_K+\lambda)$ on $|\psi>$ can be written in terms of the wave function F as follows:

$$((H_K+\lambda)F)\left(p^{(0)}, \{n_k^{(0)}\}_{|k|\leq K}\right) =$$

$$\left(\epsilon(p^{(0)}) + \sum_{|k|\leq K} \phi(k) n_k^{(0)} + \lambda\right) F\left(p^{(0)}, \{n_k^{(0)}\}\right) \tag{13}$$

$$+g \sum_{|k|\leq K} \frac{1}{2\sqrt{\phi(k)}} \left[\sqrt{n_k^{(0)}} F\left(p^{(0)} - k, \{n_l^{(0)} - \delta_{kl}\}\right) + \right.$$

$$\left. \sqrt{n_k^{(0)}+1}\; F\left(p^{(0)} + k, \{n_l^{(0)} + \delta_{kl}\}\right)\right] \tag{14}$$

We shall now write a path integral formula for $(H_K+\lambda)^{-1}$. The set of paths will be the set $\left(P(t), \{N_k(t)\}_{|k|\leq K}\right)$ such that $P(t)$ is in the lattice $\mathcal{L}$ and for each k with $|k| \leq K$, $N_k(t)$ is an integer. These paths have the following properties:

(i) at time $t=0$, they start from a given point

$$P(0) = p^{(0)} \quad N_k(0) = n_k^{(0)}, \quad |k| \leq K \tag{15}$$

(ii) the total momentum of a path is conserved, so that $\left(P(t) + \sum_{|k|\leq K} k N_k(t)\right)$ is constant.

(iii) at any integer time t , a value k such that $|k| \leq K$ is chosen at random with probability $1/(m_K\sqrt{\phi(k)})$, $m_K = \sum_{|k|\leq K} \frac{1}{\sqrt{\phi(k)}}$, and the path point at time $t+1$ is defined by:

$$\left\{ \begin{array}{l} N_l(t+1) = N_l(t) + \delta_{kl}\xi_k \\ P(t+1) = P(t) - k\,\xi_k \end{array} \right. \tag{16}$$

where $\xi_k = \pm 1$ with probability $(1/2, 1/2)$ for the chosen k.

Finally, we shall also define a functional along the path by the recursion formula

$$e(0) = \left(\epsilon(p^{(0)}) + \sum_{|l|\leq K} \phi(l) n_l^{(0)} + \lambda\right)^{-1} \tag{17}$$

$$e(t+1) = e(t) \frac{\sqrt{N_k(t) + \frac{1}{2}(\xi_k+1)}}{\epsilon(P(t+1)) + \sum_{|l|\leq K} \phi(l) N_l(t+1) + \lambda} \tag{18}$$

where, in the right hand side of (17) we have the initial conditions of the path at time 0 given by (15).

It is then clear from (8-10) and (13) that the term of order t in the perturbation series applied to the wave function F is:

$$\left[(-1)^t\left((H_{0,K}+\lambda)^{-1}W_K\right)^t(H_{0,K}+\lambda)^{-1}F\right]\left(p^{(0)},\{n_k^{(0)}\}_{|k|\leq K}\right) = \quad (19)$$
$$(-m_K g)^t \ll e(t)F\left(P(t),\{N_k(t)\}_{|k|\leq K}\right)\Big|P(0)=p^{(0)},N_k^{(0)}=n_k^{(0)} \gg$$

where $\ll \quad | \quad \gg$ denotes the conditional expectation on the paths with the initial condition (15).

If we now define a random time T by the formula

$$\mathrm{Prob}(T=t)=e^{-m_K}\frac{(m_K)^t}{t!}\quad (t \text{ integer}) \tag{20}$$

the whole perturbation series can be rewritten as

$$\left[(H_K+\lambda)^{-1}F\right]\left(p^{(0)},\,\{n_k^{(0)}\}_{|k|\leq K}\right) = \quad (21)$$
$$e^{m_K}\ll e(T)(-g)^T T!F\left(P(T),\{N_k(T)\}_{|k|\leq K}\right)\Big|P(0)=p^{(0)},N_k^{(0)}=n_k^{(0)} \gg$$

We can state the following two elementary results:

Lemma 1.1. Let us assume that the wave function F is:

$$F\left(p,\{n_k\}_{|k|\leq K}\right)=\delta(p-p^{(1)})\prod_{|k|\leq K}\delta(n_k-n_k^{(1)}) \tag{22}$$

and let us set

$$T_0=\sum_{|k|\leq K}|n_k^{(1)}-n_k^{(0)}| \tag{23}$$

Then the only non zero terms in (19) are the terms such that $t=(T_0+$ even integer) and, in particular they all have the same sign $(-1)^{T_0}$.

Theorem 1.2. *Let us assume that* $m\equiv\sum_{k\in\mathcal{L}}(\phi(k))^{-1/2}<+\infty$. *Then for a function F of type (22) or more generally in* $\mathcal{H}_{part}\otimes\mathcal{F}_B$, *we have*

$$\left[(H+\lambda)^{-1}F\right]\left(p^{(0)},\{n_k^{(0)}\}_{|k|\leq K}\right)=e^m\times$$
$$\ll e(T)(-g)^T T!F\left(P(T),\{N_k(T)\}_{|k|\leq K}\right)\Big|P(0)=p^{(0)},N_k(0)=n_k^{(0)}\gg$$

and the r.h.s. is a convergent series for $Re(\lambda)$ sufficiently large.

Remark Under the hypothesis $m<+\infty$, we have defined the propagator e^{-itH} or e^{-tH} in [5] using a similar process but in continuous time. The main reason why the hypothesis $m<+\infty$ came out was that in the continuous time case, at any finite time the expected number of modes k such that $N_k(t)$ was non zero, was finite so that at any time there was only a finite number of quanta present in the path integral. Unfortunately the hypothesis $m < +\infty$ is not physically relevant if $\phi \sim |k|$ for large k .

3. Upper bound for the matrix element of the Green function.

The problem is to obtain an upper bound for the matrix element

$$< \widetilde{\psi}_1|(H_K+\lambda)^{-1}|\widetilde{\psi}_0 >, \quad K > K_0 \tag{24}$$

where the states $|\widetilde{\psi}_j >$, $j = 0, 1$ are defined by:

$$|\widetilde{\psi}_j >= \left|p^{(j)}, \{n_k^{(j)}\}_{|k|\leq K} > , n_k^{(j)} = 0 \text{ for } |k| > K_0 \right. \tag{25}$$

The wave function of $|\widetilde{\psi}_1 >$ is:

$$\widetilde{F}_1\left(p, \{n_k\}_{|k|\leq K}\right) = \delta\left(p - p^{(1)}\right) \prod_{|k|\leq K} \delta(n - n_k^{(1)}),\ n_k^{(1)} = 0 \text{ if } K_0 < |k| \leq K \tag{26}$$

Relations (19) and (26 lead to:

$$\begin{aligned} &\left|< \widetilde{\psi}_1|\,(H_K+\lambda)^{-1}\,|\widetilde{\psi}_0 >\right| \leq \\ &\quad \sum_{t\geq 0}(m_K|g|)^t \ll e(t)\,\delta(P(t) - p^{(1)}) \\ &\quad \prod_{|k|\leq K} \delta(N_k(t) - n_k^{(1)})|P(0) = p^{(0)}, N_k(0) = n_k^{(0)} \gg \end{aligned} \tag{27}$$

For $\lambda_1 \equiv Re(\lambda) > 0$, an upper bound is found to be:

$$\left|< \widetilde{\psi}_1|(H_K+\lambda)^{-1}|\widetilde{\psi}_0 >\right| \leq \left|< \{n_k^{(1)}\}|(\widehat{H}_K+\lambda_1)^{-1}|\{n_k^{(0)}\} >\right| \tag{28}$$

where:

$$\widehat{H}_K = \sum_{|k|\leq K} \widehat{h}_k,\ \widehat{h}_k \equiv \phi(k)\, a_k^* a_k + \frac{|g|}{2}\frac{1}{\sqrt{\phi(k)}}(\, a_k^* + a_k) \tag{29}$$

Now, defining new creation and annihilation operators according to:

$$A_k = a_k + \frac{|g|}{2\phi(k)^{3/2}}, \quad A_k^* = a_k^* + \frac{|g|}{2\phi(k)^{3/2}} \tag{30}$$

allows to write

$$\widehat{H}_K = \sum_{|k|\leq K} \phi(k)\, A_k^* A_k - \frac{|g|^2}{4} \sum_{|k|\leq K} \frac{1}{\phi(k)^2} \tag{31}$$

To get rid of the extra constant, define $\lambda_{1,K}$ and $\widehat{H}'_K$ by:

$$\lambda_{1,K} \equiv \lambda_1 - \frac{|g|^2}{4} \sum_{|k|\leq K} \frac{1}{\phi(k)^2}, \qquad \widehat{H}'_K \equiv \sum_{|k|\leq K} \phi(k)\, A_k^* A_k \tag{32}$$

Then, the vacuum state for $\widehat{H}'_K$ is the product

$$\exp\left(-\frac{|g|^2}{4} \sum_{|k|\leq K} \frac{1}{\phi(k)^3}\right) \bigotimes_{|k|\leq K} \left\{ \sum_{n\geq 0} \left(\frac{-|g|}{2\phi(k)^{3/2}}\right)^n \frac{|n>}{\sqrt{n!}} \right\} \tag{33}$$

If $\sum_{|k|\leq K}(1/\phi(k)^3)$ diverges when $K \to \infty$, vector (33) is not in $\mathcal{F}_B$ and the evolution $\exp(-it\widehat{H}'_K)$ is not defined on the usual Fock space. If on the other hand:

$$\sum_{k\in\mathcal{L}} \frac{1}{\phi(k)^2} < +\infty \tag{34}$$

the operator $\widehat{H}'_K$ makes sense in the usual Fock space *without renormalization* when K goes to infinity.

To estimate the right member of (28), remark that for $K > K_0$:

$$< \{n_k^{(1)}\}|(\widehat{H}_K + \lambda_1)^{-1}|\{n_k^{(0)}\} >= \tag{35}$$
$$\int_0^\infty dt\, e^{-\lambda_1 t} \left(\prod_{|k|\leq K_0} < n_k^{(1)}|e^{-t\hat{h}_k}|n_k^{(0)} >\right) \left(\prod_{K_0<|k|\leq K} < 0|e^{-t\hat{h}_k}|0 >\right)$$

where $\widehat{h}_k$ is given by (29).

Hence, according to classical but tedious computations, we can write:

$$\left|< \{n_k^{(1)}\}|(\widehat{H}_K + \lambda_1)^{-1}|\{n_k^{(0)}\} >\right| \leq$$
$$C \int_0^\infty dt\, \exp\left[-t\left(\lambda_1 - \frac{|g|^2}{4} \sum_{|k|\leq K} \frac{1}{\phi(k)^2}\right)\right] \times \tag{36}$$
$$\exp\left[-\sum_{|k|\leq K} \frac{|g|^2}{4\phi(k)^3}(1 - e^{-t\phi(k)})\right]$$

where the constant C depends only on the sequence $\{n_k^{(j)}\}$, $|k| \leq K_0$, $j = 0, 1$.

Assume now that $\phi(k)$ behaves like $|k|$ for large values of $|k|$ and that the dimension $d \leq 2$ so that

$$\sum_{k \in \mathcal{L}} \frac{1}{\phi(k)^3} < +\infty \tag{37}$$

The above estimate is then reduced to:

$$\left| < \{n_k^{(1)}\}_{|k| \leq K_0} | (\widehat{H}_K + z)^{-1} | \{n_k^{(0)}\}_{|k| \leq K_0} > \right| \leq \tilde{C} \left(\lambda_1 - \frac{|g|^2}{4} \sum_{|k| \leq K} \frac{1}{\phi(k)^2} \right)^{-1} \tag{38}$$

This form of the bound is valid even if $\sum_{|k| \leq K} (1/\phi(k)^2)$ diverges.

Remark. It can be shown that the right-hand side of inequality (36) goes to zero when $\lim_{K \to \infty} \sum_{|k| \leq K} (1/\phi(k)^3) = \infty$.

Theorem 2.1 *If $\sum_{k \in \mathcal{L}} (1/\phi(k)^2) < \infty$, the set of matrix elements of the form $< \psi_1 | (H_K + \lambda)^{-1} | \psi_0 >$ is a sequence (in K) of holomorphic functions of λ which are uniformly bounded for $Re(\lambda) \geq (|g|^2/4) \sum_{k \in \mathcal{L}} (1/\phi(k)^2)$.*

As a consequence, the existence of at least one convergent subsequence is ensured and we are left with the task of showing that such a subsequence does not converge to zero. This will be done in the next section for λ real.

4. Lower bound for the Green function

4.1. CHOICE OF A RELEVANT SUBSET OF PATHS Ω_1.

For real λ, i.e. $\lambda \equiv \lambda_1$, we want to estimate from below the matrix element of the cut-off Green function between states $|\psi_0>$ and $|\psi_1>$, where $|\psi_j> \equiv |p^{(0)}, \{n_l^{(0)}\}_{|k| \leq K}>$, $j = 0, 1$. To this end, we use expression (19) with the following wave function:

$$F(P(t), \{N_k(t)\}_{|k| \leq K}) = \delta(P(t) - p^{(1)}) \prod_{|k| \leq K} \delta(N_k(t) - n_k^{(1)}) \tag{39}$$

By conservation of momentum the matrix elements are zero unless the following condition is satisfied:

$$p^{(1)} + \sum_{|k| \leq K} k\, n_k^{(1)} = p^{(0)} + \sum_{|k| \leq K} k\, n_k^{(0)} \tag{40}$$

As expressed in lemma 1.1, the only non-zero terms in the series (19) are of rank $T_0 + 2t$, with $T_0 = \sum_{|k| \leq K} |n_k^(0) - n_k^{(1)}|$. Thus the absolute value of the matrix element can be bounded from below by ignoring all sign problems and summing only on the indices $T_0 + 2t$, $t \geq 0$. Now each element of the summation is bounded from below by integrating only on a well chosen subset Ω_1 of the set of all paths, so that the estimations reads:

$$\begin{aligned} &\left| < \psi_1 | (H_K + \lambda)^{-1} | \psi_0 > \right| \geq \\ &\quad \sum_{t \geq 0} (m_K |g|)^{T_0 + 2t} \ll e(T_0 + 2t)\, \delta\left(P(T_0 + 2t) - p^{(1)}\right) \times \\ &\quad \prod_{|k| \leq K} \delta\left(N_k(T_0 + 2t) - n_k^{(1)}\right) \chi_{\Omega_1} | P(0) = p^{(0)}, N_k(0) = n_k^{(0)} \gg \end{aligned} \tag{41}$$

where χ_{Ω_1} is the characteristic function of subset Ω_1.

Let us call Ω_1 the subset of trajectories of the process with the following properties:

(i) at time T_0, $N_k(T_0) = n_k^{(1)}$ for all $|k| \leq K$. In particular, because of (40), on has: $P(T_0) = p^{(1)}$.

(ii) in the time interval $[0, T_0]$, all trajectories in Ω_1 are identical, so that the functional $e(t)$ is constant on Ω_1 for $t \leq T_0$. We shall define:

$$e_0(\lambda_1) \equiv e(T_0) \quad \text{which is constant on the subset } \Omega_1 \tag{42}$$

These conditions imply that the trajectories go as fast as possible to their end point.

(iii) at times $t > T_0$, $N_k(t)$ is equal either to $n_k^{(1)}$ or $n_k^{(1)} + 1$ and, in the latter case $N_k(t+1) = n_k^{(1)}$.
In particular for $t = T_0 +$ odd integers, all $N_l(t)$ are equal to $n_l^{(1)}$ except only one, say $N_k(t)$ which is $n_k^{(1)} + 1$ and as a consequence $P(t) = p^{(1)} - k$. For $t = T_0 +$ even integers, all $N_k(t)$ are equal to $n_k^{(1)}$ and $P(t) = p^{(1)}$.

Thus, in Ω_1 all the δ factors in the second member of (41) are equal to 1 and we have to compute

$$\Phi_K(t) \equiv (m_K|g|)^{T_0 + 2t} \ll e(T_0 + 2t)\, \chi_{\Omega_1} \mid P(0) = p^{(0)},\, N_k(0) = n_k^{(0)} \gg \tag{43}$$

Now let Ω_2 denote the set of trajectories starting from:

$$N_k(0) = n_k^{(1)} \quad P(0) = p^{(1)} \tag{44}$$

and having the property (iii) of Ω_1 for all t > 0 . It is clear, by the Markov property and the definition of Ω_1 on $[0, T_0]$ that

$$\Phi_K(t) = (m_K|g|)^{T_0+2t} e_0(\lambda_1) \left((2m_K)^{T_0} \prod_{|k|\leq K} \phi(k)^{\frac{|n_k^{(0)} - n_k^{(1)}|}{2}} \right)^{-1} \times$$

$$\ll e(2t)\, \chi_{\Omega_2} \,|P(0) = p^{(1)},\ N_k(0) = n_k^{(1)} \gg \qquad (45)$$

since the contribution of the interval $[0, T_0]$ consists of the constant functional $e_0(\lambda_1)$ and the weight of the individual trajectories given by:

$$(2m_K)^{-T_0} \prod_{|k|\leq K} (\sqrt{\phi(k)})^{-|n_k^{(0)} - n_k^{(1)}|}$$

4.2. LOWER BOUND FOR THE CUT-OFF GREEN FUNCTION.

Now, we have only to compute $e(2t)$ for paths in Ω_2 between 0 and $2t$. In that case, each $N_k(s)$ jumps $2t_k$ times on this time interval so that:

$$t = \sum_{|k|\leq K} t_k \qquad (46)$$

A set of $\{t_k\}_{|k|\leq K}$ satisfying (46) is called a configuration. The weight of such a configuration is exactly

$$w(\{t_k\}) = \prod_{|k|\leq K} \left(2m_K\sqrt{\phi(k)}\right)^{-2t_k} \qquad (47)$$

because the probability that $N_k(s)$ jumps is equal to $(2m_K\sqrt{\phi(k)})^{-1}$.

The number of configurations is $(t!/\prod_{|k|\leq K} t_k!)$ because each time N_k jumps by $+1$ from $n_k^{(1)}$ to $n_k^{(1)} + 1$, it immediately jumps back to $n_k^{(1)}$ by the property (iii) of the trajectories in Ω_1 or Ω_2.

The factors $(N_k(s) + \frac{1}{2}(\xi_k + 1)^{1/2}$ in $e(2t)$ are all equal to $(n_k^{(1)} + 1)^{1/2}$ and the total contribution of these factors to $e(2t)$ is

$$\prod_{|k|\leq K} (n_k^{(1)} + 1)^{t_k} \qquad (48)$$

The free field terms in the denominator of e(2t) in (18) can only have one of the following forms:

$$(\epsilon(p^{(1)}) + \sum_{|k|\leq K} \phi(k) n_k^{(1)} + \lambda_1)$$

or

$$(\epsilon(p^{(1)} - k) + \sum_{|k|\leq K} \phi(k)(n_k^{(1)} + 1) + \lambda_1)$$

Setting

$$\mu \equiv \lambda_1 + \sum_{|k|\leq K} \phi(k) n_k^{(1)} \tag{49}$$

we finally obtain:

$$m_K^{2t} \ll e(2t)\chi_{\Omega_2}|P(0) = p^{(1)}, N_k(0) = n_k^{(1)} \gg = \tag{50}$$

$$\sum_{\{t_k\}}' \frac{t!}{\prod_{|k|\leq K} t_k!} w(\{t_k\}) \prod_{|k|\leq K} (n_k^{(1)} + 1)^{t_k} \times$$

$$\left[\prod_{|k|\leq K} (\epsilon(p^{(1)} - k) + \phi(k) + \mu)(\epsilon(p^{(1)} + \mu) \right]^{-t_k}$$

where $\sum'$ is a summation over all configurations $\{t_k\}_{|k|\leq K}$ satisfying (46).

We can estimate this from below by taking account of the inequality: $(n_k^{(1)} + 1)^{t_k} \geq 1$ and we find, using (47):

$$m_K^{2t} \ll e(2t)\chi_{\Omega_2}|P(0) = p^{(1)}, N_k(0) = n_k^{(1)} \gg \quad \geq \tag{51}$$

$$\frac{1}{2^{2t}} \sum\nolimits'_{\{t_k\}} \frac{t!}{\prod_{|k|\leq K} t_k!} \prod_{|k|\leq K} (u_k)^{t_k}$$

where u_k is defined by:

$$u_k^{-1} \equiv (\epsilon(p^{(1)} - k) + \phi(k) + \mu)(\epsilon(p^{(1)}) + \mu)\phi(k) \tag{52}$$

In fact, recognizing that the lower bound in (51) involves the binomial coefficients, we can rewrite it as: $\frac{1}{2^{2t}} \left(\sum_{|k|\leq K} u_k\right)^t$ and we finally obtain from (41);

$$\left| < \psi_1 |(H_k + \lambda)^{-1}| \psi_0 > \right| \geq \tag{53}$$

$$\left(\frac{|g|}{2}\right)^{T_0} e_0(\lambda) \left[\prod_{|k|\leq K_0} \phi(k)^{(1/2)|n_k^{(0)} - n_k^{(1)}|} \right]^{-1} \left[1 - \frac{|g|^2}{4} \sum_{|k|\leq K} u_k \right]^{-1}$$

This is meaningful provided

$$\frac{|g|^2}{4} \sum_{|k|\leq K} u_k < 1 \tag{54}$$

We can always choose $p^{(1)} = 0$ by changing frames. Now if we suppose that:

$$\sum_{k\in\mathcal{L}} \frac{1}{\phi(k)(\phi(k)+\epsilon(k))} < \infty \tag{55}$$

then, for μ (i.e. λ) large enough, condition (54) can be realized. In the physical case where $\phi(k) \sim |k|$ for large k, condition (55) is realized if $\sum_{k\in\mathcal{L}}(1/\phi(k)^2 < +\infty$. We can state the result:

Theorem 3.1.*Let $|\psi_0>$ and $|\psi_1>$ represent two states $|\psi_j>=|p^{(j)}, \{n_k^{(j)}\}_{|k|\leq K}>$ satisfying the momentum conservation (40). Then for $K' > K$, the Green function $<\psi_1|(H_K+\lambda)^{-1}|\psi_0>$ has a positive lower bound, independent of K and of the states $|\psi_j>$, provided λ is large enough and the following condition is satisfied:*

$$\sum_{k\in\mathcal{L}} \frac{1}{\phi(k)(\phi(k)+\epsilon(k))} < +\infty \tag{56}$$

The bound in (53) can be computed using the fact that

$$e_0(\lambda_1) = \left(\epsilon(p^{(1)}) + \sum_{|l|\leq K} \phi(l)\, n_l^{(1)} + \lambda_1\right)^{-1} \tag{57}$$

The result is:

$$\begin{aligned} &|<\psi_1|(H_K+\lambda)^{-1}|\psi_0>| \geq \\ &C\left[\epsilon(p^{(1)}) + \sum_{|l|\leq K} \phi(l) n_l^{(1)} + \lambda_1 - \frac{|g|^2}{4}\sum_{|k|\leq K}(1/\phi(k)^2)\right]^{-1} \end{aligned} \tag{58}$$

because for large k, we have $\epsilon(p^{(1)}-k) \sim |k| \sim \phi(k)$.

5. Renormalization

5.1. COUPLING CONSTANT RENORMAIZATION.

Collecting the results of theorems 2.1 and 3.1, we can state:

Theorem 3.2. *When $\sum_{k\in\mathcal{L}}(1/\phi(k)^3)$ is convergent, the upper and lower bounds of the cut-off Green function $<\psi_1|(H_K+\lambda)^{-1}|\psi_0>$ given by (38) and (58) respectively are of the same order of magnitude even if $\sum_{k\in\mathcal{L}}(1/\phi(k)^2)$ is divergent.*

This result allows to perform a renormalization of the coupling constant. In fact, the above developments can be carried out in exactly the same way with a coupling constant $g_K = g_1 Z_K$ where Z_K depends on the cut-off in such a way that $\lim_{K\to\infty} Z_K^2 \sum_{|k|\le K}(1/\phi(k)^2)$ is finite and g_1 is the physical coupling constant. In that way, we get finite values for the Green function matrix elements when the cut-off is removed.

This result remains true even if $\sum_{k\in\mathcal{L}}(1/\phi(k)^3)$ is divergent. Indeed, a close examination of the right-hand side of (36) shows that the last term, which now reads:

$$\exp\left[-\left(\sum_{|k|\le K}\frac{1-e^{-t\phi(k)}}{4\phi(k)^3}|g_1|^2\right)\left(\sum_{|k|\le K}\frac{1}{\phi(k)^2}\right)^{-1}\right]$$

is bounded by one independently of K. As an example, we now apply this renormalization process to an estimation of the radiative correction to the propagator of the free particle.

5.2. RADIATIVE CORRECTION TO THE PROPAGATOR OF THE FREE PARTICLE

The correction to the propagator of the free particle is obtained when we consider the Green function

$$< p,0|(H_K+\lambda)^{-1}|p,0> \equiv K(p,\lambda) \tag{59}$$

where $|0>$ represents the vacuum of the Fock space. In this case, we can obtain better estimates.

- *Estimate from above* It is easy to obtain an estimate from above, replacing H_K by $H_K' \equiv \widehat{H}_K + \epsilon(p)$. In this case, we have

$$|K(p,\lambda)| \le \left[\lambda+\epsilon(p)-\frac{|g|^2}{4}\sum_{|k|\le K}\frac{1}{\phi(k)^2}\right]^{-1} \tag{60}$$

Remark that the constant $\tilde{C}$ coming from (38) is now equal to one.

- *Estimate from below.* In (53), T_0 as defined by (23) is equal to zero so that $e_1(\lambda) = 1$ in (57). From (53) we deduce the lower bound applying to the present case:

$$|K(p,\lambda)| \ge \tag{61}$$
$$\left[\lambda+\epsilon(p)-\frac{|g|^2}{4}\sum_{|k|\le K}\frac{1}{\phi(k)(\phi(k)+\epsilon(p-k)+\lambda)}\right]^{-1}$$

- *Result.* From (60) and (61), we conclude that the effective hamiltonian can be bounded between the lower bound:

$$\epsilon(p)-\frac{|g|^2}{4}\sum_{|k|\le K}\frac{1}{\phi(k)^2}$$

and the upper bound which is the negative of the root of the denominator in (61). This root is:

$$\lambda \equiv -\epsilon(p) + s(p)$$

where $s(p) \sim s_K(p)$ with

$$s_K(p) \equiv \frac{|g|^2}{4} \sum_{|k|\leq K} \frac{1}{\phi(k)(\phi(k) + \epsilon(p-k) - \epsilon(p))} \tag{62}$$

Thus the upper bound of the effective hamiltonian is $\epsilon(p) - s_K(p)$. At this point, it is possible to renormalize the propagator of the particle when $\sum_{k\in\mathcal{L}}(1/\phi(k)^2) = +\infty$. The method is to choose a running coupling constant g_K defined by:

$$g_K = g_1 \left(\sum_{|k|\leq K} \frac{1}{\phi(k)^2} \right)^{(-1/2)}$$

and to use g_K instead of g in the two preceding formulas. Then, when the cut off constant K goes to infinity we see that the propagator of the particle is bounded from below and from above. More precisely, we can state:

Theorem 4.1. *The effective hamiltonian of the particle is bounded from below by*

$$\epsilon(p) - (|g|^2/4)$$

and from above by

$$\epsilon(p) - \lim_{K\to\infty} s_K(p) \left(\sum_{|k|\leq K} (1/\phi(k)^2) \right)^{-1}$$

where s_K is defined by (62).

References

1. Maslov, V.P. and Chebotarev (1978), A.P., *Viniti Itogui Nauki*, **15**, p.5.
2. Combe, P., Høegh-Krohn, R., Rodriguez, R., Sirugue, M., and Sirugue-Collin, M., *Comm. Math. Phys.*, **77**, p. 269 (1980); *J. Math. Phys.*, **23**, p. 405, (1982).
3. Bertrand, J., and Rideau, G., (1981), Stochastic jump processes in the phase space representation of quantum mechanics, in *Mathematical problems in theoretical physics*, Lecture Notes in Physics, **153**, Springer-Verlag.
4. Bertrand, J., Gaveau,B. and Rideau,G. (1984), Poisson processes and quantum field theory: a model, in *Lecture Notes in Mathematics*, **1136**, Springer-Verlag.
5. Bertrand, J., Gaveau,B. and Rideau,G. (1985), Quantum fields and Poisson processes, I. Interaction of a cut-off boson field with a quantum particle, *Lett. Math. Phys.*, **9**, pp. 73-82.
6. Bertrand, J., Gaveau,B. and Rideau,G. (1985), Quantum fields and Poisson processes, II. Interaction of boson-boson and boson-fermion fields with a cut-off, *Lett. Math. Phys.*, **9**, pp. 353-362.

DIFFEOMORPHISM GROUP REPRESENTATIONS AND NONLINEAR QUANTUM THEORIES

GERALD A. GOLDIN
Depts. of Mathematics and Physics, Rutgers University
New Brunswick, New Jersey 08903, USA
E-mail: gagoldin@dimacs.rutgers.edu

In honor of Guy Rideau

Abstract. Quantum theories with nonlinear time-evolution equations are suggested by certain continuous unitary representations of diffeomorphism groups. Their physical interpretation is discussed in the light of nonlinear gauge transformations.

1. Introduction

In this paper I shall review briefly how nonlinear quantum theories arise in a natural way from a certain class of continuous unitary representations of the group of diffeomorphisms of physical space. I shall then offer some comments related to the interpretation of these theories, which are believed to describe dissipative quantum systems.

The group-theoretical approach to quantum theory on which this work is based was developed largely in collaboration with D. H. Sharp at Los Alamos National Laboratory (LANL); for recent reviews, see [7], [10], [11]. Most of the results on nonlinear quantum mechanics were obtained jointly with H.-D. Doebner at the Arnold Sommerfeld Institute for Mathematical Physics (ASI), Technical University of Clausthal, Germany [3], [4]; some recent calculations are also in collaboration with P. Nattermann [5]. However, I take responsibility for the views expressed here regarding the physical interpretation of nonlinear quantum theories in terms of positional measurements. I am grateful to the LANL and the ASI for hospitality during the course of the research. I would also like to thank Université Paris 7 -

J. Bertrand et al. (eds.), Modern Group Theoretical Methods in Physics, 157–166.

Denis Diderot for support during my recent visits, where additional work relating to this subject was completed.

Quantum mechanics on a spatial manifold M can be described by representating, with self-adjoint operators, an infinite-dimensional Lie algebra of local nonrelativistic densities and currents. If ψ and ψ^* are second-quantized fields satisfying canonical commutation or anticommutation relations, define the gauge-invariant operator-valued distributions $\rho_{op}(\mathbf{x}) = \psi^*(\mathbf{x})\psi(\mathbf{x})$ and $\mathbf{j}_{op}(\mathbf{x}) = (\hbar/2mi)\,[\psi^*(\mathbf{x})\,\nabla\psi(\mathbf{x}) - (\nabla\psi^*(\mathbf{x}))\psi(\mathbf{x})]$. Then let $\rho_{op}(f) = \int \rho_{op}(\mathbf{x})f(\mathbf{x})d\mathbf{x}$, and $j_{op}(\mathbf{g}) = \int \mathbf{j}_{op}(\mathbf{x})\cdot\mathbf{g}(\mathbf{x})d\mathbf{x}$, where f is a C^∞ scalar function and $\mathbf{g}$ a C^∞ vector field on the physical space; f and the components of $\mathbf{g}$ are test functions vanishing rapidly (with all derivatives) as $|\mathbf{x}| \to \infty$ in Euclidean space. These operators obey commutator brackets,

$$[\rho_{op}(f_1), \rho_{op}(f_2)] = 0, \quad [\rho_{op}(f), j_{op}(\mathbf{g})] = i\,(\hbar/m)\rho_{op}(\mathbf{g}\cdot\nabla f),$$
$$[j_{op}(\mathbf{g}_1), j_{op}(\mathbf{g}_2)] = -i\,(\hbar/m)j_{op}([\mathbf{g}_1, \mathbf{g}_2])\,, \tag{1}$$

where $[\mathbf{g}_1, \mathbf{g}_2]$ is the Lie bracket of the vector fields. The Lie algebra represented by (1) is the natural semidirect sum of the Abelian algebra of functions with the algebra *Vect*(M) of vector fields. It is the same regardless of whether ψ and ψ^* satisfy commutator or anticommutator brackets; the particle content of the theory is no longer encoded in the algebra itself, but in the choice of representation (up to unitary equivalence).

Correspondingly, one has a continuous unitary representation (CUR) of an infinite-dimensional group, the semidirect product of the additive group $\mathcal{S}$ of scalar functions with the group *Diff*(M) of diffeomorphisms of M. Let $\phi_t^{\mathbf{g}}$ be the one-parameter group of flows generated by the vector field $\mathbf{g}$. With $U(f) = \exp[i\rho_{op}(f)]$ and $V(\phi_t^{\mathbf{g}}) = \exp[(itm/\hbar)j_{op}(\mathbf{g})]$, we have

$$U(f_1)V(\phi_1)U(f_2)V(\phi_2) = U(f_1 + f_2 \circ \phi_1)V(\phi_1\phi_2)\,, \tag{2}$$

where $\phi_1\phi_2 = \phi_2 \circ \phi_1$ is the composition of the diffeomorphisms.

Now a wide variety of quantum systems have been shown to arise, or been predicted, as inequivalent unitary representations of this semidirect product group G. These include N-particle systems obeying Bose statistics, Fermi statistics, parastatistics, or–in two space dimensions–the unexpected possibility of intermediate, or "anyon" statistics [8] [13]. This fact leads to the perspective that G should be regarded as a universal group of local symmetries for quantum theory [9], and that all its inequivalent unitary representations are to be taken seriously. There is an interesting one-parameter family of distinct CUR's, all of which can be interpreted as describing a single particle in (for instance) $\mathbf{R}^3$. These are given, in the Hilbert space of square-integrable functions $\mathcal{H} = \mathcal{L}^2(\mathbf{R}^3, dx)$, by

$$U(f)\Phi(\mathbf{x}) = \exp[if(\mathbf{x})]\Phi(\mathbf{x})\,,$$

$$V(\phi)\Phi(\mathbf{x}) = \exp\left[\, i\,\Gamma \ln \mathcal{J}_\phi(\mathbf{x})\,\right] \Phi(\phi(\mathbf{x}))\sqrt{\mathcal{J}_\phi(\mathbf{x})}\,, \tag{3}$$

$\Gamma \in \mathbf{R}$, where $\mathcal{J}_\phi(\mathbf{x})$ is the Jacobian of ϕ at $\mathbf{x}$. When $\Gamma = 0$ we have ordinary quantum mechanics, and the self-adjoint generators are given as usual by

$$\rho_{op}(f)\Phi(\mathbf{x}) = f(\mathbf{x})\Phi(\mathbf{x})\,,$$

$$j_{op}^{\Gamma=0}(\mathbf{g})\Phi(\mathbf{x}) = \frac{\hbar}{2mi}\{\mathbf{g}(\mathbf{x})\cdot\nabla\Phi(\mathbf{x}) + \nabla\cdot[\mathbf{g}(\mathbf{x})\Phi(\mathbf{x})\,]\}\,. \tag{4}$$

But for $\Gamma \neq 0$ we have from (3),

$$j_{op}^{\Gamma}(\mathbf{g})\Phi(\mathbf{x}) = j_{op}^{\Gamma=0}(\mathbf{g})\Phi(\mathbf{x}) + D\rho_{op}(\nabla\cdot\mathbf{g})\Phi(\mathbf{x})\,. \tag{5}$$

Here D is a real number with the dimensions of a diffusion coefficient, related to the dimensionless parameter Γ by the formula $\Gamma = Dm/\hbar$. Then we have the problem of interpreting these new representations physically.

Doebner and I proposed the following. Let $\Psi(\mathbf{x}, t)$ be a normalized, time-evolving wave function. Consider the expectation values of the density and current operators, $\rho(\mathbf{x}, t) = (\Psi, \rho_{op}(\mathbf{x})\Psi)$, and $\mathbf{j}^D(\mathbf{x}, t) = (\Psi, \mathbf{j}_{op}^{\Gamma}(\mathbf{x})\Psi)$. (We have chosen units in which $\rho(\mathbf{x}, t)$ is the probability density, and $\mathbf{j}^D(\mathbf{x}, t)$ the probability current.) Then

$$\begin{aligned}
\rho_{op}(\mathbf{x},t) &= \overline{\Psi(\mathbf{x},t)}\Psi(\mathbf{x},t)\,,\\
\mathbf{j}^D(\mathbf{x},t) &= \frac{\hbar}{2mi}(\overline{\Psi}\nabla\Psi - \Psi\nabla\overline{\Psi}) - D\nabla(\overline{\Psi}\Psi) \qquad (6)\\
&= \mathbf{j}^{D=0}(\mathbf{x},t) - D\nabla\rho\,.
\end{aligned}$$

The additional assumption that ρ and $\mathbf{j}^D$ satisfy the equation of continuity gives us

$$\partial_t\rho \;=\; -\nabla\cdot\mathbf{j}^D \;=\; -\nabla\cdot\mathbf{j}^{D=0} + D\nabla^2\rho\,, \tag{7}$$

which is a Fokker-Planck type equation in which the real number D serves as a (quantum) diffusion coefficient. At this point in the development, it seems as though the interpretation of $D \neq 0$ as introducing diffusion will provide the essential basis for our physical understanding. But there are some surprises ahead.

Combining (6) with (7) yields the new nonlinear Schrödinger equation in the form that Doebner and I proposed it [3] [4]:

$$\begin{aligned}
i\hbar\frac{\partial}{\partial t}\Psi &= H_0\Psi + i\,I[\Psi]\Psi + R[\Psi]\Psi\,,\\
\text{with } I[\Psi] &= \frac{1}{2}\hbar D\frac{\nabla^2(\bar{\Psi}\Psi)}{\bar{\Psi}\psi} = \frac{1}{2}\hbar D\frac{\nabla^2\rho}{\rho}\,; \qquad (8)
\end{aligned}$$

where the linear operator H_0 is $-(\hbar^2/2m)\nabla^2+V(\mathbf{x})$, and where R is a real, nonlinear functional multiplying Ψ. Our assumptions up to now do not tell us the form of $R[\Psi]$; therefore we impose only the additional, reasonable conditions that the terms in R be Euclidean-invariant rational expressions, homogeneous in Ψ, with no higher than second derivatives in the numerator only. This leads to the following parameterized family of possibilities:

$$\begin{aligned} R[\Psi] &= \hbar D'\{c_1 \frac{m}{\hbar}\frac{\nabla\cdot\mathbf{j}^{D=0}}{\rho} + c_2\frac{\nabla^2\rho}{\rho} + c_3\frac{m^2}{\hbar^2}\frac{(\mathbf{j}^{D=0})^2}{\rho^2} \\ &\quad + c_4\frac{m}{\hbar}\frac{\mathbf{j}^{D=0}\cdot\nabla\rho}{\rho^2} + c_5\frac{(\nabla\rho)^2}{\rho^2}\} \\ &= \hbar D'\sum_{i=1}^{5} c_i R_i[\Psi]\,, \end{aligned} \tag{9}$$

where D' also has the dimensions of a diffusion coefficient. Our task, then, is to understand the physics of a quantum theory in which the time-evolution of the wave functions is given by a nonlinear Schrödinger equation such as this, and to establish the physical meaning of the parameters in (8) and (9.

Properties of these equations that have been previously noted include the following: (a) The time-evolution conserves probability. (b) The equation is homogeneous in that if Ψ is a solution, $\alpha\Psi$ is a solution ($\forall\alpha\in\mathbf{C}$). This property permits one to construct a *hierarchy* of N-particle equations satisfying the separation property (i.e., for which initially uncorrelated, noninteracting subsystems remain uncorrelated) [12]. (c) With $V\equiv 0$, plane waves are solutions. (d) With $V\equiv 0$ we have Euclidean- and time-translation invariance, but the sign of D appears to introduce an *arrow of time* in a fundamental way. See, however, the discussion below for a more careful interpretation of the conditions under which time-reversal invariance is broken! Distinct values of D can occur, for different particle species. (e) Extra, dissipative terms occur in the Ehrenfest relations for the time-rate of change of the expectation values $\langle -i\hbar\nabla\rangle$ and $\langle i\hbar\partial_t\rangle$. (f) When V is a potential that ordinarily would accommodate bound states, there exist stationary state solutions to (8). These typically go over smoothly to the stationary solutions of the linear Schrödinger equation as we let $D, D'\to 0$. (g) Necessary and sufficient conditions for Galileian-invariance of the equations are: $c_1+c_4=c_3=0$. The invariance of just a subfamily means that the Galileian-invariance of the Fokker-Planck equation is broken at the level of the underlying Schrödinger equation; but the nonlinear terms, though resembling a nonlinear quantum potential, do not entirely spoil the invariance. (h) For a further subfamily of these equations, characterized by $D'c_1 = D = -D'c_4$ together with $c_2+2c_5=c_3=0$, the dissipative terms in $(d/dt)\langle -i\hbar\nabla\rangle$ vanish. Ehrenfest's theorem in quantum mechanics is satisfied [4], and the equations are linearizable *via* a nonlinear transformation (see below) [1] [5] [14].

Still more general nonlinear equations in quantum mechanics, motivated by (8), are discussed in [2].

2. Nonlinear gauge transformations

Those nonlinear Schrödinger equations belonging to the subfamily obeying Ehrenfest's theorem can be transformed into linear Schrödinger equations by means of the nonlinear transformation

$$\Psi \mapsto \Psi' = N(\Psi) = |\Psi| \exp\left[\, i(\gamma \ln|\Psi| + \Lambda \arg\Psi)\,\right], \tag{10}$$

where

$$\gamma = -\frac{2mD}{\hbar}\left(1 - \frac{4m}{\hbar}D'c_2 - \frac{4m^2D^2}{\hbar^2}\right)^{-\frac{1}{2}},$$

$$\Lambda = \left(1 - \frac{4m}{\hbar}D'c_2 - \frac{4m^2D^2}{\hbar^2}\right)^{-\frac{1}{2}}, \tag{11}$$

on condition that

$$\frac{4m}{\hbar}D'c_2 < 1 - \frac{4m^2D^2}{\hbar^2}. \tag{12}$$

That is, if Ψ solves (8)-(9), with $c_2 + 2c_5 = c_3 = 0$, $D'c_1 = D = -D'c_4$, and $4\Gamma'c_2 < 1 - 4\Gamma^2$ (where we are letting $\Gamma' = D'm/\hbar$), then Ψ' solves the linear Schrödinger equation,

$$i\frac{\hbar}{\Lambda}\frac{\partial}{\partial t}\Psi' = \left[-\frac{\hbar^2}{2\Lambda^2 m}\nabla^2 + V(\mathbf{x})\right]\Psi'. \tag{13}$$

But the significance of the nonlinear transformation (10) goes beyond providing a handy solution technique. I shall argue (see below) for the interpretation of N as a *nonlinear gauge transformation* in that distinct nonlinear Schrödinger equations related by N should be regarded as describing exactly the same physics! This means that the parameters D and $D'c_j$ in (8)-(9) are not themselves susceptible to direct physical interpretation, as they are not invariant under N.

Since the argument of Ψ is only defined *modulo* integer multiples of 2π, the transformation N is not actually well-defined by (10) as a mapping having a domain whose elements are wave functions. This observation leads to some further discussion of physical equivalence among solutions to (8)-(9), which I shall not include here; for now, let us simply assume that for a given Ψ, an appropriate selection of $N(\Psi)$ can always be made.

Now N is labeled by the two real parameters γ, Λ, with $\Lambda \neq 0$. Two such transformations, having in general different values of γ and Λ, can be performed successively to yield a third; i.e.

$$N_{(\gamma_1,\Lambda_1)} \circ N_{(\gamma_2,\Lambda_2)} = N_{(\gamma_1+\Lambda_1\gamma_2,\Lambda_1\Lambda_2)}\,, \tag{14}$$

so that we have the group law of the one-dimensional affine group.

Any transformation N in this group is *local* in the sense that N depends only on the value of Ψ at a point, not on its derivatives or its values at other points. N also preserve *rays* in the Hilbert space; i.e., for any complex number α, $(\alpha\Psi)'$ belongs to the same ray as does Ψ'. And N leaves the probability density ρ invariant, while the current $\mathbf{j}$ is transformed:

$$\rho' = \rho\,,$$

$$\mathbf{j}' = \Lambda\mathbf{j} + \frac{\gamma}{2}\nabla\rho\,. \tag{15}$$

To determine just how N transforms a general nonlinear Schrödinger equation in the category under consideration, it is helpful to rewrite the entire right-hand side of (8) as a nonlinear function of the density and the current multiplying Ψ. Since

$$\nabla^2\Psi = \left(iR_1[\Psi] + \frac{1}{2}R_2[\Psi] - R_3[\Psi] - \frac{1}{4}R_5[\Psi]\right)\Psi\,, \tag{16}$$

we obtain the general form for the Schrödinger equation,

$$i\frac{\partial}{\partial t}\Psi = i\sum_{j=1}^{2}\nu_j R_j[\Psi]\Psi + \sum_{j=1}^{5}\mu_j R_j[\Psi]\Psi + \mu_0 V\Psi. \tag{17}$$

A straightforward calculation now shows that $\Psi' = N_{(\gamma,\Lambda)}(\Psi)$ continues to obey (17), but with a new set of coefficients. Denoting the transformed coefficients with primes, one obtains the following transformation rules [5]:

$$\nu_1' = \frac{\nu_1}{\Lambda}\,, \quad \nu_2' = -\frac{\gamma}{2\Lambda}\nu_1 + \nu_2\,,$$

$$\mu_1' = -\frac{\gamma}{\Lambda}\nu_1 + \mu_1\,, \quad \mu_2' = \frac{\gamma^2}{2\Lambda}\nu_1 - \gamma\nu_2 - \frac{\gamma}{2}\mu_1 + \Lambda\mu_2\,,$$

$$\mu_3' = \frac{\mu_3}{\Lambda}\,, \quad \mu_4' = -\frac{\gamma}{\Lambda}\mu_3 + \mu_4\,,$$

$$\mu_5' = \frac{\gamma^2}{4\Lambda}\mu_3 - \frac{\gamma}{2}\mu_4 + \Lambda\mu_5\,, \quad \mu_0' = \Lambda\mu_0\,. \tag{18}$$

Thus we have the fact that the eight-parameter family of equations given by (17) is invariant, as a family, under a two-parameter group of nonlinear transformations.

Now it is an elementary fact in quantum mechanics that certain linear transformations of the Hilbert space leave the physics invariant. Multiplication of all wave functions by the fixed phase $\exp[i\theta]$ does not change the values of any physical observables; the set of such operations, called "gauge transformations of the first kind", comprise the group $U(1)$. More generally, one can transform the wave functions by acting with the multiplication operator $\exp[i\theta(\mathbf{x})]$, where $\theta(\mathbf{x})$ is a real-valued, measurable function. The new wave function then satisfies a transformed Schrödinger equation, in which $-i\hbar\nabla$ is replaced by $-i\hbar\nabla - \hbar\,\mathrm{grad}\,\theta(\mathbf{x})$. But the substitution $-i\hbar\nabla \to -ih\nabla - (e/c)\,\mathbf{A}(\mathbf{x})$ describes a charged particle moving under the influence of a magnetic vector potential $\mathbf{A}(\mathbf{x})$. In the case at hand, the magnetic field $\mathbf{B} = \nabla \times \mathbf{A} \equiv 0$, so that the transformation has no observable effect. Such transformations are called "gauge transformations of the second kind".

It has been remarked by various authors that all quantum-mechanical measurements ultimately reduce to positional measurements at various times (see e.g. [6]). Adopting this view, two different time-evolution equations will not predict different physical effects as long as there always exist corresponding *time-dependent* solutions to each equation that describe identical outcomes for positional measurements; i.e., for which the time-dependent probability densities are the same.

This comment does *not* merely reflect a mathematical change of variable in Hilbert space, but embodies a physical assumption about the special nature of position measurements. I would like to say just a little bit more about this. To observe that all measurements are fundamentally positional means that when we say we "measure" momentum, energy, angular momentum, etc. in quantum mechanics, we are really making *inferences* (i.e., performing calculations) from the outcomes of position measurements at various different times. But the *process of inferring* momentum, energy, angular momentum, etc. from position outcomes (i.e., the actual choice of calculation to perform) depends on what is *assumed* about the time-evolution! For instance, the idea that the linear operator $-i\hbar\nabla$ describes momentum–that the distribution of the outcomes of a large number of momentum measurements (obtained from position measurements at different times on identically-prepared quantum states) can be predicted from it according to the usual rules of quantum mechanics–assumes that the time-evolution is given by the usual Schrödinger equation.

Thus members of the family of nonlinear Schrödinger equations discussed here, when they are related by one of the transformations $N_{(\gamma,\Lambda)}$, describe the same physics in the sense that there is indeed no physical procedure for distinguishing them. Therefore let us regard the group of nonlinear transformations $N_{(\gamma,\Lambda)}$ as a nonlinear generalization of the linearly acting $U(1)$-gauge group. It seems reasonable to call such a nonlinear transformation a "gauge transformation of the third kind".

3. Further discussion of the physical interpretation

At this point we are able to look again at the physical interpretation of the nonlinear Schrödinger equation. Since we have an 8-parameter family of equations, and a 2-parameter gauge group acting on the parameters, we expect to have (in general) six independent, gauge-invariant quantities labeling the classes of equations in the family. It will be these gauge-invariants, rather than the original parameters, that characterize the physics described by the nonlinear Schrödinger equations.

Functionally independent gauge-invariant quantities can be obtained by straightforward but fairly complicated calculations. That the following constitute one such set is verifiable by direct substitution of (18):

$$\tau_1 = \nu_2 - \frac{1}{2}\mu_1\,, \quad \tau_2 = \nu_1\mu_2 - \nu_2\mu_1\,, \quad \tau_3 = \frac{\mu_3}{\nu_1}\,,$$
$$\tau_4 = \mu_4 - \mu_1\frac{\mu_3}{\nu_1}\,, \quad \tau_5 = \nu_1\mu_5 - \nu_2\mu_4 + \nu_2^2\frac{\mu_3}{\nu_1}\,, \tag{19}$$
$$\tau_0 = \nu_1\mu_0\,.$$

To begin the process of interpreting the quantities τ_j, consider the effect of time reversal on the nonlinear Schrödinger equation. Sending $t \to -t$ is equivalent to setting new coefficients $\nu_j^{\mathbf{T}} = -\nu_j\,(j = 1,2)$ and $\mu_j^{\mathbf{T}} = -\mu_j\,(j = 1,\ldots,5)$, where the superscript $\mathbf{T}$ denotes time reversal. Now $N_{(\gamma,\Lambda)}(\Psi) = \overline{\Psi}$ when $\gamma = 0$ and $\Lambda = -1$. From (18) we then see that if Ψ solves Eq. (17), $\overline{\Psi}$ solves the time-reversed equation as long as $\nu_2 = \mu_1 = \mu_4 = 0$. But in terms of the τ's, we calculate that $\tau_j^{\mathbf{T}} = -\tau_j$ for $j = 1$ or $j = 4$, while $\tau_j^{\mathbf{T}} = +\tau_j$ for $j = 2, 3$ or 5. Thus nonzero values for *either* of the two parameters τ_1 or τ_4 introduce an "arrow of time" into quantum mechanics.

For the usual, linear Schrödinger equation, we have $\nu_1 = -\hbar/2m, \nu_2 = 0$, $\mu_1 = 0$, $\mu_2 = -\hbar/4m$, $\mu_3 = \hbar/2m$, $\mu_4 = 0$, $\mu_5 = \hbar/8m$, and $\mu_0 = 1/\hbar$. The corresponding gauge-invariants are: $\tau_1 = 0$, $\tau_2 = \hbar^2/8m^2$, $\tau_3 = -1$, $\tau_4 = 0$, $\tau_5 = -\hbar^2/8m^2$, and $\tau_0 = -1/2m$. If $V(\mathbf{x}) \equiv 0$, however, note that μ_0 and τ_0 are not determined. We see that the quotient $\hbar/m$ can be obtained from the gauge invariants, while m and $\hbar$ are obtained individually only for nonzero potentials.

We also observe that neither the original diffusion coefficient D nor its dimensionless counterpart $\Gamma = Dm/\hbar$, whose introduction led to all of the above, are actually physical quantities in the sense of being gauge-invariant! When we write (8) in the form (17), we have

$$\nu_1 = -\frac{\hbar}{2m}, \quad \nu_2 = \frac{1}{2}D, \quad \mu_1 = D'c_1, \quad \mu_2 = -\frac{\hbar}{4m} + D'c_2,$$

$$\mu_3 = \frac{\hbar}{2m} + D'c_3, \quad \mu_4 = D'c_4, \quad \mu_5 = \frac{\hbar}{8m} + D'c_5, \tag{20}$$

$$\mu_0 = \frac{1}{\hbar}.$$

As ν_2 is proportional to D, we see that $D \neq 0$ contributes to a possibly nonzero value for τ_1. But $\mu_1 = D'c_1$ also contributes to τ_1. Thus the condition that $D \neq 0$ is, by itself, neither necessary nor sufficient to break the time-reversal invariance. This is a direct consequence of the fact that the value of D–even its sign–changes under nonlinear gauge transformations. In fact, the particular equation studied by Ushveridze [15] [16] corresponds to the values $D = 0$ and $D'c_1 \neq 0$ (with the remaining $c_j = 0$) in (8); thus this equation is also to be understood as breaking time-reversal invariance.

Since the Galileian invariance of (8) is equivalent to imposing the conditions $c_1 + c_4 = c_3 = 0$ [4], the coefficients in this case take the values $\nu_1 = -\hbar/2m$, $\nu_2 = D/2$, $\mu_1 = c_1 D'$, $\mu_2 = -\hbar/4m + c_2 D'$, $\mu_3 = \hbar/2m$, $\mu_4 = -c_1 D'$, and $\mu_5 = \hbar/8m + c_5 D'$, where $\hbar/m$, D, $c_1 D'$, $c_2 D'$, and $c_5 D'$ are freely chosen. Now we obtain values for the gauge invariants that fix $\tau_3 = -1$ and $\tau_4 = 0$, with the other values permitted to vary. The condition of Galileian invariance still permits some irreversibility, as one can have $\tau_1 \neq 0$ in a Galileian-invariant theory.

The additional conditions characterizing the equations obeying Ehrenfest's theorem are: $D'c_1 = D$, together with $c_2 + 2c_5 = 0$. The first of these conditions sets $\tau_1 = 0$, so that we have only time-reversal invariant quantum mechanics. The second condition, in combination with the first, sets $\tau_5 = -(1/2)\tau_2$. Thus we can write $\tau_2 = \hbar^2/8m^2 - \kappa$, where $\kappa = (\hbar/2m)D'c_2 + (1/2)D^2$. The case $\kappa = 0$ is the linear Schrödinger equation. As long as $\kappa < \hbar^2/8m^2$ so that $\tau_2 > 0$, which is condition (12), the equation is linearizable by means of a gauge transformation, changing only the effective value of $\hbar/m$.

In conclusion, the results reported here are all consistent with the interpretation that the general, nonlinear Schrödinger equation that Doebner and I proposed can describe dissipative processes in quantum mechanics in a fundamental way. A group of nonlinear transformations is to be interpreted as a gauge group for the theory, with the gauge-invariant parameters characterizing the physical nature of the dissipation. Nonzero values for two of our parameters break the time-reversal invariance, and for specified conditions on the parameters we have Galileian invariance. For still more restrictive conditions on the parameters the equation is linearizable. In that case alone, the physics is in principle indistinguishable from ordinary quantum mechanics.

References

1. Auberson, G. and Sabatier, P. C. (1994) On a Class of Homogemeous Nonlinear Schrödinger Equations, *J. Math. Phys.***35**, 4028.
2. Dodonov, V. V. and Mizrahi, S. S. (1995) Uniform Nonlinear Evolution Equations for Pure and Mixed Quantum States, *Annals of Physics* **237**, 226-268.
3. Doebner, H.-D. and Goldin, G. A. (1992) On a General Nonlinear Schrödinger Equation Admitting Diffusion Currents, *Phys. Lett. A* **162**, 397-401.
4. Doebner, H.-D. and Goldin, G. A. (1994) Properties of Nonlinear Schrödinger Equations Associated with Diffeomorphism Group Representations, *J. Phys. A: Math. Gen.* **27**, 1771-1780.
5. Doebner, H.-D., Goldin, G. A., and Nattermann, P. (1995) A Family of Nonlinear Schrödinger Equations: Linearizing Transformations, and Resulting Structure, to be published.
6. Feynman, R. P. and Hibbs, A. R. (1965) "Quantum Mechanics and Path Integrals", McGraw-Hill, New York, p. 96.
7. Goldin, G. A. (1992) The Diffeomorphism Group Approach to Nonlinear Quantum Systems, *Int. J. Mod. Phys. B* **6**, 1905-1916.
8. Goldin, G. A., Menikoff, R., and Sharp, D. H. (1981) Representations of a Local Current Algebra in Non-Simply Connected Space and the Aharonov-Bohm Effect, *Journal of Mathematical Physics* **22**, 1664-1668.
9. Goldin, G. A., Menikoff, R., and Sharp, D. H. (1983) Diffeomorphism Groups, Gauge Groups, and Quantum Theory, *Physical Review Letters* **51**, 2246-2249.
10. Goldin, G. A. and Sharp, D. H. (1989) Diffeomorphism Groups and Local Symmetries: Some Applications in Quantum Physics, in B. Gruber and F. Iachello, eds., "Symmetries in Science III", Plenum, New York, 181-205.
11. Goldin, G. A. and Sharp, D. H. (1991) The Diffeomorphism Group Approach to Anyons, *Int. J. Mod. Phys. B* **5**, 2625-2640.
12. Goldin, G. A. and Svetlichny, G. (1994) Nonlinear Schrödinger Equations and the Separation Property, *J. Math. Phys.* **35**, 3322-3332.
13. Leinaas, J. M. and Myrheim, J. (1977) On the Theory of Identical Particles, *Nuovo Cimento* **37B**, 1.
14. Nattermann, P. (1994) Solutions of the General Doebner-Goldin Equation via Nonlinear Transformations, in "Proceedings of the XXVI Symposium on Mathematical Physics, Torun, December 7-10, 1993", Nicolas Copernicus University Press, Torun, 47.
15. Ushveridze, A. G. (1994) Dissipative Quantum Mechanics: A Special Doebner-Goldin Equation, its Properties and Exact Solutions, *Phys. Lett. A* **185**, 123-127.
16. Ushveridze, A. G. (1994) The Special Doebner-Goldin Equation as a Fundamental Equation of Dissipative Quantum Mechanics, *Phys. Lett. A* **185**, 128-132.

CONDITIONAL SYMMETRIES FOR 1ST ORDER SYSTEMS OF PDES IN THE CONTEXT OF THE CLAIRIN METHOD

ALFRED M. GRUNDLAND
Centre de recherches mathématiques, Université de Montréal, C.P. 6128, succursale Centre-ville, Montréal, Québec H3C 3J7, Canada.

AND

GUY RIDEAU
Laboratoire de Physique Théorique et Mathématique, Université Paris, Denis Diderot, 2 place Jussieu, F–75251, Paris, Cedex 05, France.

Abstract. The existence of p-dimensional Abelian Lie algebra of conditional symmetries admitted by 1st order systems of PDEs in p independent variables is discussed in detail. The link between Bäcklund transformations and conditional symmetries is investigated. The obtained results are applied to several examples of integrable and nonintegrable systems.

1. Introduction

In this paper we study conditional symmetries for 1st order PDEs with many independent and dependent variables. The notion of conditional symmetries evolved in the process of extending Lie's classical theory of symmetries of PDEs, which started with the work of G. Bluman and J. Cole [1]. This "nonclassical" approach, later developed by D. Levi and P. Winternitz [2] as well as others (e.g. P. Olver and P. Rosenau [3]), consists basically in modifying the original system by adding to it a certain differential constraint of 1st order for which a symmetry criterion is identically satisfied. The overdetermined system of equations obtained in this way admits, in some cases, a larger class of Lie point symmetry groups and, consequently, can provide new classes of solutions of the original system. One can go a

J. Bertrand et al. (eds.), Modern Group Theoretical Methods in Physics, 167–178.

step further and add multiple differential constraints compatible with our basic system, looking for still larger classes of solutions. The term "conditional symmetries" relates to this particular case [4].

Our study of conditional symmetries is based on the paper [5] where the existence of conditional symmetries of PDEs of an arbitrary order was discussed. Here we concentrate on a system of 1st order PDEs appended with p-differential constraints of 1st order, where p is the number of independent variables in the original system. Our choice of the number of differential constraints is motivated by our main aim which is the construction of the Bäcklund transformation (B.T.) for the original system. It was M. Clairin in 1903 [6] who first observed that the B.T. for a given scalar equation in two independent variables can usually be defined by two differential constraints. Our approach can be seen as a modification of the Clairin method for the case of many independent variables. We use it in conjunction with the symmetry reduction method (SRM) and are able to propose a new, systematic and relatively simple way of constructing B.T. given by 1st order differential equations which in the following we will call 1st order B.T.

In Section 2 we determine the conditions necessary for the existence of the conditional symmetries for 1st order PDEs. They are expressed by nonlinear system of PDEs and functional equations, for which we postulate the solutions in the form of the formal power series in dependent variables, with nonconstant coefficients. In further analysis we show that this postulate allows us to distinguish the cases for which we are able to construct 1st order B.T.

In Section 3 we apply our theoretical results to the examples of integrable systems which possess B.T. (AKNS equations, generalized Liouville equation and Burgers equations). We consider as well some nonintegrable systems for which no B.T. exists. In this case we can construct special classes of solutions.

2. The existence of conditional symmetries for 1st order systems of PDEs and their associated 1st order Bäcklund transformations

Consider a system of m PDEs in p independent variables $x = (x^1, \ldots, x^p)$ and q dependent variables $u = (u^1, \ldots, u^q)$ of the form

$$\Delta^s(x^i, u^\alpha, u_i^\alpha) = 0, \quad s = 1, \ldots, m, \tag{1}$$

where u_i^α denotes the 1st order partial derivatives of u^α, i.e. $u_i^\alpha = (\partial u^\alpha)/(\partial x^i)$.

We denote by $U = \mathbb{R}^q$ and $X = \mathbb{R}^p$ the differential manifolds representing the spaces of dependent and independent variables, respectively. $J^1 = J^1(X \times U)$ denotes the 1-jet space over X and $(x^i, u^\alpha, u_i^\alpha)$ are the local coordinates on J^1.

Consider a set of p vector fields Y_k, defined on $J^0 = X \times U$

$$Y_k = \xi_k^i(x,u)\partial_i + \varphi_k^\alpha(x,u)\partial_\alpha, \quad k = 1,\dots,p \tag{2}$$

where $\partial_i = \partial/\partial x^i$ and $\partial_\alpha = \partial/\partial u^\alpha$. We adopt here the summation convention over repeated indices. The 1st prolongation of Y_k are the vector fields

$$\mathrm{pr}^{(1)} Y_k = Y_k + (D_j Q_k^\alpha + \xi_k^i u_{ij}^\alpha)\partial_{u_j^\alpha}, \text{ where } u_{ij}^\alpha = \frac{\partial^2 u^\alpha}{\partial x^i \partial x^j} \tag{3}$$

defined on J^2. D_j denote the total derivative operators in X restricted to J^1

$$D_j = \partial_j + u_j^\alpha \partial_\alpha, \quad j = 1,\dots,p \tag{4}$$

and Q_k^α denote the characteristics of the vector fields Y_k which are defined by

$$Q_k^\alpha(x^i, u^\alpha, u_i^\alpha) = \varphi_k^\alpha - \xi_k^i u_i^\alpha. \tag{5}$$

Let us augment the original system (1) by requiring that the characteristics (5) be equal to zero. Thus we obtain an overdetermined system

$$\Delta^s(x^i, u^\alpha, u_i^\alpha) = 0, \tag{6a}$$

$$Q_k^\alpha(x^i, u^\alpha, u_i^\alpha) = 0. \tag{6b}$$

Let us identify the systems of equations $\Delta^s : J^1 \to \mathbb{R}^m$ and $Q_k^\alpha : J^1 \to \mathbb{R}^{q\times p}$ with the subvarieties of the solutions space $\mathcal{S}_\Delta = \{(x^i, u^\alpha, u_i^\alpha):$ $\Delta^s(x^i, u^\alpha, u_i^\alpha) = 0, s = 1,\dots,m\}$ and $\mathcal{S}_Q = \{(x^i, u^\alpha, u_i^\alpha): Q_k^\beta(x^i, u^\alpha, u_i^\alpha) = 0;$ $k = 1,\dots,p; \alpha = 1,\dots,q\}$, respectively.

Definition 2.1 ([4]) Vector fields Y_k are called conditional symmetries iff they are tangent to the intersection of the subvarieties $\mathcal{S}_\Delta$ and $\mathcal{S}_q$, i.e. $\mathcal{S} = \mathcal{S}_\Delta \cap \mathcal{S}_q$.

The infinitesimal criterion for the existence of the symmetry group of the overdetermined system (6a)–(6b) is given by

$$\mathrm{pr}^{(1)} Y_k(\Delta^s) = 0 \tag{7a}$$

and

$$\mathrm{pr}^{(1)} Y_k(Q_i^\alpha) = 0 \tag{7b}$$

for each $k = 1, \ldots, p$, whenever equations (6a)–(6b) are satisfied.

Note that in order for the set of constraints (6b) to be consistent they have to be chosen in such a way that (7b) is identically satisfied [2].

Let us assume that the initial system of PDEs (1) admits a p dimensional Lie module $\mathcal{G}$ of conditional symmetries in the sense of the Definition 2.1 with the generators given by (2) such that

$$[Y_k, Y_\ell] = \tau^r_{k\ell} Y_r, \quad r = 1, \ldots, p \tag{8}$$

where $\tau^r_{k\ell}$ depend on x and u. Note that the equations (2), (6a)–(6b) and (7a)–(7b) are invariant under point transformations.

Let us suppose, that $\det \xi^i_k \neq 0$. Then the vector fields Y_k can be written in the form

$$Z_r = (\xi^{-1})^i_r Y_i = \partial_r + \phi^\alpha_r \partial_\alpha; \quad r = 1, \ldots, p; \tag{9}$$

where $\phi^\alpha_r = (\xi^{-1})^i_r \varphi^\alpha_i$. The commutators of Z_k, Z_j are

$$[Z_k, Z_j] = (\phi^\alpha_{j,k} + \phi^\alpha_{j,u^\beta}\phi^\beta_k - \phi^\alpha_{k,j} - \phi^\alpha_{k,u^\beta}\phi^\beta_j)\partial_\alpha.$$

So, Z_k form a Lie algebra only when

$$\phi^\alpha_{j,k} + \phi^\alpha_{j,u^\beta}\phi^\beta_k - \phi^\alpha_{k,j} - \phi^\alpha_{k,u^\beta}\phi^\beta_j = 0. \tag{10}$$

This means that $Z_1, \ldots, Z_p$ span a p-dimensional Abelian algebra, i.e. $[Z_k, Z_j] = 0$ for $1 \leq k, j \leq p$. Taking the form of the vector fields (9) into account we can express the overdetermined system (6a)–(6b) as follows

$$\Delta^s(x^i, u^\alpha, u^\alpha_i) = 0 \tag{11a}$$

and

$$u^\alpha_i - \phi^\alpha_i(x, u) = 0. \tag{11b}$$

Substituting (11b) into (11a) we get

$$\Delta^s(x^i, u^\alpha, \phi^\alpha_i) = 0. \tag{12}$$

At this point a clear analogy between our approach and the Clairin method can be seen, since the form of the constraints (11b) at which we arrive here is identical as that introduced by M. Clairin [6]. Its main advantage lies in the fact that the constraints (11b) constitute a system of 1st order ODEs. If the first condition (10) on the function ϕ^α_i is satisfied then (11b) can be solved by successive integration. The second condition (12) on the function ϕ^α_i guarantees the compatibility of the constraints (11b) with the original system (1). This means that having the general solution of (11b) we can then proceed to the construction of the solution of the initial system (1).

Definition 2.2 A Lie algebra L spanned by vector fields $X_1, \ldots, X_p$ is called a conditional symmetry algebra of the system $\Delta^s = 0$ if the following conditions:

(i) $[X_k, X_j] = C^\ell_{kj} X_\ell$, where C^ℓ_{kj} are structure constants

and

(ii) $\mathrm{pr}^{(1)} X_k(\Delta^s) = 0$ modulo $\Delta^s = 0$ and $Q^\alpha_i = 0$

are satisfied.

Then under the above assumptions we have the following theorem:

Theorem 2.3 *Consider p smooth vector fields*

$$Z_k = \partial_k + \phi^\alpha_k \partial_\alpha; \quad k = 1, \ldots, p.$$

If the functions ϕ^α_k *are chosen so as to satisfy*

(i) $\phi^\alpha_{k,j} - \phi^\alpha_{j,k} + \phi^\alpha_{k,u^\beta}\phi^\beta_j - \phi^\alpha_{j,u^\beta}\phi^\beta_k = 0,$

(ii) $\Delta^s(x^i, u^\alpha, \phi^\alpha_i) = 0, \quad s = 1, \ldots, m,$

on some neighbourhood of $(x_0, u_0) \in \mathcal{S}$, *then the first order system*

$$\Delta^s(x^i, u^\alpha, u^\alpha_i) = 0$$

admits a p-dimensional Abelian Lie algebra L of conditional symmetries(in the sense of Definition (2.2) *with generators* Z_k, $1 \leq k \leq p$.

Proof. Generators Z_k constitute conditional symmetries for the system (1) if they are Lie point symmetries of the overdetermined system (11a)–(11b). Indeed, the infinitesimal symmetry criterion (7a)–(7b) for (11a) is identically satisfied by virtue of the assumption (i), i.e.

$$\mathrm{pr}^{(1)} Z_k(Q^\alpha_j) = \phi^\alpha_{j,k} + \phi^\beta_k \phi^\alpha_{j,u^\beta} - \phi^\alpha_{k,j} - \phi^\beta_j \phi^\alpha_{k,u^\beta} \equiv 0.$$

Applying the invariance criterion (7a)–(7b) to (11b) we obtain

$$\mathrm{pr}^{(1)} Z_k(\Delta^s) = \partial_k \Delta^s + \phi^\alpha_k \partial_\alpha \Delta^s + (\phi^\alpha_{k,j} + \phi^\beta_j \phi^\alpha_{k,u^\beta})\partial_{u^\alpha_j}\Delta^s. \tag{13}$$

On the other hand, carrying out the differentiations in (ii) gives us

$$D_j \Delta^s = \partial_j \Delta^s + \phi^\alpha_{k,j}\partial_{u^\alpha_k}\Delta^s + \phi^\alpha_j(\partial_\alpha \Delta^s + \phi^\beta_{k,u^\alpha}\partial_{u^\beta_k}\Delta^s) = 0, \tag{14}$$

where $D_j = \partial_j + \phi^\alpha_j \partial_\alpha$ are the total derivatives (4) whenever (11b) holds. After substituting (14) into (13) and taking into account the assumption (i) we obtain that the equations (13) are identically equal to zero. □

Remark 2.4 (1) If the conditions (i) and (ii) are satisfied, then the graph of the solution of (11a)–(11b) is invariant under the vector fields Z_k, $1 \leq k \leq p$.

(2) If the set of solutions of the so-called "determining equations" obtained by applying the symmetry criterion to the overdetermined system (11a)–(11b) is larger than the set obtained by applying this criterion to the initial system (1) then the system (11a)–(11b) admits a larger symmetry group than the original system (1). So, new reductions for the system (1) can be obtained.

In our approach the construction of solutions to the original system (1) requires us to solve the systems (10) and (12) for ϕ_k^α as functions of x and u. These systems are, in general, nonlinear. So we cannot expect to solve them in a closed form, except in some particular cases. But nevertheless, as we show it in the next section, there exist many physically interesting systems of PDEs for which particular solutions of (10) and (12) lead to 1st order Bäcklund transformations or to solutions depending on some arbitrary functions. These particular solutions of (10) and (12) are obtained by expanding each function ϕ_i^α into a polynomial in the dependent variables u^α (up to a point transformation). This polynomial is reduced often to a trinomial, what means that the equations (11b) become a matrix Riccati system which can be expected to possess the Painlevé property.

We now show that the conditions (10) and (12) for the existence of conditional symmetries can be used to construct Bäcklund transformations. This construction consists of the following steps.

Substituting the assumed polynomial forms of the functions ϕ_k^α into the system (10) and (12) and next requiring that the coefficients of the successive powers of u^α in the so obtained equations vanish, we obtain a system of 1st order PDEs for the coefficients of the polynomials a_{ki}^α. Denote this system by

$$E^r(x, a_{ki}^{\alpha(1)}) = 0, \quad r = 1, \ldots, \ell \leq m. \tag{15}$$

Solving(15) leads to two separate cases.

In the first case the set of solutions of (15) is parametrized by some functions, say v^α, which satisfy certain PDEs of 1st order

$$F^t(x, v^{(1)}) = 0, \quad t = 1, \ldots, n. \tag{16}$$

Then there exists a 1st order B.T. between the sets of solutions of the initial system (1) and the system (16). For any particular solution of (16) the system (11b) defines a specific transformation between the system (16) and (1). In other words, from each solution of (16) we can construct a solution of (1) and vice versa. Note that in this case the original system (1) admits infinitely many compatible constraints (11b) which are parametrized by v^α.

In the second case, the set of solutions of (15) is uniquely defined (with the freedom of some constants only). Then a B.T. does not exist and only particular solution to (1) can be constructed by applying the symmetry reduction technique.

3. Examples of applications

Now we present several examples which illustrate the above theoretical considerations concerning the construction of the conditionally invariant solutions. Our study starts with a class of completely integrable equations [8], [9] and we construct the Bäcklund transformations for them. Next we consider some equations which are not completely integrable but nevertheless some classes of nontrivial conditionally invariant solutions are obtained.

Due to the limited space we can present here only the main results, without going into the details.

3.1. THE NONLINEAR WAVE EQUATION

Consider an overdetermined system of PDEs in two independent variables obtained by adding the 1st order differential constraints

$$u_x = \varphi(x,t,u), \quad u_y = \psi(x,t,u) \tag{17}$$

to the nonlinear wave equation

$$u_{xt} = b(u). \tag{18}$$

The equations (17) correspond to the requirement that the characteristics of the vector fields

$$Z_1 = \partial_x + \varphi(x,t,u)\partial_u, \quad Z_2 = \partial_t + \psi(x,t,u)\partial_u \tag{19}$$

be equal to zero. Assume that the C^2 functions φ and ψ are quadratic in u, i.e.

$$\begin{aligned} \varphi &= a_2(x,t)u^2 + a_1(x,t)u + a_0(x,t), \\ \psi &= b_2(x,t)u^2 + b_1(x,t)u + b_0(x,t). \end{aligned} \tag{20}$$

We were able to reconstruct the well-known result [7] that the only equations of the type (18), up to an arbitrary point transformation, which admit a 1st order B.T. of Riccati type are such that

$$b(u) = \{u, e^u, \sin u, \sinh u\}. \tag{21}$$

The system of the equations (18) with (17), up to an invertible transformation of the dependent variable $u \mapsto v = \eta(u)$, can be written as an overdetermined system for v. The compatibility conditions for this system lead us to the differential equations of the form (10) and (12) for φ and ψ as functions of x, t and v. After the transformation $u \mapsto v = \eta(u)$ we substitute (20) into equations for φ and ψ and next we require that the coefficients of the successive powers of v in so obtained equations vanish. Solving them we obtain the well-known condition [7] $d/du(b^{-1}(d^2b)/(du^2)) = 0$, from which we get (21).

Using our approach we looked for the B.T. of Riccati type for the equation (18) with the right hand side of the form (21) and we reconstructed easily the known results (see e.g. [8, 9]) in all four cases.

3.2. THE CUBIC SCHRÖDINGER EQUATION

$$iu_t + u_{xx} = \varepsilon u|u|^2, \quad \varepsilon = \pm 1. \tag{22}$$

Assume that (22) admits 2-dimensional Abelian Lie algebra spanned by the vector fields of the form (9), i.e.

$$\begin{aligned} Z_1 &= \partial_x + \varphi(x,t,u,\overline{u})\partial_u + \overline{\varphi}(x,t,u,\overline{u})\partial_{\overline{u}}, \\ Z_2 &= \partial_t + \psi(x,t,u,\overline{u})\partial_u + \overline{\psi}(x,t,u,\overline{u})\partial_{\overline{u}}. \end{aligned} \tag{23}$$

The characteristic equations of Z_k are

$$u_x = \varphi, \quad u_t = \psi, \tag{24a}$$

$$\overline{u}_x = \overline{\varphi}, \quad \overline{u}_t = \overline{\psi} \tag{24b}$$

The conditions (24a)–(24b) imply that $\overline{\varphi}$ and $\overline{\psi}$ are the complex conjugate of φ and ψ, respectively. We look for the solutions of the overdetermined system formed by (22) and (24a) invariant under the vector fields (23). Under the assumption that the functions φ and ψ are in the form of finite power series we are able to reconstruct a 1st order auto B.T. [10]

$$\begin{aligned} u_x &= -\frac{1}{2}vu^2 - \lambda u + \frac{\varepsilon}{2}\overline{v}, \text{ where } iv_t + v_{xx} = \varepsilon v|v|^2, \\ u_t &= -\frac{i}{2}(\lambda v + v_x)u^2 + \frac{i}{2}(\varepsilon|v|^2 - 2\lambda^2)u + i\frac{\varepsilon}{2}(\lambda\overline{v} - \overline{v}_x), \end{aligned} \tag{25}$$

where λ is an arbitrary complex parameter. Note that both equations (25) are Riccati equations for the function u.

3.3. THE BURGERS EQUATIONS

The hierarchy of matrix Burgers type equations and the associated B.T. were found in [11]. As in illustration we consider here the first and second equation from this hierarchy

$$U_t = U_{xx} - 2UU_x, \tag{26}$$

and

$$U_t = U_{xxx} + 3(U^2U_x - (UU_x)_x), \tag{27}$$

respectively, where U is a square matrix. We append each of these equations with a 1st order constraint of the form

$$U_x = AU^2 + UBU + U^2C + DU + UE + F \tag{28}$$

where $A, B, \ldots, F$ are matrix functions of x, t. Substituting (28) into (26) and (27) we obtain the forms of the second constraints for each of these basic equations. Following the procedure outlined in Section 2 we obtain auto-B.T. for (26)

$$\begin{aligned} U_x &= U^2 - U(\lambda I + V) + \lambda V, \text{ where } V_t = V_{xx} - 2VV_x, \\ U_t &= (-U^2 + \lambda(V + V_x))(\lambda I + V) + U((\lambda I + V)^2 - V_x) \end{aligned} \tag{29}$$

and for (27)

$$\begin{aligned} U_x &= U^2 - U(\lambda I + V) + \lambda V \\ U_t &= -6U^4 - 6U^3Q + 6U^2QU - 4UQU^2 + 3(UQ)^2 \\ &\quad + 4\lambda VU^2 - 6\lambda UVU + 5\lambda U^2V - U^2V_x \\ &\quad + (2UQ - \lambda U + QU)V_x - UQ^3 - 4\lambda VUQ + 3\lambda UVQ \\ &\quad - 2\lambda UQV - \lambda(2V + QV_x) + \lambda V(Q^2 + \lambda V) \end{aligned} \tag{30}$$

where $V_t = V_{xxx} + 3(V^2V_x - (VV_x)_x)$, $Q = (\lambda I + V)$, λ is the Bäcklund parameter and we denote the identity matrix by I.

Note that both equations (29) are of the Riccati form for the matrix U. But it is not the case for (30), where only one of them is a Riccati type equation; the second one contains the 4th order polynomial. These results are consistent with [11].

The similar analysis was done by the authors for the KdV and the MKdV equations and associated B.T. were constructed.

3.4. THE GENERALIZED LIOUVILLE EQUATION

The generalization of the Liouville equation for $(n+1)$-dimension was formulated by P. Santini [12] and has the form

$$\frac{\partial}{\partial t}\left(\sum_{j=1}^{n}[u_{x_jx_j}-\frac{1}{2}(u_{x_j})^2]\right)=0. \tag{31}$$

We look for the solutions of (31) invariant under $(n+1)$ vector fields

$$Z_0=\partial_t+\varphi^{(0)}(x,t,u)\partial_u, \quad Z_j=\partial_{x_j}+\varphi^{(j)}(x,t,u)\partial_u, \quad j=1,\ldots,n. \tag{32}$$

According to the postulates of our method we require that the functions $\varphi^{(0)}$ and $\varphi^{(j)}$ can be expressed as finite Fourier series in u. In this case they are represented by trinomials in u. A 1st order auto-B.T. for this case was constructed for the first time by P. Santini and the author [13]

$$(u-v)_{x_j}=w_{x_j}e^{(u-v)/2}, \quad (u-v)_t=w_te^{(u-v)/2}, \tag{33}$$

where the function v satisfies the nonlinear equation (31) and the function w satisfies the linear equation

$$\sum_{j=1}^{n}(w_{x_jx_j}-v_{x_j}w_{x_j})=0. \tag{34}$$

Note that after the transformations $u=2\ln f$ and $v=2\ln g$ in the equation (33) we can rewrite (33) in the Riccati form. An explicit solution of (31) in $(2+1)$-dimension can be easily constructed from the constant solution

$$u=v_0+2\ln|f(x_1+ix_2,t)+g(x_1-ix_2,t)|, \quad v_0\in\mathbb{R} \tag{35}$$

where f and g are arbitrary functions of their arguments.

3.5. EXAMPLES OF NONINTEGRABLE EQUATIONS

Let us consider the double Sine-Gordon equation

$$u_{xt}=\frac{1}{2}\sin 2u+\lambda\sin u. \tag{36}$$

We look for the solutions of (36) invariant under the vector fields (19). According to the Theorem 1 the conditions (10) and (12), in this case take the form

$$\varphi_t+\psi\varphi_u=\frac{1}{2}\sin 2u+\lambda\sin u, \quad \psi_x+\varphi\psi_u=\frac{1}{2}\sin 2u+\lambda\sin u. \tag{37}$$

We look for solutions of (37) in the form of finite order polynomials in u. In this case, like previously, they are trinomials in u. From our further analysis it results that the differential constraints compatible with the system (36) can only have the freedom of some constants

$$(38)\qquad u_x = \alpha(1 + 2\lambda u + u^2), \quad u_t = \frac{\alpha}{\beta}(1 + 2\lambda u + u^2), \qquad \alpha, \beta \in \mathbb{R}.$$

The solution of (38) and, consequently, of (36) has the form of a traveling wave (see e.g. [14])

$$(39)\qquad u = -\lambda + c \tanh\left(cx - \frac{t}{c\beta} - x_0\right), \quad c^2 = \alpha.$$

Let us consider now the Klein-Gordon equation u^n, i.e.

$$(40)\qquad u_{xt} = u^n.$$

The considerations similar to those in the above example lead us to the system of constraints compatible with (40), i.e.

$$(41)\qquad u_x = (t - t_0)u^n, \quad u_t = (t - t_0)^{-1}u^{n-1}.$$

The solution of (41) has the form

$$(42)\qquad u = \left[(n-1)^2(t_0 - t)^n(x_0 - x)^n\right]^{1/n-1}$$

4. Final remarks

We have demonstrated the link between the Bäcklund transformations and the symmetry reduction method in its "nonclassical" version. This link enabled us to propose an algorithm for the construction of 1st order B.T. for PDEs with many independent and dependent variables. This new approach is tested in Section 3. It proved to be effective, as we were able to reproduce easily the known results which were obtained usually by much more complicated procedures. Its potential for providing new meaningful results was shown in the case of the generalized Liouville equation.

The conditional symmetry method presented here is limited by several assumptions. For example we study only first order differential constraints. The question arises whether our approach can be extended to higher order differential constraints and if this may provide new classes of solutions. Other requirement of the method, worth investigating is the condition $\det \xi^i_k = 0$ (see (9)). We can expect that relaxing it can broaden the applicability of our approach. These and other questions will be addressed in future work.

5. Acknowledgements

The authors have benefited from helpful discussions with D. Levi and P. Winternitz on matters related to this work. This research was supported in part by the Natural Sciences and Engineering Research Council of Canada and the Fonds FCAR du Québec.

References

1. G.W. Bluman and J.D. Cole (1969) The general similarity solution of the heat equation, *J. Math. Mech.* **18**, pp. 1025–1042.
2. D. Levi and P. Winternitz (1989) Nonclassical symmetry reduction: Example of the Boussinesq equation, *J. Phys. A: Math. Gen.*, **22**, pp. 2915–2924.
3. P.J. Olver and Ph. Rosenau (1987) Group invariant solutions of differential equations, *SIAM J. Appl. Math.* **47**, pp. 263–278.
4. P. J. Olver, E. M. Vorob'ev (1995) Nonclassical and conditional symmetries, *CRC Handbook of Lie Group Analysis*, **Vol. 3**, chapt. XI, CRC press, London (N. H. Ibragimovv, ed).
5. A M. Grundland, J. Tafel (1995), On the existence of nonclassical symmetries of partial differential equations, *J. Math. Phys.* (accepted).
6. M. J. Clairin (1903) Sur quelques équations aux dérivées partielles du second ordre, *Ann. Fac. Sci.*, Univ. Toulouse (2), Ser. 5, 437–459.
7. W. F. Shadwick (1978) The Bäcklund problem for the equation $z_{x^1 x^2} = f(z)$, *J. Math. phys* **19**, no. 11, 2312–2317.
8. C. Rogers, W. F. Shadwick (1982) Bäcklund transformations and their applications, *Academic press*, New York.
9. M. J. Ablowitz, P. A. Clarkson (1991) Solitons, nonlinear evolution equations and inverse scattering, *Cambridge University Press*, Cambridge.
10. J. Harnad, P. Winternitz (1982), Pseudopotentials and Lie symmetries for the generalized nonlinear Schrödinger equation, *J. Math. Phys.* **23**, no. 4, 517–525.
11. D. Levi, D. Ragnisco, M. Bruschi (1983) Continuous and discrete matrix Burgers hierarchies, *Il Nuovo Cimento* **74**, no. 1, 33–51.
12. P. Santini, private communication.
13. A. M. Grundland, P. Santini, The conditional symmetries for the generalized Liouville equation (in preparation).
14. R.K. Bullough, P.J. Caudrey (1980), Solitons, *Topics in Current Physic* bf 17, Springer Verlag, Berlin.

QUASISTATIONARY APPROXIMATION AND GAUGE INVARIANCE IN CLASSICAL ELECTRODYNAMICS

ANDRÉ HESLOT
Laboratoire de Physique Théorique et Mathématique,
Université Paris 7 - Denis Diderot,
2, Place Jussieu, F-75251 Paris Cedex 05 - France.

Abstract. The quasistationary approximation of classical electrodynamics is given a precise and consistent formulation in terms of fields. Its physical meaning and relationship with gauge invariance are discussed.

Contribution in honor of Guy Rideau

1. Introduction

Many applications of classical electrodynamics (particularly, to the study of electric circuits) make use of the *quasistationary approximation*. By this, we mean that the fields are calculated as in the stationary (i.e., static) case, except that the electric field is corrected to take account of the Faraday induction law : in other words, the emission of radiation and propagation phenomena are neglected. The approximation is valid as soon as the system under consideration is much smaller than the wavelength associated with its field, and this explains its practical importance.

However, it seems that the foundations of this approximation have not received much attention : as far as we know, the presentations of the subject are always incomplete or contradictory. The purpose of the present paper is to give a proper formulation of the quasistationary approximation, and to clarify its physical meaning. Our main result is that it can be ultimately formulated as a simplified version of classical electrodynamics in which the gauge invariance is partly lost, *but which is by itself logically consistent.*

J. Bertrand et al. (eds.), Modern Group Theoretical Methods in Physics, 179–188.

2. Survey of classical electrodynamics

Classical electrodynamics studies the interactions between the electromagnetic field $(\mathbf{E}, \mathbf{B})$ on the one hand, and the charged particles on the other hand. Its basic equations divide thus in two groups :

• Maxwell equations, which rule the field and, in particular, express how it is influenced by charged particles, through the charge and current densities ρ and $\mathbf{j}$:

$$\begin{aligned} \operatorname{curl} \mathbf{E} &= -\partial \mathbf{B}/\partial t, & (1)\\ \operatorname{div} \mathbf{E} &= \rho/\epsilon_0, & (2)\\ \operatorname{curl} \mathbf{B} &= (\mathbf{j}/\epsilon_0 + \partial \mathbf{E}/\partial t)/c^2, & (3)\\ \operatorname{div} \mathbf{B} &= 0, & (4) \end{aligned}$$

where ϵ_0 and c stand as usual for the permittivity and velocity of light in empty space.

• The equations of motion for the charged particles : the influence of the field on a particle with charge q and velocity $\mathbf{v}$ is described by the Lorentz force,

$$\mathbf{F} = q(\mathbf{E} + \mathbf{v} \times \mathbf{B}). \tag{5}$$

In most applications to macroscopic systems, e.g., electric circuits, the Lorentz force is incorporated in phenomenological laws such as the generalized Ohm law.

As is well known, the $\partial \mathbf{E}/\partial t$ term (displacement current) in equation (3) was introduced by Maxwell in order to solve the contradiction between Ampère law $\operatorname{curl} \mathbf{B} = \mathbf{j}/\epsilon_0 c^2$ and the local conservation law of electric charge, i.e.,

$$\operatorname{div} \mathbf{j} = -\partial \rho/\partial t. \tag{6}$$

It is this $\partial \mathbf{E}/\partial t$ term, together with the $\partial \mathbf{B}/\partial t$ term in equation (1) (Faraday law of induction), which is responsible for the mutual generation of $\mathbf{E}$ and $\mathbf{B}$ and thus, for the phenomena of radiation and propagation of the electromagnetic field.

3. Solution of Maxwell equations

The general solution of Maxwell equations (1) and (4) is

$$\begin{aligned} \mathbf{E} &= -\nabla U - \partial \mathbf{A}/\partial t, & (7)\\ \mathbf{B} &= \operatorname{curl} \mathbf{A}. & (8) \end{aligned}$$

The non-uniqueness of the potentials U and $\mathbf{A}$, for given $\mathbf{E}$ and $\mathbf{B}$, is the gauge invariance property of classical electrodynamics. It means in particular that *there is no preferred decomposition of the electric field* $\mathbf{E}$ *in an electrostatic-like part* $-\nabla U$ *and an induction part* $-\partial\mathbf{A}/\partial t$.

The gauge invariance allows to impose a supplementary condition on the potentials. We choose the Lorentz gauge condition,

$$\operatorname{div}\mathbf{A} + \frac{1}{c^2}\frac{\partial U}{\partial t} = 0. \tag{9}$$

Then the two remaining Maxwell equations (2) and (3) read

$$\nabla^2 U - \frac{1}{c^2}\frac{\partial^2 U}{\partial t^2} = -\frac{\rho}{\epsilon_0}, \tag{10}$$

$$\nabla^2 \mathbf{A} - \frac{1}{c^2}\frac{\partial^2 \mathbf{A}}{\partial t^2} = -\frac{\mathbf{j}}{\epsilon_0 c^2}. \tag{11}$$

Suppose that the charged particles described by the densities ρ and $\mathbf{j}$ are contained in a finite volume V. Then the general solution of (10) and (11) can be written under the form of retarded potentials,

$$U(t) = \frac{1}{4\pi\epsilon_0}\int_V \frac{\rho(t-r/c)}{r}\,d\tau + (\text{ext. rad}), \tag{12}$$

$$\mathbf{A}(t) = \frac{1}{4\pi\epsilon_0 c^2}\int_V \frac{\mathbf{j}(t-r/c)}{r}\,d\tau + (\text{ext. rad}), \tag{13}$$

where for simplicity only the time arguments have been displayed, and $r = |\mathbf{r}|$, where $\mathbf{r}$ is the vector from the volume element $d\tau$ (at which ρ and $\mathbf{j}$ are considered) to the point at which U and $\mathbf{A}$ are evaluated. The expression (ext. rad.) stands for external radiation. It represents the general solution of the homogeneous equations associated with (10) and (11) : physically, this free solution is interpreted as produced by charged particles situated outside the volume V, and will be omitted in what follows. An essential feature of (12) and (13) is the retardation time r/c, which means that electromagnetic effects propagate with the finite velocity c.

From (12) and (13), the electromagnetic field $(\mathbf{E},\mathbf{B})$ can be calculated, using (7) and (8). This gives :

$$\mathbf{E}(t) = \frac{1}{4\pi\epsilon_0}\int_V \rho(t-r/c)\frac{\mathbf{r}}{r^3}\,d\tau + \frac{1}{4\pi\epsilon_0}\int_V \frac{1}{c}\frac{\partial\rho}{\partial t}(t-r/c)\frac{\mathbf{r}}{r^2}\,d\tau - \frac{1}{4\pi\epsilon_0 c^2}\int_V \frac{1}{r}\frac{\partial\mathbf{j}}{\partial t}(t-r/c)\,d\tau, \tag{14}$$

$$\mathbf{B}(t) = \frac{1}{4\pi\epsilon_0 c^2}\int_V \mathbf{j}(t - r/c) \times \frac{\mathbf{r}}{r^3}\, d\tau + \frac{1}{4\pi\epsilon_0 c^2}\int_V \frac{1}{c}\frac{\partial \mathbf{j}}{\partial t}(t - r/c) \times \frac{\mathbf{r}}{r^2}\, d\tau. \quad (15)$$

In the stationary case, i.e., ρ and $\mathbf{j}$ independent of time, formulas (14) and (15) reduce to

$$\mathbf{E} = \frac{1}{4\pi\epsilon_0}\int_V \rho \frac{\mathbf{r}}{r^3}\, d\tau, \quad (16)$$

$$\mathbf{B} = \frac{1}{4\pi\epsilon_0 c^2}\int_V \mathbf{j} \times \frac{\mathbf{r}}{r^3}\, d\tau. \quad (17)$$

(16) and (17) are the Coulomb and Biot-Savart laws respectively : thus (14) and (15) can be regarded as generalizations of these laws to nonstationary situations. Surprisingly enough, these formulas were not well known until recently, even though they prove very convenient when discussing the structure of the field in the general case (see [1, 2, 3, 4]).

4. Presentation of the quasistationary approximation (QSA)

Suppose we have to study a system of charges contained in a volume V. It may be for instance an electric circuit, for which we wish to calculate the electromagnetic field it produces. This field is given by (14) and (15). However, these formulas do not solve the problem, since the charge and current densities ρ and $\mathbf{j}$ depend themselves on the field through the Lorentz force (5). No general exact solution is known to this problem of mutual interaction between charged particles and field : this comes from the complexity of formulas (14) and (15), due mainly to the retardation term r/c in the time arguments of the densities ρ and $\mathbf{j}$.

However, if the volume V is small enough, it should be possible to replace (14) and (15), *inside and in the vicinity of* V, by approximate formulas obtained by neglecting partially the effect of the retardation times. We shall explain this more precisely in the next section. Let us say for the moment that the expression of the field becomes then simple enough so that the problem of its interaction with the charged particles in V can often be solved with a good precision. The densities ρ and $\mathbf{j}$ which correspond to the conditions on the system can then be determined, which allows to calculate finally the field at any point in space using the exact formulas (14) and (15).

Of course, the condition of validity of the approximation is that the variation of the densities ρ and $\mathbf{j}$ must be small during the time a/c, where a represents the linear dimensions of the volume V : thus the situation must be close enough to a stationary one, and this is the reason for which this approximation is called the *quasistationary approximation* (QSA). The condition of validity can be also written $a/c \ll T$, where T is the characteristic time of the system, or $a \ll \lambda$, where $\lambda = cT$ is the corresponding wavelength of the electromagnetic field.

Most textbooks pretend that QSA amounts to saying that the $\partial \mathbf{E}/\partial t$ term (displacement current) in Maxwell equation (3) can be neglected. Generally, no detailed analysis is given (nor seems even necessary), because of the order of the exposition, i.e., Electro- and Magnetostatics/Induction phenomena and electric circuits/Displacement current and Maxwell equations (see e.g. [5, 6, 7]). Sometimes, it is specified that it is that term which is responsible for the radiation phenomena, and thus for the retardation time we want to neglect (see [8]). Whatever the argument, it is not satisfactory to neglect the displacement current, *since one is left then with equations which contradict the local conservation law of electric charge* (6).

In fact, the correct method will be to expand the field, given by (14) and (15), as a power series in the retardation time r/c, and to retain the first few terms of the series (compare with [9]).

5. The fields in QSA

Let us consider first the magnetic field, which is easier to discuss. Expanding (15) as a power series in the retardation time r/c, one gets

$$\begin{aligned} \mathbf{B} &= \frac{1}{4\pi\epsilon_0 c^2}\int_V \left[\mathbf{j} - \frac{r}{c}\frac{\partial \mathbf{j}}{\partial t} + \frac{r^2}{2c^2}\frac{\partial^2 \mathbf{j}}{\partial t^2} + \dots\right] \times \frac{\mathbf{r}}{r^3}\, d\tau \\ &\quad + \frac{1}{4\pi\epsilon_0 c^2}\int_V \frac{1}{c}\left[\frac{\partial \mathbf{j}}{\partial t} - \frac{r}{c}\frac{\partial^2 \mathbf{j}}{\partial t^2} + \dots\right] \times \frac{\mathbf{r}}{r^2}\, d\tau, \end{aligned}$$

that is,

$$\mathbf{B} = \frac{1}{4\pi\epsilon_0 c^2}\int_V \left[\mathbf{j} - \frac{r^2}{2c^2}\frac{\partial^2 \mathbf{j}}{\partial t^2} + \dots\right] \times \frac{\mathbf{r}}{r^3}\, d\tau, \qquad (18)$$

where the time dependence has been dropped since all quantities refer now to the same time. In QSA, we keep only the term of order zero, i.e., $\mathbf{B}$ is simply given by the Biot-Savart formula (17).

Consider now the electric field. Expanding (14) as a power series in the retardation time r/c, one gets

$$\begin{aligned}\mathbf{E} &= \frac{1}{4\pi\epsilon_0}\int_V \left[\rho - \frac{r}{c}\frac{\partial\rho}{\partial t} + \frac{r^2}{2c^2}\frac{\partial^2\rho}{\partial t^2} + \ldots\right]\frac{\mathbf{r}}{r^3}\,d\tau \\ &\quad + \frac{1}{4\pi\epsilon_0}\int_V \frac{1}{c}\left[\frac{\partial\rho}{\partial t} - \frac{r}{c}\frac{\partial^2\rho}{\partial t^2} + \ldots\right]\frac{\mathbf{r}}{r^2}\,d\tau \\ &\quad - \frac{1}{4\pi\epsilon_0 c^2}\int_V \frac{1}{r}\left[\frac{\partial\mathbf{j}}{\partial t} - \frac{r}{c}\frac{\partial^2\mathbf{j}}{\partial t^2} + \ldots\right]d\tau,\end{aligned}$$

that is,

$$\begin{aligned}\mathbf{E} &= \frac{1}{4\pi\epsilon_0}\int_V \left[\rho - \frac{r^2}{2c^2}\frac{\partial^2\rho}{\partial t^2} + \ldots\right]\frac{\mathbf{r}}{r^3}\,d\tau \\ &\quad - \frac{1}{4\pi\epsilon_0 c^2}\int_V \frac{1}{r}\left[\frac{\partial\mathbf{j}}{\partial t} - \frac{r}{c}\frac{\partial^2\mathbf{j}}{\partial t^2} + \ldots\right]d\tau. \qquad (19)\end{aligned}$$

In the second integral, only the first term will be kept in QSA,

$$\mathbf{E}_{m1} = -\frac{1}{4\pi\epsilon_0 c^2}\int_V \frac{1}{r}\frac{\partial\mathbf{j}}{\partial t}\,d\tau. \qquad (20)$$

Concerning the first integral, the situation is more complicated : under the conditions of validity of QSA, the second term,

$$\mathbf{E}_{m2} = -\frac{1}{4\pi\epsilon_0}\int_V \frac{1}{2c^2}\frac{\partial^2\rho}{\partial t^2}\frac{\mathbf{r}}{r}\,d\tau \qquad (21)$$

is small compared with the first, which simply represents the Coulomb formula (16),

$$\mathbf{E}_s = \frac{1}{4\pi\epsilon_0}\int_V \rho\frac{\mathbf{r}}{r^3}\,d\tau, \qquad (22)$$

and this should normally lead us to neglect $\mathbf{E}_{m2}$. But it is not difficult to show, using the local conservation law (6) of electric charge, that we can also write

$$\mathbf{E}_{m2} = -\frac{1}{4\pi\epsilon_0}\int_V \frac{1}{2c^2}\left[\frac{1}{r}\frac{\partial\mathbf{j}}{\partial t} - \left(\frac{\partial\mathbf{j}}{\partial t}\cdot\mathbf{r}\right)\frac{\mathbf{r}}{r^3}\right]d\tau, \qquad (23)$$

By comparison with (20), this means that, except under special circumstances, $\mathbf{E}_{m2}$ and $\mathbf{E}_{m1}$ are of the same order of magnitude. Similarly, it can be shown that the magnetic force $q\mathbf{v}\times\mathbf{B}$ on a charged particle in V

is generally of the same order of magnitude as $q\mathbf{E}_{m2}$ and $q\mathbf{E}_{m1}$. *Thus we must keep or neglect together* $\mathbf{E}_{m1}$, $\mathbf{E}_{m2}$, *and* $\mathbf{B}$. Since we do not want to keep only the Coulomb part $\mathbf{E}_s$ of the field, which is often washed out at the macroscopic level by the tendency of matter to electric neutrality, we finally retain, for the fields in QSA,

$$\begin{aligned} \mathbf{E} &\simeq \frac{1}{4\pi\epsilon_0}\int_V \rho\frac{\mathbf{r}}{r^3}\,d\tau \\ &\qquad -\frac{1}{4\pi\epsilon_0 c^2}\int_V \frac{1}{r}\frac{\partial \mathbf{j}}{\partial t}\,d\tau - \frac{1}{4\pi\epsilon_0}\int_V \frac{1}{2c^2}\frac{\partial^2\rho}{\partial t^2}\frac{\mathbf{r}}{r}\,d\tau, \qquad (24) \\ \mathbf{B} &\simeq \frac{1}{4\pi\epsilon_0 c^2}\int_V \mathbf{j}\times\frac{\mathbf{r}}{r^3}\,d\tau. \qquad (25) \end{aligned}$$

Notice that these expressions can be formally obtained from (14) and (15) through an expansion to second order in $1/c$.

6. The potentials in QSA

Let us define the new quantities

$$\begin{aligned} U &= \frac{1}{4\pi\epsilon_0}\int_V \frac{\rho}{r}\,d\tau, \qquad (26) \\ \mathbf{A} &= \frac{1}{4\pi\epsilon_0 c^2}\int_V \frac{\mathbf{j}}{r}\,d\tau + \frac{1}{4\pi\epsilon_0 c^2}\int_V \frac{1}{2}\frac{\partial\rho}{\partial t}\frac{\mathbf{r}}{r}\,d\tau. \qquad (27) \end{aligned}$$

Then it is easy to check that the fields in QSA are related to U and $\mathbf{A}$ by formulas (7) and (8). These U and $\mathbf{A}$ play thus the role of potentials in QSA. More precisely, introducing the components $\mathbf{E}_s$ and $\mathbf{E}_m = \mathbf{E}_{m1}+\mathbf{E}_{m2}$ of the electric field $\mathbf{E}$ (formulas (20), (21) and (22)), we have

$$\begin{aligned} \mathbf{E}_s &= -\nabla U, \qquad (28) \\ \mathbf{E}_m &= -\partial\mathbf{A}/\partial t, \qquad (29) \\ \mathbf{B} &= \text{curl}\,\mathbf{A}. \qquad (30) \end{aligned}$$

Notice that U and $\mathbf{A}$ are *not* the retarded potentials in which the retardation time would be neglected, because then there would not be the second term in (27). Moreover, this second term accounts for the component $\mathbf{E}_{m2}$ of $\mathbf{E}$, which originally came from the scalar potential. The transition to these new potentials can be considered thus as a gauge transformation : in fact, with the help of (6), it can be shown that

$$\text{div}\,\mathbf{A} = 0, \qquad (31)$$

i.e., the new potentials satisfy the gauge condition which is usual in the static case, instead of the Lorentz gauge condition (9).

7. The equations of QSA

The QSA fields and potentials have been given explicitly in the last two sections. But it is also useful, and as we shall see very instructive, to exhibit equations which are satisfied by these fields and potentials.

The simplest case is that of U : its expression (26) is the same as in the static case, therefore it satisfies Poisson equation,

$$\Delta U = -\rho/\epsilon_0. \tag{32}$$

Said another way, in the electric field $\mathbf{E} = -\nabla U - \partial\mathbf{A}/\partial t$, the Coulomb part $\mathbf{E}_s = -\nabla U$ (formula (22)) obeys the laws of Electrostatics, i.e.,

$$\operatorname{curl}\mathbf{E}_s = \mathbf{0}, \tag{33}$$

$$\operatorname{div}\mathbf{E}_s = \rho/\epsilon_0. \tag{34}$$

The case of the magnetic field is more subtle : $\mathbf{B}$ is given by the same expression as in the static case, namely Biot-Savart formula, but it cannot obey the laws of Magnetostatics, because Ampère law $\operatorname{curl}\mathbf{B} = \mathbf{j}/\epsilon_0 c^2$ can be satisfied only if $\operatorname{div}\mathbf{j} = 0$, which does not hold in general. By direct calculation from (25), one obtains in fact

$$\operatorname{curl}\mathbf{B} = (\mathbf{j}/\epsilon_0 + \partial\mathbf{E}_s/\partial t)/c^2, \tag{35}$$

$$\operatorname{div}\mathbf{B} = 0, \tag{36}$$

which is consistent with the local conservation law of electric charge (6) because of equation (34). As anticipated, $\mathbf{B}$ does not obey the laws of Magnetostatics. However, there is an essential simplification in comparison with the exact Maxwell equations (3) and (4), because $\partial\mathbf{E}/\partial t$ is replaced by $\partial\mathbf{E}_s/\partial t$, and we just saw that $\mathbf{E}_s$ can be calculated from Electrostatics independently of $\mathbf{B}$, whereas the electric field $\mathbf{E}$ in the exact theory depends on $\mathbf{B}$ through the Faraday law (1). *Thus the mutual generation of* $\mathbf{E}$ *and* $\mathbf{B}$, *which is responsible for the phenomena of radiation and propagation, disappears in QSA* (compare with [10], where QSA is called the Faraday approximation; see also [11]). Notice that (35) and (36) may be considered as the equations of Magnetostatics, if we include in $\mathbf{j}$ the term $\epsilon_0\partial\mathbf{E}_s/\partial t$, which plays here the role of the displacement current : in fact, it is only in QSA that this concept of displacement current really makes sense, as a source of **B** which does not depend on **B** itself. Finally these results can also be formulated as an equation for **A** : from (30), (31) and (35), one gets

$$\Delta \mathbf{A} = -(\mathbf{j}/\epsilon_0 + \partial \mathbf{E}_s/\partial t)/c^2, \tag{37}$$

which is again the equation of Magnetostatics if we include in **j** the displacement current.

Once $\mathbf{E}_s$ and **B** are known, the last step is to determine the component $\mathbf{E}_m = -\partial \mathbf{A}/\partial t$ of the electric field **E**. One immediately obtains, from (29), (30) and (31),

$$\operatorname{curl} \mathbf{E}_m = -\partial \mathbf{B}/\partial t, \tag{38}$$

$$\operatorname{div} \mathbf{E}_m = 0. \tag{39}$$

As could have been anticipated, $\mathbf{E}_m$ describes thus the induction phenomena.

Of course, the preceding equations for the three fields $\mathbf{E}_s$, **B** and $\mathbf{E}_m$ (or for the potentials), have to be completed with the Lorentz force (5), which we may write

$$\mathbf{F} = q(\mathbf{E}_s + \mathbf{E}_m + \mathbf{v} \times \mathbf{B}). \tag{40}$$

8. Discussion

Summarizing the results of the preceding section, the important point is the natural decomposition of the electric field **E** in QSA in an electrostatic part $\mathbf{E}_s$ and an induction part $\mathbf{E}_m$. As we said in section 3, such a decomposition has no physical meaning in the exact theory because of gauge invariance. In that sense, we may say that *the gauge invariance is partially lost in QSA*. In particular, the scalar potential U determined by (28) is fixed up to an additive constant only (which represents what is left of gauge invariance as far as U is concerned), and thus the difference of potential between two points is uniquely defined : this difference represents the voltage across the two points, as used in the theory of electric circuits.

Another point of interest is that the equations of the preceding section represent a simplified version of classical electrodynamics which is of course approximate, *but which is by itself logically consistent*, at least when no change of inertial frame is considered. In particular, this allows to study energetic questions directly from these equations, as we shall see now. From Lorentz force law (5) or (40), the time derivative of the kinetic energy $\mathcal{E}$ of the charged particles contained in the volume V is

$$\frac{d\mathcal{E}}{dt} = \int_V \mathbf{j}.\mathbf{E}\, d\tau. \tag{41}$$

But we have in QSA the decomposition $\mathbf{E} = \mathbf{E}_s + \mathbf{E}_m$ and thus,

$$\frac{d\mathcal{E}}{dt} = \int_V \mathbf{j}.\mathbf{E}_s\, d\tau + \int_V \mathbf{j}.\mathbf{E}_m\, d\tau. \tag{42}$$

Using the QSA equations (35) and (33), and taking account of the behaviour of the fields at infinity, it can be shown that

$$\int_V \mathbf{j}.\mathbf{E}_s\, d\tau = -\frac{d}{dt}\int_{\mathbb{R}^3} \frac{\epsilon_0 E_s^2}{2}\, d\tau. \tag{43}$$

In the same way, using (35) and (38), one obtains

$$\int_V \mathbf{j}.\mathbf{E}_m\, d\tau = -\frac{d}{dt}\int_{\mathbb{R}^3} \frac{\epsilon_0 B^2 c^2}{2}\, d\tau. \tag{44}$$

These results lead to assign to the fields $\mathbf{E}_s$ and $\mathbf{B}$ the energy densities $\epsilon_0 E_s^2/2$ and $\epsilon_0 B^2c^2/2$ respectively. Of course, there is no radiated power here, i.e., the corresponding surface integrals vanish.

References

1. Tran-Cong Ton (1991) On the time-dependant, generalized Coulomb and Biot-Savart laws, *Am. J. Phys.* **59**, pp. 520–528.
2. Jefimenko, O.D. (1993) Force exerted on a stationary charge by a moving electric current or by a moving magnet, *Am. J. Phys.* **61**, pp. 218–222.
3. Jefimenko, O.D. (1994) Direct calculation of the electric and magnetic fields of an electric point charge moving with constant velocity, *Am. J. Phys.* **62**, pp. 79–85.
4. Lorrain, P., Corson, D.R. and Lorrain, F. (1988) *Electromagnetic Fields and Waves*, W. H. Freeman, New York, 3rd edition, chapter 37, section 37-6.
5. Slater, J.C. and Frank, N.H. (1947) *Electromagnetism*, McGraw-Hill, chapter VII, section 4.
6. Panofsky, W.K.H. and Phillips, M. (1962) *Classical Electricity and Magnetism*, Addison-Wesley, chapter 10, section 10-1.
7. Reitz, J.R., Milford, F.J. and Christy, R.W. (1992) *Foundations of electromagnetic theory*, 4th edition, Addison Wesley, chapter 16, section 1.
8. Landau, L.D. and Lifshitz, E.M. (1960) *Electrodynamics of continuous media*, Pergamon Press, chapter VII, section 45.
9. Fano, R.M., Lan Jen Chu and Adler, R.B. (1960) *Electromagnetic fields, energy and forces*, John Wiley and Sons, chapter 6, section 6-5.
10. Soodak, H. and Tiersten, M.S. (1994) Dynamic interpretation of Maxwell's equations, *Am. J. Phys.* **62**, pp. 907–913.
11. Zapolsky, H.S. (1987) Does charge conservation imply the displacement current ? *Am. J. Phys.* **55**, p. 1140.

GRADED POISSON BRACKETS AND FIELD THEORY

Y. KOSMANN-SCHWARZBACH
U.R.A. 169 du C.N.R.S., Centre de Mathématiques
Ecole Polytechnique, F-91128 Palaiseau, France

Abstract. An introduction to Gerstenhaber algebras, and in particular exact Gerstenhaber algebras, also called BV-algebras, motivated by recent developments in field theory and not so recent results in differential and Poisson geometry.

1. Introduction.

Our aim is first to point out the role of the graded Poisson brackets in field theory, then to formally introduce the Gerstenhaber algebras and describe many examples of those algebras drawn from differential geometry as well as from field theory. In many cases, there exists an additional structure, an operator, usually denoted Δ, or b_0, or ∂, that generates a graded bracket which measures to what extent this operator fails to be a derivation of the associative product. In this case, the Gerstenhaber algebra is called an exact Gerstenhaber algebra, or a Batalin-Vilkovisky algebra (BV-algebra), because such operators are basic to the Batalin-Vilkovisky quantization of field theories.

2. Graded Poisson brackets (antibrackets) in classical and quantum field theory.

The classical BRST operator, $\widehat{S}$, can be expressed as a graded bracket with a fixed element S in the graded Lie algebra of functions on a phase space extended with the ghosts and the antighosts. By the graded Jacobi identity, the fundamental property of $\widehat{S} = [S, .]$,

$$\widehat{S}^2 = 0 ,$$

J. Bertrand et al. (eds.), Modern Group Theoretical Methods in Physics, 189–196.

is equivalent to the classical master equation,

$$[S, S] = 0 .$$

In the quantization theory of Batalin and Vilkovisky, there is a quantum correction to the operator $\widehat{S}$, and the quantum BRST-BV operator, acting on functions of even and odd variables, is of the form $\widehat{S}_\hbar = [S, .] - \hbar\Delta$, where Δ is an operator of order 2 and square 0. (See example 4.3.). Witten (1990) clarified the relation between the graded bracket [,], and the differential operator Δ. The fact that Δ does not act as a derivation of the associative product is the origin of the graded bracket, which is defined by

$$[F, G] = \Delta(FG) - (\Delta F)G - (-1)^{|F|}F(\Delta G) , \tag{2.1}$$

where $|F|$ is the degree of F. Witten shows that the basic properties (graded Jacobi identity and graded Leibniz rule) can be derived from this formula for the graded bracket, which he calls the antibracket and denotes by { , }. Then, the condition

$$\widehat{S}_\hbar^2 = 0$$

is equivalent to the closure condition (so that the action can be integrated using the Feynman path integral), *i. e.*,

$$\Delta(e^{-\frac{S}{\hbar}}) = 0 ,$$

or to the quantum master equation,

$$\frac{1}{2}[S, S] - \hbar\Delta S = 0 .$$

It is very striking that, several years earlier, Koszul (1985) had studied such structures, which we shall describe more precisely in the next sections, on abstract graded algebras and applied the results to differential and Poisson geometry.

More recently, it was shown (Zwiebach (1993), Lian and Zuckerman (1993), Getzler (1994)) that such structures also appear in string theory. In fact, on the BRST cohomology of the state-space of a topological chiral algebra there exist an associative, graded commutative multiplication, a graded Poisson bracket and an operator b_0 of square 0 which satisfy the same relation (2.1) as above, up to an overall sign factor. The nonchiral case has been treated by Penkava and Schwarz (1991). For the case when the central charge is 1 , the chiral cohomology has been computed by Lian and Zuckerman (1993).

We shall now give a definition of Gerstenhaber algebras, offer examples from differential and Poisson geometry, and then suggest further applications of this theory.

3. Gerstenhaber algebras.

The first instance of a graded Lie algebra structure on a graded vector space equipped with an associative, graded commutative product, one of the gradations being shifted by 1, and the two laws satisfying a compatibility condition in the form of a graded Leibniz law, seems to have appeared in Gerstenhaber (1963).

Definition. Let A be an associative, graded commutative algebra over a commutative field of characteristic 0. A *Gerstenhaber algebra structure* on A is a graded Lie algebra structure, $[\ ,\ ]$, on the vector space A with gradation shifted by 1, the two laws being related by the graded Leibniz rule. More precisely, for all a, b and c in A,

(*i*) $$[b,a] = -(-1)^{(|a|-1)(|b|-1)}[a,b]\ ,$$

(*ii*) $$(-1)^{(|a|-1)(|c|-1)}[a,[b,c]] + (-1)^{(|b|-1)(|a|-1)}[b,[c,a]] + (-1)^{(|c|-1)(|b|-1)}[c,[a,b]] = 0\ ,$$

(*iii*) $$[a, b\wedge c] = [a,b]\wedge c + (-1)^{(|a|-1)|b|}b\wedge [a,c]\ ,$$

where $|a|$ is the degree of $a \in A$, and $\wedge$ is the associative product.

Relation (i) is the graded skewsymmetry, (ii) is the graded Jacobi identity, while (iii) is the graded Leibniz rule. Gerstenhaber algebras are also called G-algebras (Gerstenhaber and Schack (1992) or Schouten algebras (Kosmann-Schwarzbach and Magri (1990), and see example 3.2 for the justification of this terminology), and their brackets are variously called *graded Poisson brackets* or *odd Poisson brackets* or *antibrackets*, or sometimes *graded Lie brackets* or just *graded brackets.* There are many examples of Gerstenhaber algebras.

Example 3.1. *The Hochschild cohomology of an associative algebra.* Gerstenhaber (1963) showed the existence, on the Hochschild cohomology space of any associative algebra, A, of a graded Lie bracket and a graded commutative product satisfying the preceding properties. The bracket is obtained by skewsymmetrizing the (nonassociative) composition product of cochains, while the graded commutative product is the cup-product defined in terms of the ring structure of A.

Example 3.2. *The Schouten bracket of multivectors.* In differential geometry, the fundamental example of a Gerstenhaber algebra is that of the exterior algebra of multivectors on a smooth manifold, M, equipped with the Schouten bracket. By multivectors we mean smooth fields of skewsymmetric contravariant tensors. The Schouten bracket (see *e. g.*, Tulczyjew

(1974)) is the unique extension of the Lie bracket of vector fields into a Gerstenhaber algebra bracket [,], such that

$$[f,g] = 0 , \quad [X,f] = X.f ,$$

for $f, g \in C^\infty(M)$, and X any vector field on M. In view of the Hochschild-Kostant-Rosenberg theorem, the Schouten bracket appears as a special case of the graded bracket of example 3.1.

The Schouten bracket is a very important tool in Poisson geometry as was shown by Flato, Lichnerowicz and Sternheimer (1975), Lichnerowicz (1977) and many more recent papers and monographs. In fact, a bivector P on M is a Poisson bivector if and only if $[P,P] = 0$. Moreover, $[P,f]$ is the opposite of the Hamiltonian vector field with Hamiltonian, f.

Example 3.3. *The algebra of functions on an odd symplectic supermanifold.* Gerstenhaber algebras also arise as algebras of functions on supermanifolds of type $(n|n)$ equipped with an odd symplectic (or Poisson) structure. See Schwarz (1993), Getzler (1994), Alexandrov, *et al.* (1995). In fact, the exterior algebra of multivectors on a manifold M of example 3.2 can be considered to be the algebra of functions on T^*M considered as an odd symplectic manifold of type $(n|n)$, where n is the dimension of M.

Example 3.4. *The Koszul bracket of differential forms on a Poisson manifold.* On any Poisson manifold, (M,P), the exterior algebra of differential forms is a Gerstenhaber algebra when it is equipped with the Koszul bracket, $[\ ,\ ]_P$, which is uniquely defined by

$$[f,g]_P = 0 \quad [\alpha,f]_P = (\mathcal{P}\alpha).f , \quad [df,dg]_P = d\{f,g\}_P .$$

Here $f,\ g \in C^\infty(M)$, α is any differential 1-form on M, d is the de Rham differential, $\mathcal{P}$ is the linear bundle map from the cotangent bundle to the tangent bundle of M defined by $\langle\beta,\mathcal{P}\alpha\rangle = \mathcal{P}(\alpha,\beta)$, for the differential 1-forms α and β, and

$$\{f,g\}_P = \mathcal{P}(df).g = \langle dg, \mathcal{P}(df)\rangle = P(df,dg)$$

is the Poisson bracket on $C^\infty(M)$ defined by the Poisson structure, P. In particular, $[df,g]_P = \{f,g\}_P$. See Koszul (1985), although the restriction of this bracket to differential one-forms was discovered independently earlier by several authors.

Example 3.5. *The Gerstenhaber algebra of a Lie algebroid.* Examples 3.2 and 3.4 are special cases of the following general construction (Kosmann-Schwarzbach and Magri (1990), Mackenzie and Xu (1994)). A Lie algebroid is a vector bundle, E, over a manifold M equipped with a Lie bracket, [,], on the vector space of sections of E, and with a vector-bundle morphism

a from E to TM, the tangent bundle of M, called the *anchor*, such that a maps brackets of sections of E to Lie brackets of vector fields on M, and

$$[x, fy] = f[x, y] + (a(x).f)y$$

where $f \in C^\infty(M)$, and x and y are any sections of E. The space of sections of ΛE, the exterior algebra bundle of E, is a Gerstenhaber algebra. There is a more general, entirely algebraic construction based on Palais pairs (see Gerstenhaber and Schack (1992)), variously called pseudo-Lie algebras, (K, R)-Lie algebras, Lie-Rinehart algebras, differential Lie algebras, and diverse other names.

4. Exact Gerstenhaber algebras.

The Gerstenhaber algebras which possess an operator of degree -1 which is related to the graded bracket in the following sense play an important role both in geometry and in field theory.

Definition. A linear operator of degree -1, $\partial : A \to A$, *generates* the graded bracket $[\ ,\ ]$ on A if

$$(4.1) \qquad [a, b] = (-1)^{|a|}(\partial(a \wedge b) - \partial a \wedge b - (-1)^{|a|} a \wedge \partial b) ,$$

for all a and $b \in A$. A Gerstenhaber algebra is called *exact* if there exists an operator of degree -1, of square 0 which generates the graded bracket.

It is natural to ask what conditions on ∂, a linear operator of degree -1 and square 0 on a graded associative algebra $(A, \wedge)$, will imply that formula (4.1) defines a graded Lie bracket satisfying the graded Leibniz rule. Koszul (1985) showed that it is sufficient that ∂ be of order 2 and vanish on the unit. It is easy to prove that, conversely, if ∂ generates a Gerstenhaber bracket and vanishes on the unit, it is necessarily of order at most 2 (and exactly 2 if the bracket is not identically 0).

Example 4.1. *The Koszul bracket on a Poisson manifold.* On a Poisson manifold, the Koszul bracket is generated by the Poisson homology operator,

$$\partial_P = [i_P, d] ,$$

the graded commutator of the interior product of forms by the Poisson bivector and the de Rham differential.

Example 4.2. *The Schouten bracket on an orientable manifold.* On an orientable manifold, M, of dimension n, any nonvanishing section of $\Lambda^n TM$ defines an operator that generates the Schouten bracket (Koszul (1985)).

When the manifold is symplectic, the Schouten bracket is generated by $\partial = [i_\omega, d_P]$, the graded commutator of the interior product of multivectors by the symplectic form, and the corresponding Lichnerowicz-Poisson cohomology operator. (Here P denotes the bivector inverse to the nondegenerate 2-form ω.) This property (Kosmann-Schwarzbach and Magri (1990)) is, of course, dual to that of the Koszul bracket of example 4.2.

Example 4.3. *The Schouten bracket on a vector space.* The Schouten bracket of fields of multivectors on a vector space with global coordinates $x_i, i = 1, \cdots, n$, is generated, in particular, by the operator

$$\Delta = \sum_{i=1}^{n} \frac{\partial}{\partial x^i} \frac{\partial}{\partial \xi_i} ,$$

where a field of multivectors is viewed as a function of the even variables $x_i, i = 1, \cdots, n$, and of the odd variables

$$\xi_i = \frac{\partial}{\partial x^i}, i = 1, \cdots, n .$$

This is the operator that appears in Witten (1990).

Example 4.4. *The algebraic Schouten bracket on the exterior algebra of a Lie algebra.* If g is the Lie algebra of a Lie group G, the restriction of the Schouten bracket to the left-invariant multivectors on G defines a graded bracket on Λg called the algebraic Schouten bracket, which is generated by the homology operator of the Lie algebra g.

Example 4.5. *The exact Gerstenhaber algebra of a symplectic supermanifold with a volume element.* Let $\mathcal{M}$ be a supermanifold of type $(n|n)$ equipped with

(i) an odd Poisson structure, and
(ii) a volume element (Berezinian), μ.

Then the algebra of functions on $\mathcal{M}$ is an exact Gerstenhaber algebra whose graded bracket is generated by the operator ∂ defined by

$$\partial f = \frac{1}{2} \mathrm{div}_\mu(H_f) ,$$

where H_f is the Hamiltonian vector field associated with the function f, and div_μ denotes the divergence of a vector field with respect to the volume element μ. This is the geometric setting for BV quantization. See Schwarz (1993), Getzler (1994), Alexandrov, *et al.* (1995). Example 4.3 is a special case of this situation.

Example 4.6. *The exact Gerstenhaber algebra structure on a chiral topological algebra.* On the BRST cohomology of a super chiral algebra (obtained from a chiral algebra with central charge 26 by adding ghosts or from a superconformal field theory in dimension 2), there exist both an associative product and an operator b_0 defining an exact Gerstenhaber algebra structure. See Lian and Zuckerman (1993), Penkava and Schwarz (1991).

Example 4.7. *The dual exact Gerstenhaber algebras of a Lie bialgebra.* Any Lie bialgebra g defines a pair of exact Gerstenhaber algebras in duality, $(\Lambda g, \Lambda g^*)$.

Moreover, the Lie algebra cohomology operator of g (resp., g^*) is a derivation of the graded bracket on Λg^* (resp., Λg). These properties were expressed using various languages by various authors. The Lie bialgebras appear as a special case of a much more general situation, that of the *Lie bialgebroids*, introduced by Mackenzie and Xu (1994).

The relationship between Lie bialgebroids and Gerstenhaber algebras was studied in Kosmann-Schwarzbach (1994), where a new proof of the self-duality of the notion of Lie bialgebroid was given. In that paper, we showed that a Lie bialgebroid gives rise to a pair of Gerstenhaber algebras in duality, each equipped with a derivation of the bracket. We also showed that any triangular Lie bialgebroid in the sense of Mackenzie and Xu (1994) gives rise to a pair of Gerstenhaber algebras in duality, one of which is exact. Examples of Lie bialgebroids also arise in the theory of Poisson-Nijenhuis structures, which are closely related to the bihamiltonian structures that play a very important role in the theory of integrable systems.

References

1. Alexandrov, M., Kontsevich, M., Schwarz, A. and Zaboronsky, O. (1995) The Geometry of the master equation and topological quantum field theory, preprint hep-th.
2. Flato, M., Lichnerowicz, A. and Sternheimer, D. (1975) Algèbres de Lie attachées à une variété canonique, *J. Math. pures et appl.*, 54, 445-480.
3. Gerstenhaber, M. (1963) The Cohomology structure of an associative ring, *Ann. of Math.*, 78, 267-288.
4. Gerstenhaber, M. and Schack, S. (1992) Algebras, bialgebras, quantum groups and algebraic deformations, *Deformation theory and quantum groups with applications to mathematical physics*, Gerstenhaber, M. and Stasheff, J, eds., *Contemporary Mathematics* 134, Amer. Math. Soc., Providence, R.I., 51-92.
5. Getzler, E. (1994) Batalin-Vilkovisky algebras and two-dimensional topological field theory, *Comm. Math. Phys.*, 159, 265-285.
6. Kosmann-Schwarzbach, Y. (1994) Exact Gerstenhaber algebras and Lie bialgebroids, *Acta Appl. Math.*, to appear.
7. Kosmann-Schwarzbach, Y. and Magri, F. (1990) Poisson-Nijenhuis structures, *Ann. Inst. Henri Poincaré*, Phys. Théor., 53, 35-81.
8. Koszul, J.-L. (1985) Crochet de Schouten-Nijenhuis et cohomologie, *Elie Cartan et les mathématiques d'aujourd'hui, Astérisque*, hors série, Soc. Math. Fr., 257-271.
9. Lian, B. H. and Zuckerman, G. (1993) New perspectives on the BRST-algebraic structure of string theory, *Comm. Math. Phys.*, 154, 613-646.
10. Lichnerowicz, A. (1977) Les variétés de Poisson et leurs algèbres de Lie associées, *J. Diff. Geom.*, 12, 253-300.
11. Mackenzie, K. and Xu, P. (1994) Lie bialgebroids and Poisson groupoids, *Duke Math. J.*, 73, 415-452.
12. Penkava, M. and Schwarz, A. (1991) *On some algebraic structures arising in string*

theory, preprint, hep-th/912071.
13. Schwarz, A. (1993) Geometry of Batalin-Vilkovisky quantization, *Comm. Math. Phys.*, 155, 249-260.
14. Tulczyjew, W. (1974) The graded Lie algebra of multivector fields and the generalized Lie derivative of forms, *Bull. Acad. Pol. Sc.*, 22, 937-942.
15. Witten, E. (1990) A note on the antibracket formalism, *Modern Physics Letters*, A 5, 487-494.
16. Zwiebach, B. (1993) Closed string theory: quantum action and the B-V master equation, *Nucl. Phys.*, B 390, 33-152.

THE BASIC LEMMA IN THE THEORY OF FORMAL GROUPS

CARLOS MORENO
L.P.T.M. Univ. ParisVII
2 Place Jussieu, F-75251 Paris
Dpto. de Física Teórica, Univ. Complutense,
E-28040 Madrid

AND

FELIX GALVEZ
I.F.P. "Luis Buñuel"
E-28933 Móstoles, Madrid
Dpto. de Física Teórica, Univ. Complutense,
E-28040 Madrid

To Guy Rideau

Abstract.[1]
We prove the theorem by M. Lazard about the cohomology of the classical analyseur in characteristic zero in the framework of the basic notions of the simplicial homotopy theory outlined by Drinfeld. In the end we must compute the cohomology of the quotient simplicial set $I^r/\partial I^r$, where I^r is the rth cartesian product of the standard 1-simplex I and ∂I^r is its boundary. It is zero but in degree r in which case it is of dimension 1. Lazard theorem plays a fundamental role in the theory of quasi-Hopf quantized universal enveloping-algebras.

1. Introduction

In [2] M. Lazard introduced the notion of *analyseur*. It is an algebraic structure where the notion of *formal group law* can be developed. Lazard defines also the cohomology complex of an *analyseur*. Given a formal group law to

[1]Lecture to the RIDEAU COLLOQUIUM, University "Paris VII (Denis Diderot)", Paris, 9-10 January 1995.

J. Bertrand et al. (eds.), Modern Group Theoretical Methods in Physics, 197–210.

the order l, some cohomology classes of the above complex are obstructions to the existence of a formal group law to the order $l+1$.
The particular case of interest for us is *the classical analyseur with n-parameters over the field* $\mathbb{R}$. It is defined in essence as the direct limit

$$\ldots \overset{\circ}{\mathbb{R}}[x^1,\ldots,x^p] \longrightarrow \overset{\circ}{\mathbb{R}}[x^1,\ldots,x^p,x^{p+1}]\ldots$$

where $x^a = (x_1^a,\ldots,x_n^a)$, $a = 1, 2, 3, \ldots,$ jointly with the algorithm of substitutions [13].
The cohomology of this analyseur is computed in [2] and in degree p is isomorphic to the *p-exterior power of a real vector space of dimension n*. This result is the basic lemma in the theory of formal group laws in characteristic zero.
If $\mathfrak{g}$ is a Lie algebra the cohomology of the cogebra $U(\mathfrak{g})$ is defined in [3]. It is easy to prove, see below, that the cohomology complex of the classical analyseur with n-parameters over $\mathbb{R}$, see (2) below, is isomorphic to the cohomology complex of the cogebra $U(\mathfrak{g})$ for *any* Lie algebra $\mathfrak{g}$ of dimension n, and therefore its cohomology in degree p is $\wedge^p\mathfrak{g}$.
Given an invariant star product [4][5] to the order l, some cohomology classes of the above complex are obstructions to the existence of an invariant star product to the order $l+1$. This is because the invariant Hochschild cohomology complex on the Lie group is isomorphic to the cohomology complex of the cogebra $U(\mathfrak{g})$ [6], and thus isomorphic to the cohomology complex of the classical analyseur.
Drinfeld obtains triangular quantum groups from invariant star products on Lie groups endowed with an invariant symplectic structure, and obtains these invariant star products, *essentially*, from the Campbell-Hausdorff formal group law of the Lie group [1] [6] [7] [8]. Therefore the cohomology that allows the existence of formal Hausdorff group law is the same (isomorphic) as the cohomology that allows the existence of triangular quantum groups. Drinfeld has outlined a way to compute the cohomology of the cogebra $U(\mathfrak{g})$ in the framework of the basic notions of the *simplicial homotopy theory* [11]. To develop this computation we prove here an isomorphism theorem by Drinfeld, Theorem 4 below, and we continue to extract the simplicial contents from Lazard proof.

2. Formal group laws and the cohomology complex of M.Lazard

2.1.

Definition 1

Let A be a commutative ring and let $\{f_a(x_1{}^1,\ldots,x_n{}^1,x_1{}^2,\ldots,x_n{}^2)\} \equiv f(x^1,x^2)$, *where* $a = 1,2,\ldots,n$, $x^1 = (x_1{}^1,\ldots,x_n{}^1), x^2 = (x_1{}^2,\ldots,x_n{}^2)$, *be a set of n formal power series with coefficients in A and without constant*

terms. With $r \in \mathbb{N}$, let $(\Gamma^r f) = (f(f(x^1,x^2),x^3) - f(x^1,f(x^2,x^3)))^r$ be the component of total degree r of the formal power series $(\Gamma f)(x^1,x^2,x^3) = (f(f(x^1,x^2),x^3) - f(x^1,f(x^2,x^3)))$.
1) We will say that the set $f(x^1,x^2)$ defines a formal group law to the order r if a) $f^1(x^1,x^2) = x^1 + x^2$, and b) $\left(\Gamma^l f\right)(x^1,x^2,x^3) = 0$, $l = 1,\ldots,r$.
2) We will say that the set $f(x^1,x^2)$ defines a formal group law if a) is satisfied and b) is satisfied for all $r \in \mathbb{N}$.

Definition 2 *Two formal group laws $f(x^1,x^2)$ and $g(x^1,x^2)$ will be equivalent if there are n formal power series $\{\phi_a(x^1),\ a = 1,\ldots,n\} \equiv \phi(x^1)$, without constant terms (hence invertible for the composition law) such that $f(\phi(x^1),\phi(x^2)) = \phi\left(g(x^1,x^2)\right)$.*

2.2.

We will consider here a very particular case in the work of M. Lazard. We consider only the case of *l'analyseur classique à n paramètres sur le corps* $\mathbb{R}$. Let us suppose that $f(x^1,x^2)$ defines a formal group law to the order r. Then $(\Gamma_{r+1} f)(x^1,x^2,x^3)$ is a 3-cocycle in a cohomology complex introduced in [2].

Definition 3 *Let $\overset{\circ}{\mathbb{R}}\left[x^1,\ldots,x^p\right]$ be the space of polynomials $F(x^1,\ldots,x^p)$ in $p\cdot n$ indeterminates, without constant term. Let d_p be the operator defined by the expression,*

$$d_p F(x^1,\ldots,x^p,x^{p+1}) = F(x^2,\ldots,x^{p+1}) + (-1)^{p+1} F(x^1,\ldots,x^p) + \sum_{i=1}^{p} (-1)^i F(x^1,\ldots,x^{i-1},x^i + x^{i+1},x^{i+2},\ldots,x^p,x^{p+1}). \tag{1}$$

Then $d_p \circ d_{p-1} = 0$, and Lazard complex is the following

$$\ldots \overset{\circ}{\mathbb{R}}\left[x^1,\ldots,x^{p-1}\right] \xrightarrow{d_{p-1}} \overset{\circ}{\mathbb{R}}\left[x^1,\ldots,x^p\right] \xrightarrow{d_p} \overset{\circ}{\mathbb{R}}\left[x^1,\ldots,x^{p+1}\right] \ldots \tag{2}$$

It is clear from the definition of d_p that if $F^r(x^1,\ldots,x^p)$ is a polynomial of total degree r then $d_p F^r(x^1,\ldots,x^{p+1})$ is also a polynomial of total degree r. Complex (2) is therefore the direct sum of its homogeneous components

$$\ldots \overset{\circ}{\mathbb{R}}{}^r\left[x^1,\ldots,x^{p-1}\right] \xrightarrow{d_{p-1}} \overset{\circ}{\mathbb{R}}{}^r\left[x^1,\ldots,x^p\right] \xrightarrow{d_p} \overset{\circ}{\mathbb{R}}{}^r\left[x^1,\ldots,x^{p+1}\right] \ldots \tag{3}$$

Among the results obtained in [2] a very particular one is the following

Theorem 1 (Lazard) *The cohomology of the complex (3) is* $H^p(\overset{\circ}{\mathbb{R}^r}[*]) = 0$, *if* $p \neq r$, *and*

$$H^r(\overset{\circ}{\mathbb{R}^r}[*]) \approx \mathbb{R} - lin\left\{\sum_{\sigma\in\mathfrak{S}_r}(-1)^\sigma x^1_{i_{\sigma(1)}}\cdots x^r_{i_{\sigma(r)}};\quad \forall i_a \in (1,\ldots,n)\right\},$$

where $\mathfrak{S}_r$ *is the symmetric group acting on the set* $\{1,\ldots,r\}$.

This theorem allows to prove, for example

Theorem 2 *1) A set of* n *formal power series of the form* $f(x^1,x^2) = x^1 + x^2 + f^2(x^1,x^2)$, *is a formal group law to the order* 2 *if and only if the homogeneous term of degree* 2, $f^2(x^1,x^2)$, *is bilinear.*
2) It can be extended to a group law to the order 3 *if and only if the bilinear form* $f^2(x^1,x^2) - f^2(x^2,x^1)$ *is a Lie bracket.*
3) A formal group law to the order $r \geq 3$ *can be extended to a formal group law to any order.*

Another similar Theorem about the equivalence of two formal group laws can be also proven [2].

2.3.

Let $\mathfrak{g}$ be a real n-dimensional Lie algebra and $(\mathbf{e}_1,\ldots,\mathbf{e}_n)$ a basis of $\mathfrak{g}$. Let $S(\mathfrak{g}) = \bigoplus_{m\geq 0} S^m(\mathfrak{g})$ be the graded bigebra structure of the symmetric tensors (zero characteristic) on $\mathfrak{g}$. The following mapping Φ is a 0-graded bigebra isomorphism: $S(\mathfrak{g}) \ni \mathbf{e}_1^{\alpha(1)}\ldots\mathbf{e}_n^{\alpha(n)} \xrightarrow{\Phi} (x_1^1)^{\alpha(1)}\ldots(x_n^1)^{\alpha(n)} \in \mathbb{R}[x^1]$ where $\alpha(l) \in \mathbb{N}$ and $\Phi(1) = 1$. In fact, using the canonical graded algebra isomorphism $\mathbb{R}[x^1] \otimes \overset{p}{\ldots} \otimes \mathbb{R}[x^1] \simeq \mathbb{R}[x^1,\ldots,x^p]$, we obtain (p=2)

$$\Delta\left(\mathbf{e}_1^{\alpha(1)}\ldots\mathbf{e}_n^{\alpha(n)}\right) = (\mathbf{e}_1\otimes 1 + 1\otimes\mathbf{e}_1)^{\alpha(1)}\ldots(\mathbf{e}_n\otimes 1 +$$

$$+1\otimes\mathbf{e}_n)^{\alpha(n)} \longrightarrow (x_1^1 + x_1^2)^{\alpha(1)}\ldots(x_n^1 + x_n^2)^{\alpha(n)} = \Delta\left((x_1^1)^{\alpha(1)}\ldots(x_n^1)^{\alpha(n)}\right).$$

where Δ is the coproduct of $S(\mathfrak{g})$ or of $\mathbb{R}[x^1]$. We have then $\Delta\circ\Phi^{-1} = \Phi^{-1}\circ\Delta$. The isomorphism Φ induces an isomorphism of graded algebras: $S(\mathfrak{g})\otimes \overset{p}{\ldots} \otimes S(\mathfrak{g}) \xrightarrow{\Phi} \mathbb{R}[x^1,\ldots,x^p]$. Therefore Φ^{-1} is an isomorphism from the complex (2) to the complex

$$\ldots S(\mathfrak{g})\otimes \overset{p-1}{\ldots} \otimes S(\mathfrak{g}) \xrightarrow{d_{p-1}} S(\mathfrak{g})\otimes \overset{p}{\ldots} \otimes S(\mathfrak{g}) \xrightarrow{d_p} S(\mathfrak{g})\otimes \overset{p+1}{\ldots} \otimes S(\mathfrak{g})\ldots \quad (4)$$

$$d_p(s_1 \otimes \ldots \otimes s_p) = 1 \otimes s_1 \otimes \ldots \otimes s_p + \tag{5}$$

$$\sum_{i=1}^{p}(-1)^i s_1 \otimes \ldots \otimes \Delta s_i \otimes \ldots \otimes s_p + (-1)^{p+1} s_1 \otimes \ldots \otimes s_p \otimes 1,$$

and from the complex (3) to the r-homogeneous component of the complex (4), that is to say to

$$\ldots \left(S(\mathfrak{g}) \otimes \overset{p-1}{\ldots} \otimes S(\mathfrak{g})\right)^r \xrightarrow{d_{p-1}} (S(\mathfrak{g}) \otimes \overset{p}{\ldots} \otimes S(\mathfrak{g}))^r \xrightarrow{d_p} \left(S(\mathfrak{g}) \otimes \overset{p+1}{\ldots} \otimes S(\mathfrak{g})\right)^r \ldots \tag{6}$$

where the differential is the restriction of that in (5). The cogebra structure of $U(\mathfrak{g})$ and $S(\mathfrak{g})$ are isomorphic through the symmetrization mapping η. Clearly complex (6) is thus isomorphic to the r-homogeneous component of the complex

$$\ldots U(\mathfrak{g}) \otimes \overset{p-1}{\ldots} \otimes U(\mathfrak{g}) \xrightarrow{d_{p-1}} U(\mathfrak{g}) \otimes \overset{p}{\ldots} \otimes U(\mathfrak{g}) \xrightarrow{d_p} U(\mathfrak{g}) \otimes \overset{p+1}{\ldots} \otimes U(\mathfrak{g}) \ldots \tag{7}$$

$$d_p(u_1 \otimes \ldots \otimes u_p) =$$

$$= 1 \otimes u_1 \otimes \ldots \otimes u_p + \sum_{i=1}^{p}(-1)^i u_1 \otimes \ldots \otimes \Delta u_i \otimes \ldots \otimes u_p + (-1)^{p+1} u_1 \otimes \ldots \otimes u_p \otimes 1. \tag{8}$$

That is to say complex (6) is isomorphic to the following complex

$$\ldots \left(U(\mathfrak{g}) \otimes \overset{p-1}{\ldots} \otimes U(\mathfrak{g})\right)^r \xrightarrow{d_{p-1}} (U(\mathfrak{g}) \otimes \overset{p}{\ldots} \otimes U(\mathfrak{g}))^r \xrightarrow{d_p} \left(U(\mathfrak{g}) \otimes \overset{p+1}{\ldots} \otimes U(\mathfrak{g})\right)^r \ldots \tag{9}$$

where the differential is the restriction of that in (8). Here Δ is the coproduct of $U(\mathfrak{g})$, and $U^r(\mathfrak{g}) = \eta(S^r(g))$. Complexes (2) and (3) are thus respectively isomorphic to complexes (7) and (8).
As far as we know complex (8) was explicitly written for the first time by P. Cartier in [3]. From Theorem 1, we obviously get

Theorem 3 (Lazard) *The cohomology of the complex (9) is* $H^p\left(\left(U(\mathfrak{g})^{\otimes *}\right)^r\right) = 0$, *if* $p \neq r$ *and* $H^r\left(\left(U(\mathfrak{g})^{\otimes *}\right)^r\right) = \mathfrak{g} \wedge \overset{r}{\ldots} \wedge \mathfrak{g}$.

2.4.

The graded bigebra $U(\mathfrak{g})$ is isomorphic to the graded bigebra of the invariant differential operators on every connected Lie group G with Lie algebra

$\mathfrak{g}$. The elements of $U(\mathfrak{g})\otimes \overset{p}{\ldots} \otimes U(\mathfrak{g})$ can therefore be identified with the invariant p-multidifferential operators on G. That is to say with the multidifferential p-linear mappings on the associative algebra $C^\infty(G)$ with values in $C^\infty(G)$. Complex (7) is thus isomorphic to the invariant differential cohomology complex of the usual $C^\infty(G)$-bimodule $C^\infty(G)$.This complex is then a subcomplex of the cohomology complex associated to the bar resolution (or standard homology complex) of the $C^\infty(G)$ -bimodule $C^\infty(G)$, which in turn is the associated homology complex of a simplicial $\mathbb{R}$-vector space, $(\beta_*(C^\infty(G); C^\infty(G)), d^*, s^*)$ [9]. Complex (7) is therefore the associated cohomology complex of a cosimplicial vector space [11] defined by the following "face" and "degeneracy" operators:

$$\begin{aligned} \partial^i_{p+1} &: U^{\otimes p}(\mathfrak{g}) \longrightarrow U^{\otimes(p+1)}(\mathfrak{g}), \quad i = 0, 1, \ldots, (p+1) \\ \sigma^j_{p-1} &: U^{\otimes p}(\mathfrak{g}) \longrightarrow U^{\otimes(p-1)}(\mathfrak{g}), \quad j = 0, 1, \ldots, (p-1) \end{aligned}$$

and

$$\begin{aligned} \partial^0_{p+1}(u_1 \otimes \ldots \otimes u_p) &= (1 \otimes u_1 \otimes \ldots \otimes u_p) \\ \partial^l_{p+1}(u_1 \otimes \ldots \otimes u_p) &= (u_1 \otimes \ldots \Delta u_l \otimes \ldots \otimes u_p) \\ \partial^{p+1}_{p+1}(u_1 \otimes \ldots \otimes u_p) &= (u_1 \otimes \ldots \otimes u_p \otimes 1) \\ \sigma^j_{p-1}(u_1 \otimes \ldots \otimes u_p) &= \epsilon(u_{j+1})(u_1 \otimes \ldots \otimes \hat{u}_i \ldots \otimes u_p) \end{aligned}$$

Where $l = 1, \ldots, p$; $j = 0, \ldots, (p-1)$ and ϵ is the counit of $U(\mathfrak{g})$.
Using only the cogebra structure of $U(\mathfrak{g})$ we can directly verify commutation relations characterizing a cosimplicial vector space.These mappings are such that if we define $\partial_{p+1} \equiv \sum_{i=0}^{p+1}(-1)^i \partial^i_{p+1}$, we have $\partial_{p+1} \circ \partial_p = 0$ and $d_p = \partial_{p+1}$. (d_p of (8) and we will write $\partial^i_{p+1} \equiv d^i_p$, $i \in \{0, \ldots, (p+1)\}$. Thus complex (2) is isomorphic to the associated cohomology complex of the cosimplicial vector space $U^{\otimes *}(\mathfrak{g})$.

3. Drinfeld framework to Lazard theorem

3.1.

Let Δ be the category of monotonic maps $\mu : [p] \longrightarrow [q]$, $[p] = \{0, 1, \ldots, p\}$, $p \in \mathbb{N}$[9] [11]. Let ∂^i_p, and σ^i_p, $(i = 0, \ldots, p)$ be respectively the usual monomorphisms and epimorphisms. Let Δ° be the opposite category of Δ and let I be the functor $I : \Delta^\circ \longrightarrow$ **Set** defined by $I[p] = \Delta([p], [1]) = \left\{\mu^p_0, \mu^p_1, \ldots, \mu^p_{p+1}\right\}$, where $i \in \{0, 1, \ldots, p, (p+1)\}$ and,

$$\mu^p_i : \begin{pmatrix} 0 & 1 & \ldots & p-i & p-i+1 & \ldots & p-1 & p \\ 0 & 0 & \ldots & 0 & 1 & \ldots & 1 & 1 \end{pmatrix}$$

$$d_l^p : \Delta([p],[1]) \longrightarrow \Delta([p-1],[1]), \quad s_l^p : \Delta([p],[1]) \longrightarrow \Delta([p+1],[1]), \quad (10)$$

$$d_l^p \mu_i^p = \mu_i^p \circ \partial_p^l, \quad s_l^p \mu_i^p = \mu_i^p \circ \sigma_p^l,$$

$l = 0, 1, \ldots, p$. We can obtain $d_{p-l}^p \mu_i^p = \mu_i^{p-1}, \quad 0 \le i \le l$ and $d_{p-l}^p \mu_i^p = \mu_{i-1}^{p-1}, \quad l < i \le (p+1)$. Also $s_{p-l}^p \mu_i^p = \mu_i^{p+1}, \quad 0 \le i \le l$ and $s_{p-l}^p \mu_i^p = \mu_{i+1}^{p+1}, \quad l < i \le (p+1)$. Therefore functor I defines a simplicial set usually called the standard 1-simplex.

3.2.

Consider now the r-cartesian product, I^r of the simplicial set I. It is the functor, $I^r : \Delta^\circ \longrightarrow \mathbf{Set}; \quad \mathbf{I^r[p]} = (\mathbf{I[p]} \times \overset{r}{\ldots} \times \mathbf{I[p]})$.
If $\left(\mu_{i_1}^p, \ldots, \mu_{i_r}^p\right) \in (I[p] \times \overset{r}{\ldots} \times I[p]), i_a \in \{0, \ldots, p+1\}$, and if we define $d_i^p, s_i^p, \; i = 0, \ldots, p$,, (again the same symbols!) acting diagonally on the elements of $I^r[p]$, I^r is a (covariant) functor.

3.3.

The boundary ∂I^r of I^r is the subfunctor of I^r defined by

$$\partial I^r[p] = \cup_{i-1}^r (I \times \ldots \times \overset{i}{\partial I} \times \ldots \times I)[p] \equiv \cup_{i=1}^r (I[p] \times \ldots \times \overset{i}{\partial I[p]} \times \ldots \times I[p]),$$

where $\partial I[p] = \{\mu_0^p, \mu_{p+1}^p\}$, is the boundary of the standard 1-simplex . Let us write $\phi_p^r : \partial I^r[p] \longrightarrow I^r[p]$, for the inclusions. Let $I/\partial I$ be the corresponding quotient simplicial set and $C_*(I^r, \partial I^r) \equiv C_*(I^r/\partial I^r) \equiv C_*(I^r)/C_*(\partial I^r)$, the corresponding homology complex.

3.4.

Let $C^*(I^r)$ and $C^*(\partial I^r)$ be respectively the associated cohomology complexes of the complexes $C_*(I^r)$ and $C_*(\partial I^r)$. Face and degeneracy operators are defined by

$$d_{p-1}^i \equiv \partial_p^i \equiv (d_i^p)^*, \; i = 0, 1, \ldots, p; \quad \text{and } s_{p+1}^i \equiv \sigma_p^i \equiv (s_i^p)^*, \quad i = 0, 1, \ldots, p.$$

If $\{\mu_{p-1}^i;\; i = 0, 1, \ldots, p\}$ is the dual basis of the basis $\{\mu_i^{p-1};\; i = 0, 1, \ldots, p\}$, we can obtain the following fundamental relations

$$d_{p-1}^i \mu_{p-1}^k = \mu_p^k, \; 0 \le k < i$$

$$d^i_{p-1}\mu^k_{p-1} = \mu^{k+1}_p,\ i < k \leq p \tag{11}$$
$$d^i_{p-1}\mu^i_{p-1} = \mu^i_p + \mu^{i+1}_p.$$

$$\begin{aligned} s^i_{p+1}\mu^t_{p+1} &= \mu^t_p,\ 0 \leq t \leq (p-i), \\ s^i_{p+1}\mu^t_{p+1} &= \mu^{t-1}_p,\ (p-i+1) < t \leq p, \\ s^i_{p+1}\mu^t_{p+1} &= 0,\ t = (p-i+1) \end{aligned} \tag{12}$$

3.5.

Dualizing the injection ϕ^r_*, we get a morphism of complexes $C^*(I^r) \xrightarrow{\phi^*_r} C^*(\partial I^r)$. Drinfeld considers the subcomplex $Ker\,\phi^*_r$. It is defined through

$$Ker\ \phi^p_r = \mathbb{R} - lin\{(\mu^{i_1}_p \otimes \ldots \otimes \mu^{i_r}_p),\ i_a \in (1, \ldots, p)\}.$$

This space is generated by all zero p-cochains when acting on p-chains $(\mu^p_{j_1}, \ldots, \mu^p_{j_r})$ *where at least one* j_a *is* 0 *or* $p+1$. A standard argument proves that $C^p(I^r/\partial I^r)$ is canonically isomorphic to the orthogonal, in $C^p(I^r)$, to $C_p(\partial I^r)$ and this is just $Ker\,\phi_p{}^r$, which is then isomorphic to the associated cohomology complex of a simplicial set, and we will identify from now on these two complexes.

3.6.

Let r be fixed and let $\mathbb{R}(\mathfrak{S}_r)$ be the group algebra over $\mathbb{R}$. The group $\mathfrak{S}_r$ *acts* on the right on the vector space $\mathfrak{g}\otimes \overset{r}{\ldots} \otimes\mathfrak{g}$ as a group of automorphisms. We consider $\mathfrak{g}\otimes \overset{r}{\ldots} \otimes\mathfrak{g}$ then as a right $\mathbb{R}(\mathfrak{S}_r)$-module. Similarly the group $\mathfrak{S}_r$ *acts* on the left on the vector space $Ker\,\phi^r_p$ as a group of automorphisms. We consider this space then as a left $\mathbb{R}(\mathfrak{S}_r)$-module. We can therefore define the vector space $(\mathfrak{g}\otimes \overset{r}{\ldots} \otimes\mathfrak{g})\otimes_{\mathfrak{S}_r} Ker\,\phi^r_p$ as the quotient of the vector space $(\mathfrak{g}\otimes \overset{r}{\ldots} \otimes\mathfrak{g}) \otimes Ker\,\phi^r_p$, by the usual relations.

Lemma 1 *Let* $(\mu^{i_1}_p \otimes \ldots \otimes \mu^{i_r}_p)$ *be an element of the basis chosen in* $Ker\,\phi^r_p$. *We then have* $(d^i_p \circ \sigma)(\mu^{i_1}_p \otimes \ldots \otimes \mu^{i_r}_p) = (\sigma \circ d^i_p)(\mu^{i_1}_p \otimes \ldots \otimes \mu^{i_r}_p)$, $i = 0, 1, \ldots, p+1$ *and similar equalities for operators* $s^j_p, j = 0, 1, \ldots, p-1$.

Proof

It is clear from the definitions and the action of d^i_p and s^j_p given in (11),(12) or because of the analogous commutations relations for d^{p+1}_i and s^{p-1}_j.

From this Lemma and relations (11),(12) we easily conclude that the action of d^i_p and s^j_p is also diagonal. This Lemma also allows us to consider the following mappings, noted again d^i_p and s^j_p,

$$d_p^i : (\mathfrak{g}\otimes \overset{r}{\ldots} \otimes\mathfrak{g}) \otimes_{\mathfrak{S}_r} Ker\, \phi_p^r \longrightarrow (\mathfrak{g}\otimes \overset{r}{\ldots} \otimes\mathfrak{g}) \otimes_{\mathfrak{S}_r} Ker\, \phi_{p+1}^r.$$

$$d_p^i \left((x_1 \otimes \ldots \otimes x_r) \otimes_{\mathfrak{S}_r} (\mu_p^{i_1}, \ldots, \mu_p^{i_r})\right) = (x_1\otimes\ldots\otimes x_r)\otimes_{\mathfrak{S}_r} d_p^i(\mu_p^{i_1}, \ldots, \mu_p^{i_r}). \tag{13}$$

And similarly for mappings s_p^j. Clearly these mappings are well defined.

Theorem 4 *Let ψ_p^r be defined as follows*

$$(S(\mathfrak{g})\otimes \overset{p}{\ldots} \otimes S(\mathfrak{g}))^r \xrightarrow{\psi_p^r} (\mathfrak{g}\otimes \overset{r}{\ldots} \otimes\mathfrak{g}) \otimes_{\mathfrak{S}_r} Ker\, \phi_p^r,$$

where

$$h_1^1 \ldots h_{r_1}^1 \otimes\ldots\otimes h_1^p \ldots h_{r_p}^p \xrightarrow{\psi_p^r} h_1^1\otimes \overset{r}{\ldots} \otimes h_{r_p}^p \otimes_{\mathfrak{S}_r} \left(\mu_p^1(r_1), \mu_p^2(r_2), \ldots, \mu_p^p(r_p)\right), \tag{14}$$

$\mu_p^a(r_a) \equiv \mu_p^a\otimes \overset{r_a}{\ldots} \otimes\mu_p^a$ and $r_1 + \ldots + r_p = r$. These mappings are vector space isomorphisms such that $d_p^i \circ \psi_p^r = \psi_{p+1}^r \circ d_p^i$, $(i = 0, 1, \ldots, p+1)$. Thus they define an isomorphism from the complex (6) to the complex

$$(\mathfrak{g}\otimes \overset{r}{\ldots} \otimes\mathfrak{g}) \otimes_{\mathfrak{S}_r} Ker\, \phi_p^r, \qquad d_p = \sum_{i=0}^{p+1}(-1)^i d_p^i,$$

where d_p^i is defined as in (13), and therefore the respective cohomologies are isomorphic.

Proof

a) It must be understood in (14) that $r_a = 0$ in case the block $h_1^a \ldots h_{r_a}^a$, $a = 1, \ldots, p$, is 1 (hence of degree 0), and then that, in the right hand side of (14), the corresponding blocks $\ldots \otimes h_1^a \otimes \ldots \otimes h_{r_a}^a \otimes \ldots$ and $\mu_p^a(r_a)$ are missing.

b) In the left hand side of (14), we have

$$d_p^a (\ldots \otimes h_1^a \ldots h_{r_a}^a \otimes \ldots) = \ldots \otimes \Delta(h_1^a \ldots h_{r_a}^a) \otimes \ldots =$$

$$= \ldots \otimes (h_1^a \otimes 1 + 1 \otimes h_1^a)\ldots(h_{r_a}^a \otimes 1 + 1 \otimes h_{r_a}^a) \otimes \ldots,$$

In the right hand side of (14), we have

$$d_p^a \left[\left(h_1^1 \otimes \ldots \otimes h_{r_p}^p\right) \otimes_{\mathfrak{S}_r} \left(\ldots \otimes \mu_p^a \otimes \ldots \otimes \mu_p^a \otimes \ldots\right)\right] =$$

$$= \left(h_1^1 \otimes \ldots \otimes h_{r_p}^p\right) \otimes_{\mathfrak{S}_r} \left(\ldots \otimes d_p^a\mu_p^a \otimes \ldots \otimes d_p^a\mu_p^a \otimes \ldots\right) =$$

$$= \left(h_1^1 \otimes \ldots \otimes h_{r_p}^p\right) \otimes_{\mathfrak{S}_r} \left(\ldots \otimes \left(\mu_{p+1}^a + \mu_{p+1}^{a+1}\right) \otimes \overset{r_a}{\ldots} \otimes \left(\mu_{p+1}^a + \mu_{p+1}^{a+1}\right) \otimes \ldots\right).$$

The action of isomorphisms ψ_p^r is now clear.

3.7.

The cohomology of the complex $\left((\mathfrak{g} \otimes \overset{r}{\ldots} \otimes \mathfrak{g}) \otimes_{\mathfrak{S}_r} Ker\ \phi_*^r;\ d_*\right)$, is isomorphic to the cohomology of the complex $\left((\mathfrak{g} \otimes \overset{r}{\ldots} \otimes \mathfrak{g}) \otimes_{\mathfrak{S}_r} C^*(I^r, \partial I^r);\ d_*\right)$. The set of classes $(e_{i_1} \otimes \ldots \otimes e_{i_r}) \otimes_{\mathfrak{S}_r} \left(\mu_p^{j_1} \otimes \ldots \otimes \mu_p^{j_p}\right)$, where $1 \le i_1 \le i_2 \le \ldots \le n$, and $1 \le j_k \le p$, some $\mu_p^{j_k}$ can be missing, is a set of generators of the space $(\mathfrak{g} \otimes \overset{r}{\ldots} \otimes \mathfrak{g}) \otimes_{\mathfrak{S}_r} C^p(I^r, \partial I^r)$, and then every element of this space is a sum of elements of the form $(e_{i_1} \otimes \ldots \otimes e_{i_r}) \otimes_{\mathfrak{S}_r} \alpha^k$, where two elements in the sum have *different (nonordered)* set $\{i_1, \ldots, i_r\}$, and where α^k is an element of $C^p(I^r, \partial I^r)$. The following is now clear

Proposition 1 *a) The elements of the form $(e_{i_1} \otimes \ldots \otimes e_{i_r}) \otimes_{\mathfrak{S}_r} \alpha^k$, where $1 \le i_1 \le i_2 \le \ldots \le n$ runs all different sets and where α^k runs the elements in a basis of the space of cocycles in $C^p(I^r, \partial I^r)$, are a set of generators of the space of cocycles in $(\mathfrak{g} \otimes \overset{r}{\ldots} \otimes \mathfrak{g}) \otimes_{\mathfrak{S}_r} C^p(I^r, \partial I^r)$.*
b) The elements of the form $(e_{i_1} \otimes \ldots \otimes e_{i_r}) \otimes_{\mathfrak{S}_r} \beta^l$, where $1 \le i_1 \le i_2 \le \ldots \le n$, runs all different sets and where β^l runs the elements in a basis of the space of coboundaries in $C^p(I^r, \partial I^r)$, are a set of generators of the space of coboundaries in $(\mathfrak{g} \otimes \overset{r}{\ldots} \otimes \mathfrak{g}) \otimes_{\mathfrak{S}_r} C^p(I^r, \partial I^r)$.
c) The elements of the form $(e_{i_1} \otimes \ldots \otimes e_{i_r}) \otimes_{\mathfrak{S}_r} \gamma^m$, where $1 \le i_1 \le i_2 \le \ldots \le n$,, runs all different sets and where γ^m runs the classes in a basis of the space of cohomology $H^p(C^*(I^r, \partial I^r))$ are a set of generators of the space of cohomology $H^p\left((\mathfrak{g} \otimes \overset{r}{\ldots} \otimes \mathfrak{g}) \otimes_{\mathfrak{S}_r} C^*(I^r, \partial I^r)\right)$.
Thus the computation of the cohomology of complex (7) has been reduced to the computation of the cohomology $H^*(C^*(I^r, \partial I^r))$.

3.8.

To obtain the cohomology of the complex $C^*(I^r, \partial I^r)$ we can work with its normalized (sub)-complex $\bar{C}^*(I^r, \partial I^r)$. A *degenerate p-chain* [9] in the homology complex $C_*(I^r, \partial I^r)$ is one in the image of some degeneracy operator s_i^p. From the way these operators act we see that the space of degenerate p-chains is generated by elements $\left[\mu_{i_1}^p \otimes \ldots \otimes \mu_{i_r}^p\right]$ in which *at least one of the elements in the set $\{\mu_1^p, \ldots, \mu_p^p\}$ is missing. A normalized p-cochain* in $C^*(I^r, \partial I^r)$ is one resulting zero when acting on a degenerate p-chain. Thus we have

Proposition 2 *1) The space $\bar{C}^p(I^r, \partial I^r)$ of normalized p-cochains in $C^p(I^r, \partial I^r)$ is generated by the elements $(\mu_p^{i_1}, \ldots, \mu_p^{i_r})$ in which none of the elements in the set $\{\mu_p^1, \ldots, \mu_p^p\}$ is missing.*
2) If $p > r$ there are no non-zero normalized p-cochains and then $H^p(\bar{C}^*(I^r, \partial I^r)) = 0$,

$$H^p\left(\overset{\circ}{\mathbb{R}^r}[*]\right) \approx H^p\left(\left(S^{\otimes(*)}(\mathfrak{g})\right)^r\right) \approx H^p\left(\left(U^{\otimes(*)}(\mathfrak{g})\right)^r\right) = 0.$$

To obtain the rest of the spaces of the cohomology $H^*(\bar{C}^*(I^r, \partial I^r))$ we go further into the proof that Lazard has developed to compute the cohomology (Theorem 1.) of the complex (3). For this we need the following result.

3.9.

Lemma 2 *For $i = 1, \ldots, (p-1)$ let $\sigma_i, \sigma_0 \in \mathfrak{S}_p$ be respectively the transposition $(i, i+1) \to (i+1, i)$ and the unit element of the group. Let $\mathbb{Z}(\mathfrak{S}_p)$ be the group algebra of the symmetric group $\mathfrak{S}_p$ over the ring $\mathbb{Z}$. Consider the element of $\mathbb{Z}(\mathfrak{S}_p)$, $\mathbf{A} = \sum_\sigma \epsilon(\sigma)\sigma$ where $\sigma \in \mathfrak{S}_p$ and $\epsilon(\sigma)$ the signature σ. Then,*
1) For every i, $\mathbf{A}$ is in the twosided ideal generated by $(\sigma_0 - \sigma_i)$.
2) The element $(n! \cdot \sigma_0 - \mathbf{A})$ is in the twosided ideal generated by the set of elements $(\sigma_0 + \sigma_i)$, $i - 1, \ldots, (p-1)$

Proof
2) Every element $\sigma \in \mathfrak{S}_p$ is a product of a *minimal* number of *transpositions* σ_i. This number is ≥ 1 but for σ_0 and is even if $\epsilon(\sigma) = +1$ and odd if $\epsilon(\sigma) = -1$. From $\sigma_i = (\sigma_0 + \sigma_i) - \sigma_0$, we see that σ is a product of elements $(\sigma_0 + \sigma_k) - \sigma_0$, with k in some set $\mathcal{K} \subseteq \{1, \ldots, p-1\}$
From this we obtain

$$n! \cdot \sigma_0 - \mathbf{A} = (n! - 1) \cdot \sigma_0 - \sum_{\sigma \neq \sigma_0} \epsilon(\sigma) \cdot \sigma = \ldots = [\text{sum of products of} (\sigma_0 + \sigma_i)].$$

4. The cohomology of the complex $H^*(\bar{C}^*(I^r, \partial I^r))$.

4.1.

Lemma 3 *Let us introduce the following mappings*

$$\omega_p^i : C^p(I^r, \partial I^r) \longrightarrow C^{p+1}(I^r, \partial I^r), \quad \omega_p^i = d_p^i - N_p^i - N_p^{i+1},$$

where $i = 1, \ldots, p$ and

$$N_p^i : \begin{pmatrix} \mu_p^{i_1} & \cdots & \mu_p^i & \cdots & \mu_p^{i_r} \\ \mu_{p+1}^{i_1} & \cdots & \mu_{p+1}^{i+1} & \cdots & \mu_{p+1}^{i_r+1} \end{pmatrix},$$

where $i_1 \leq i_2 \leq \ldots i_r$.
Then $\forall \sigma \in \mathfrak{S}_r$, $N_p^i \circ \sigma = \sigma \circ N_p^i$, *and* $\sum_{i=1}^p (-1)^i \omega_p^i = d_p$

Proof
We obtain

$$\sum_{i-1}^{p} (-1)^i \omega_p^i =$$

$$\sum_{i=1}^{p} (-1)^i d_p^i - \left(-N_p^1 + (-1)^p N_p^{p+1}\right) = \sum_{i=1}^{p} (-1)^i d_p^i + d_p^0 + (-1)^{(p+1)} d_p^{p+1}.$$

4.2.

The coboundary of a r *normalized* cochain is a $r+1$ *normalized* cochain, hence zero. We can write

Lemma 4 *Every element of* $\bar{C}^r(I^r, \partial I^r)$ *is a cocycle.*

A straightforward computation allow us to get

Lemma 5 *Consider the element* $\beta_i = \left(\mu_{r-1}^1, \ldots, \mu_{r-1}^{i-1}, \mu_{r-1}^i(2), \mu_{r-1}^{i+1}, \ldots, \mu_{r-1}^{r-1}\right)$.
Let $i, j = 1, \ldots, r-1$ *be such that* $i \neq j$ *. Then:*
1) $\omega_{r-1}^j \beta_i = 0$,
2) $(\sigma_0 - \sigma_i)\, \omega_{r-1}^i \beta_i = 0$.

4.3.

Every element $\alpha \in \bar{C}^{r-1}(I^r, \partial I^r)$ is of the form $\alpha = \sum_{i\sigma} t_{i\sigma} \sigma(\beta_i)$ where $t_{i\sigma} \in \mathbb{R}$. From Lemmae 3, and 5, we get $d_{r-1}\alpha = \sum_i t_{i\sigma} d_{r-1}(\sigma(\beta_i)) = \sum_i (-1)^i t_{i\sigma} \sigma(\omega_{r-1}^i \beta_i)$, and then $\mathbf{A}(d_{r-1}\alpha) = \sum_i (-1)^i \epsilon(\sigma) t_{i\sigma} \mathbf{A}(\omega_{r-1}^i \beta_i) = 0$. We have proved

Lemma 6 *The skewsymmetrized of every coboundary in* $\bar{C}^r(I^r, \partial I^r)$ *is zero. That is* $\mathbf{A}(d_{r-1}\alpha) = 0$, $\forall \alpha \in \bar{C}^{r-1}(I^r, \partial I^r)$.

4.4.

Lemma 7 *Let us write* $\alpha_0 = (\mu_r^1, \ldots, \mu_r^i, \ldots, \mu_r^r)$ *in* $\bar{C}^r(I^r, \partial I^r)$ *and consider* β^i *as before. Then*

1) $(\sigma_0 + \sigma_i)\alpha_0 = d_{p-1}((-1)^i \beta_i)$. *Thus* $\sigma(\alpha_0) - \epsilon(\sigma)\alpha_0$ *is a coboundary.*
2) For any α *in* $\bar{C}^r(I^r, \partial I^r)$, $(n!\,\sigma_0 - \mathbf{A})\alpha = d_{r-1}(\sigma(\beta))$, *for some* $\beta \in \bar{C}^{r-1}(I^r, \partial I^r)$.

Proof

2) From Lemma(2) we have
$(n!\,\sigma_0 - \mathbf{A})\alpha_0 = \sum_i \rho_i \cdot (\sigma_0 + \sigma_i) \cdot \alpha_0 = d_{r-1}(\sum_i (-1)^i \cdot \rho_i \cdot \beta_i)$.From $\sigma\mathbf{A} = \mathbf{A}\sigma$ we obviously get $(n!\,\sigma_0 - \mathbf{A})\sigma(\alpha_0) = d_{r-1}(\sum_i \sigma((-1)^i \cdot \rho_i \cdot \beta_i))$. Every α in $\bar{C}^r(I^r, \partial I^r)$ is a linear combination of elements $\sigma(\alpha_0)$, hence the result. From the above Lemmae and Proposition 2 the following is now clear

Theorem 5 *The vector space* $H^r(\bar{C}^*(I^r, \partial I^r))$ *is of dimension* 1 *with basis the class* $[\mathbf{A}\alpha_0]$.

Theorem 6 *The following mapping is a vector space isomorphism*

$$\begin{array}{rcl} H^r\left((\mathfrak{g}\otimes \overset{r}{\ldots} \otimes \mathfrak{g}) \otimes_{\mathfrak{S}_r} \bar{C}^*(I^r, \partial I^r)\right) & \longrightarrow & \Lambda^r \mathfrak{g} \\ (e_{l_1} \otimes \ldots \otimes e_{l_r}) \otimes_{\mathfrak{S}_r} \mathbf{A}\alpha_0 & \longrightarrow & e_{l_1} \wedge e_{l_2} \wedge \ldots \wedge e_{l_r} \end{array}$$

Therefore $H^r\left(\overset{\circ}{\mathbb{R}^r}\,[*]\right) \approx H^r\left(\left(S^{\otimes(*)}(\mathfrak{g})\right)^r\right) \approx H^r\left(\left(U^{\otimes(*)}(\mathfrak{g})\right)^r\right) \approx \Lambda^r \mathfrak{g}$.

4.5.

To compute $H^p(\bar{C}^*(I^r, \partial I^r))$, when $r > p$, we proceed as follows. We have in this case

$$\bar{C}^p(I^r, \partial I^r) = lin - \{\left(\mu_p^1(r_1), \ldots, \mu_p^p(r_p)\right),\ r_1 + \ldots + r_p = r > p,\ r_a \geq 1\}.$$

We consider the complex $(\bar{C}^*(I^r, \partial I^r), d_*)$ as a *differential graded vector space, with differential of degree 1*, [10], and we define as follows an *homogeneous filtration*, that is compatible with the graduation, $G^l(\bar{C}^*(I^r, \partial I^r))$, $l \in \mathbb{Z}$. Let the following vector spaces be

$$G^{-l,p} = lin\left\{(\mu_p^1(r_1), \ldots, \mu_p^p(r_p)) \in \bar{C}^p(I^r, \partial I^r),\ |\ r_p \leq l\right\},$$

where $l > 0$, and $G^{l,p} = 0$, if $l \geq 0$. Then

$$\begin{array}{l} \ldots G^{l+1,p} \subseteq G^{l,p} \ldots G^{1,p} \subseteq G^{0,p} \subseteq G^{-1,p} \ldots \\ \quad \ldots \subseteq G^{-l+1,p} \subseteq G^{-l,p} \ldots \subseteq G^{p-r-1,p} \subseteq G^{p-r-2,p} \ldots . \end{array}$$

This is a stationary sequence in both senses: $G^{l,p} = 0$, if $l \geq 0$ and $G^{-l,p} = G^{p-r-1,p}$ if $l \geq (r + 1 - p)$. From the action (23) of operators d_p^i, $i = 0, 1, \ldots, p, (p+1)$ we see that $d_p(G^{l,p}) \subseteq G^{l,p+1}$, $\forall l \in \mathbb{Z}$ and $\forall p \in \mathbb{N}$.

The filtration is thus compatible with the differential. Also define $G^{-l} = \oplus_p G^{-l,p}$. It is clear that

$$G^{-l} = \oplus_q \left(G^{-l} \cap C^{-l+q}(I^r, \partial I^r)\right) \quad \text{and} \quad G^{-l,-l+p} = G^{-l} \cap C^{-l+p}(I^r, \partial I^r),$$

$$C^p(I^r, \partial I^r) = \cup_l G^{-l,p} \quad \text{and} \quad C^*(I^r, \partial I^r) = \cup_l G^{-l}.$$

We see that $C^*(I^r, \partial I^r)$, d_*, $G^*)$ is a *filtered differential vector space*, [10]. We then have a decreasing filtration of the graded vector space $\bar{C}^*(I^r, \partial I^r)$ by the subspaces G^l. It is also *regular* as a consequence of $G^l = 0$ for l big enough. The study of the *spectral sequence* $G_m^{p,q}$ *of the complex* $(\bar{C}^*(I^r, \partial I^r)$, $d_*)$ *corresponding to this filtration* allows us to obtain the result $H^p(\bar{C}^*(I^r, \partial I^r) = 0$, if $p < r$. See [12] for details and further developments. To be able to directly apply some results in [10], Ch.XV it is more convenient to work with the filtration $F^{-l,p} \equiv G^{-l,-l+p}$.

Acknowledgements The authors wish to express their gratitude to professors R. F. Alvarez-Estrada, M. Flato, A. Galindo, A. Lichnerowicz and D. Sternheimer for illuminating discussions and friendly encouragement.

References

1. V.G.Drinfeld, *On constant, quasiclassical solutions of the Yang-Baxter quantum equation*, Sov. Math. Dokl. **28** (1983), p. 667-671.
2. M. Lazard, *Lois des groupes et analyseurs*, Ann. Ec. Norm. Sup., **72** (1955), p. 299-400.
3. P. Cartier, *Hyperalgèbres et groupes de Lie formels*, Séminaire Sophus Lie, 2ème année 1955/56, Faculté des Sciences de Paris.
4. F. Bayen, M. Flato, C. Fronsdal, A. Lichnerowicz, and D. Sternheimer, *Deformation Theory and Quantization. I and II*, Ann. of Phys., **111**, 61-151 (1978)
5. A. Lichnerowicz, *Déformations d'algèbres associées à une variété symplectique (Les $*_\nu$-Produits)*, Ann. Inst. Fourier, **32**, 157-209 (1982)
6. C. Moreno et L. Valero. *Produits star invariants et équation de Yang-Baxter quantique constante*, Actes des Journées Relativistes (24-29 avril 1990, Aussois, France)
7. C. Moreno and L. Valero. *Star products and quantum groups*, Physics and Manifolds. Proceedings of the International Colloquium in honour of *Yvonne Choquet-Bruhat*, Paris, June 3-5, 1992. Kluwer Academic Publishers, 1994.
8. L. Valero. Thesis, Universidad Complutense, Madrid, 1995.
9. S. Mac Lane. *Homology*, Springer-Verlag. New York 1975.
10. H. Cartan and S. Eilenberg. *Homological Algebra*, Princeton University Press, Princeton, 1956.
11. P. Gabriel and M. Zisman. *Calculus of fractions and homotopy theory*, Springer-Verlag, New York 1967.
12. F. Galvez. Thesis to appear.
13. N. Bourbaki. *Algèbre, Chapitres 4 à 7*, Masson, Paris, 1981.

SPECTRA AND GENERALIZED EIGENFUNCTIONS OF THE ONE- AND TWO-MODE SQUEEZING OPERATORS IN QUANTUM OPTICS

BENGT NAGEL
Department of Physics, Division Theoretical Physics
Royal Institute of Technology
S-100 44 Stockholm Sweden

1. Introduction and Background Material

1.1 INTRODUCTION AND SUMMARY

In recent years the concept of squeezed state has become central in quantum optics both from the theoretical and experimental point of view [1]. In the simplest one-mode case a squeezed state is defined here as a *displaced squeezed vacuum* of the form $|\alpha,\varsigma\rangle$ = $D(\alpha)S(\varsigma)|0\rangle$ obtained by applying the *squeezing operator* $S(\varsigma)=\exp[(\varsigma^* a^2-\varsigma a^{+2})/2]$ [2] and the *displacement operator* $D(\alpha)=\exp(\alpha a^+ - \alpha^* a)$ on the vacuum state $|0\rangle$ (i e the ground state of the one-dimensional harmonic oscillator). $\varsigma = re^{i2\theta}$ and $\alpha=|\alpha|e^{i\beta}$ are complex numbers, and $a=(q+ip)/\sqrt{2}$, $a^+=(q-ip)/\sqrt{2}$ in terms of the normalized coordinate and momentum operators q and p, where $[q,p]=iI$. The term "squeezing" comes from the fact that for $\cos 2\theta > \tanh r\ (>0)$ the dispersion Δq of the coordinate operator in the squeezed state is smaller than the vacuum state value $1/\sqrt{2}$: the uncertainty of the value of q is squeezed compared to the vacuum value. If one performs a rotation in the qp-plane ("phase space") to new canonically conjugate operators $q_\theta = \cos\theta\, q + \sin\theta\, p$, $p_\theta = -\sin\theta\, q + \cos\theta\, p$ the dispersions Δq_θ and Δp_θ will be equal to $e^{-r}/\sqrt{2}$ and $e^{r}/\sqrt{2}$, respectively, and the correlation $\Delta(q_\theta p_\theta)$ [3] is zero. This implies that the variance matrix of the original set $q\,p$ (a symmetric 2x2 matrix with the variances $(\Delta q)^2$ and $(\Delta p)^2$ as diagonal elements and $\Delta\,(qp)$ as off diagonal elements) has its determinant equal to the minimum allowed value 1/4 . This can actually be taken as a characteristic property of squeezed states: they are the states (amongst a priori even mixed states) that give equality in the relation $(\Delta q)^2(\Delta p)^2-[\Delta(qp)]^2 \geq 1/4$ (sometimes called the Schrödinger-Robertson uncertainty relation). The coherent states (ζ =0) are characterized by having the variance matrix diagonal with diagonal elements equal to 1/2. An alternative characterization of queezed states uses the fact that $|\alpha,\varsigma\rangle$ is the (up to a phase factor unique) normalized vector annihilated by the transformed annihilation operator $b=\cosh r\cdot(a-\alpha)+e^{i2\theta}\sinh r\cdot(a^+-\alpha^*)$.

J. Bertrand et al. (eds.), Modern Group Theoretical Methods in Physics, 211–220.

The definition of squeezed states we prefer is as the coherent states (with vacuum as isotropy vector) of a representation (*rep*) of a group HxM(2,**R**), which is a semi-direct product of the 3-dimensional Heisenberg-Weyl group and the 3-dimensional metaplectic group, a two-fold covering of the group SU(1,1).

In experimental realizations of squeezed states the relevant state often corresponds to a two-mode squeezed state generated by applying the two-mode squeezing operator $S_2(\varsigma) = \exp(\varsigma^* a_1 a_2 - \varsigma a_1{}^+ a_2{}^+)$ on the vacuum state.

The purpose of this paper is to derive the spectrum and the (generalized) eigenfunctions, in three different forms, of the squeezing generators $(a^{+2} - a^2)/4i$ and $(a_1^+ a_2^+ - a_1 a_2)/2i$, and also of another quadratic (in a and a^+) generator introduced in the next section. Using these results the spectrum and eigenfunctions of the most general second degree polynomial hermitian Hamiltonian (not necessarily bounded below) can be obtained.

The problem of deriving the eigenfunctions of the squeezing generators has been studied before [4], but the results were not complete, in particular the doubling of the spectrum in the one-mode case was not obtained.

In the next section the group theoretical background of the squeezed states will be quickly presented, and after that the three different state representations will be given: the photon number (harmonic oscillator excitation number) representation (the *n-representation*), the ordinary configuration space representation (the *q-representation*) and the Fock-Bargmann holomorphic function representation (the *z-representation*). The last representation has the advantage that the generalized eigenfunctions will be holomorphic functions (but of course not normalizable), whereas is the q-representation we expect to get tempered distributions, and in the n-representation tempered sequences (growth bounded by some power of n). In general the starting point will be the n-representation, where the method of an earlier paper [5] can be used.

1.2. SQUEEZED STATES AS COHERENT STATES OF HxM(2, R)

The presentation given here is essentially well-known [6], so we only sketch the derivations. The condition for a linear inhomogeneous transformation of $\{a, a^+\}$

$$b = \mu a + \nu a^+ + \alpha,\ b^+ = \nu^* a + \mu^* a^+ + \alpha^* \tag{1}$$

to give a new set $\{b, b^+\}$ satisfying the canonical commutation relations (CCR) is $|\mu|^2 - |\nu|^2 = 1$, which means that the homogeneous part of the transformation is an element of the group SU(1,1). It is known that any two irreducible representations (*irreps*) of the CCR are unitarily equivalent. Now the original rep by $\{a, a^+\}$ is clearly irreducible, and it is not hard to convince oneself that this also holds for the rep by $\{b, b^+\}$ - one can show that there is a unique "new vacuum" $|0\rangle_b$ with $b|0\rangle_b = 0$. This means that one gets a set of unitary operators $\{V(\alpha, g): \alpha \in \mathbf{C}, g \in \mathrm{SU}(1,1)\}$, determined up to phase factors, such that

$$b = V(\alpha, g)^+ a V(\alpha, g). \tag{2}$$

Composing two successive transformations and using irreducibility one realizes that one has a rep up to a phase factor of the group ISU(1,1), "inhomogeneous SU(1,1)" with the composition $(\alpha',g')\cdot(\alpha,g)=(\alpha'+g'\alpha,g'g)$, where $g'\alpha=\mu'\alpha+\nu'\alpha^*$. It is then known that the obstruction against choosing the phase factors of $V(\alpha,g)$ in such a way that one gets a true rep is present already in the 2-dimensional translation subgroup $\{\alpha\}$: since a and a^+ don't commute one has to make a (central) extension of this group to the 3-dimensional Heisenberg-Weyl group $\{(\alpha,s):\alpha\in\mathbf{C},\ s\in\mathbf{R}\}$ with the composition $(\alpha',s')\cdot(\alpha,s)=(\alpha'+\alpha,s'+s+\mathrm{Im}[\alpha'\alpha^*])$. Furthermore it turns out that the resulting rep of SU(1,1) is two-valued, so that if one introduces the two-fold covering of SU(1,1), called the metaplectic group M(2,**R**), one obtains a true unitary rep of the group HxM(2,**R**). Here already the restriction to the group H is irreducible, whereas the restriction to M(2,**R**) gives the sum of two irreps. Corresponding to a factorization of a general group element in the form

$$(\alpha,g,s)=(\alpha,e,o)\cdot(0,g,0)\cdot(0,e,s) \tag{3}$$

we can write a general representing unitary operator in the product form

$$U(\alpha,g,s)=D(\alpha)U(g)e^{is},\ \text{where}\ D(\alpha)=\exp(\alpha a^+-\alpha^*a). \tag{4}$$

The standard definition of a set of coherent states of a given rep of a group G starts by choosing a particular isotropy vector of the rep which is invariant (up to a phase factor, i e gives a one-dimensional rep) under a suitably chosen subgroup H of the given group. The set of corresponding coherent states is the set of states (ignoring differences in phase factors) obtained by applying all unitary operators in the rep on this vector. The set can then be parametrized by the elements of the homogeneous space G/H.

In our case we choose the vacuum state as isotropy vector (as a more general case one could choose an arbitrary number state $|n\rangle$ to get *displaced squeezed number states*), and we then want to factorize the SU(1,1)-representing operator as

$$U(g)=S(\varsigma)\exp(i2\varphi J_o) \tag{5}$$

where as earlier $S(\varsigma)=\exp[(\varsigma^*a^2-\varsigma a^{+2})/2]$ and $J_o=(a^+a+aa^+)/4=(N+1/2)/2$ is the generator of a U(1) subgroup of SU(1,1). As

$$S(\varsigma)^+aS(\varsigma)=\cosh r\cdot a-e^{i2\theta}\sinh r\cdot a^+, \tag{6a}$$

$$\exp(-i2\varphi J_o)a\exp(i2\varphi J_o)=e^{i\varphi}a, \tag{6b}$$

we get by comparing with $U(g)^+aU(g)=\mu a+\nu a^+$ the relations

$$\mu=e^{i\varphi}\cosh r,\ \ \nu=-e^{i(2\theta-\varphi)}\sinh r. \tag{7}$$

J_o together with $J_1=(a^{+2}+a^2)/4$ and $J_2=(a^{+2}-a^2)/4i$ generate a rep

$$[J_o,J_1]=iJ_2,\ \ [J_o,J_2]=-iJ_1,\ \ [J_1,J_2]=-iJ_o \tag{8}$$

of the Lie algebra su(1,1) with Casimir operator $C_2=J_o^{\ 2}-J_1^{\ 2}-J_2^{\ 2}=-3/16$. We get two irreps of the two-fold covering of SU(1,1) belonging to the discrete series D_k^+ with $k=1/4$ and 3/4 ($C_2=k(k-1)$ in an irrep), corresponding to the spectrum Sp $J_o=\{1/4,5/4,\ldots\}$ (spanned by the even number states), and {3/4, 7/4,..} (spanned by the odd number states). We shall also study the generator $K_+=J_o+J_1=(a+a^+)^2/4$. J_o, J_2, and K_+ generate reps of the three different types of subgroups of SU(1,1): elliptic, hyperbolic, and parabolic, respectively. Since two irreps are involved, from general results it follows that the spectrum of J_2 is twice the whole real line, and for K_+ twice the positive half-axis. It remains to give explicit expressions for the corresponding generalized eigenfunctions. As mentioned earlier this will be done in three different forms, which we now proceed to define.

1.3. THREE DIFFERENT REPRESENTATIONS OF OSCILLATOR STATES

The connection between the three different representations introduced earlier, the n-, q- and z-representations, can be made via the ON bases correspondences as follows:

$$\{|n\rangle\} \leftrightarrow \{u_n(q) = N_n H_n(q) \cdot e^{-q^2/2}\} \leftrightarrow \{z^n / \sqrt{n!}\}, \text{ where } N_n = (\sqrt{\pi}\, 2^n n!)^{-1/2} \tag{9}$$

$$|\varphi\rangle = \sum a_n |n\rangle \leftrightarrow \varphi(q) = \sum a_n u_n(q) \leftrightarrow \phi(z) = \sum a_n z^n / \sqrt{n!} = e^{|z|^2/2} \langle z^* | \varphi\rangle \tag{10}$$

where $\langle \alpha | \varphi\rangle$ is the so-called coherent state representation;

$$\|\varphi\|^2 = \sum |a_n|^2 = \int |\varphi(q)|^2\, dq = \pi^{-1} \iint |\phi(z)|^2 e^{-|z|^2} d^2 z. \tag{11}$$

The generators of the representation of our group has the forms

$$\begin{aligned}
&a \leftrightarrow (q + d/dq)/\sqrt{2} \leftrightarrow d/dz, \quad a^+ \leftrightarrow (q - d/dq)/\sqrt{2} \leftrightarrow z;\\
&J_o:\ (a^+a + aa^+)/4 \leftrightarrow (-d^2/dq^2 + q^2)/4 = H_{osc}/2\hbar\omega \leftrightarrow (zd/dz + 1/2)/2;\\
&J_1:\ (a^{+2} + a^2)/4 \leftrightarrow (d^2/dq^2 + q^2)/4 \leftrightarrow (d^2/dz^2 + z^2)/4;\\
&J_2:\ (a^{+2} - a^2)/4i \leftrightarrow i(qd/dq + 1/2)/2 \leftrightarrow i(d^2/dz^2 - z^2)/4;\\
&K_+:\ (a^+ + a)^2/4 \leftrightarrow q^2/2 \leftrightarrow (d^2/dz^2 + 2zd/dz - z^2 + 1)/4.
\end{aligned} \tag{12}$$

We shall need formulas taking us from $\varphi(q)$ to $\phi(z)$ and vice versa (Bargmann and inverse Bargmann transforms). These are easily obtained using the integrals for a_n:

$$a_n = \int u_n(q)\varphi(q)dq = \pi^{-1} \iint e^{-|z|^2} (z^{*n} / \sqrt{n!})\, \phi(z) d^2 z \tag{13}$$

and the generating function for the Hermite polynomials

$$\exp(2yt - t^2) = \sum H_n(y) t^n / n!. \tag{14}$$

The result is

$$\phi(z) = \int K(z,q)\varphi(q)dq; \quad \varphi(q) = \pi^{-1} \iint K(z^*,q)\phi(z) d^2 z, \tag{15}$$

where the kernel

$$K(z,q) = (1/\pi)^{1/4} \exp(-z^2/2 - q^2/2 + \sqrt{2} zq) \tag{16}$$

2. Generalized Eigenfunctions of the Squeezing Generator J_2

We start by deriving the eigenfunctions in the n-representation. Since the eigenfunctions belong to a continuous spectrum, the sequence $\{a_n\}$ we obtain will not be square summable, only power bounded (i e a tempered sequence, dual to rapidly decreasing test sequences). See e g [5] for details about the mathematical background. So we have to solve the eigenvalue equation

$$J_2 |\lambda\rangle = [(a^{+2} - a^2)/4i] |\lambda\rangle = \lambda |\lambda\rangle \tag{17}$$

With $|\lambda\rangle = \sum f_n(\lambda) |n\rangle$ this leads to the recursion relation

$$\sqrt{n(n-1)} f_{n-2}(\lambda) - \sqrt{(n+1)(n+2)} f_{n+2}(\lambda) = i4\lambda f_n(\lambda), \tag{18}$$

evidently connecting only the even or the odd number states, corresponding to a doubling of the spectrum. To get rid of the square root we put ($N(\lambda)$ is a normalization constant)

$$f_n(\lambda) = N(\lambda) \sqrt{\Gamma([n+1]/2] / \Gamma(n/2+1)}\, a_n \tag{19}$$

to get the relation

$$(n+1)a_{n+2} + i4\lambda a_n - na_{n-2} = 0 \tag{20}$$

We first treat the case of even numbers n and put $n=2m$ and $b_m = a_{2m}$, so that

$$(m+1/2)b_{m+1} + i2\lambda b_m - mb_{m-1} = 0 \tag{21}$$

This recursion relation can be solved by the standard Laplace method introducing the expression $b_m = \int t^m y(t)dt$ in the relation, which gives after partial integrations

$$\int t^m[-d(t^2y)/dt + ty/2 + i2\lambda y + dy/dt]dt - \mathrm{I}\, t^m(1-t^2)y(t) = 0 \tag{22}$$

where I.. gives the contributions from the ends of the integration path. Taking this path to be from -1 to 1 (so that the corresponding contribution vanishes) and solving the simple first order differential equation for *y(t)* we get after an integration variable change leading to an integral from 0 to 1, and using the standard integral representation for the hypergeometric function ${}_2F_1$ (which for convenience we denote by F) the result

$$b_m = (-1)^m 2^{-1/2} \frac{\Gamma(1/4 - i\lambda)\Gamma(1/4 + i\lambda)}{\Gamma(1/2)} F(-m, 1/4 - i\lambda, 1/2; 2) \tag{23}$$

One could also have found this result directly by comparing the recursion relation with a suitable contiguity relation for the hypergeometric functions [7].

A transformation formula for transforming $z \to z/(z-1)$ in F gives

$$F(-m, 1/4 - i\lambda, 1/2; 2) = (-1)^m F(-m, 1/4 + i\lambda, 1/2; 2) \tag{24}$$

Together with the relation, valid for real λ ,

$$F(-m, 1/4 - i\lambda, 1/2; 2)^* = F(-m, 1/4 + i\lambda, 1/2; 2) \tag{25}$$

this implies, using the series expansion of F, that F is for even m a real, even polynomial of degree m in λ, whereas for odd m it is i times a real, odd polynomial of degree m.

With a change of definition of the normalization factor we can write

$$f_{2m}(\lambda) = N(\lambda)\sqrt{(1/2)_m / m!}F(-m, 1/4 + i\lambda, 1/2; 2), \quad m = 0, 1, .. \tag{26}$$

using the notation $(a)_m = \Gamma(a+m)/\Gamma(a)$.

Treating the odd n case in a similar way we obtain

$$f_{2m+1}(\lambda) = N'(\lambda)\sqrt{(3/2)_m / m!}F(-m, 3/4 + i\lambda, 3/2; 2), \quad m = 0, 1, .. \tag{27}$$

We now have to prove the (generalized) orthonormality and completeness relations of these generalized eigenfunctions, and in particular determine the normalization constants. Although this can be done using (26) and (27) and suitable integral representations it is simpler to observe that we have obtained special cases of the Pollaczek polynomials [8]

$$P_m(\lambda, b) = \sqrt{(2b)_m / m!}F(-m, b + i\lambda, 2b; 2), \quad m = 0, 1, .. \tag{28}$$

which for every $b > 0$ form a complete orthonormal set of polynomials on the real line with the weight function

$$\rho_b(\lambda) = 2^{2b-1}|\Gamma(b + i\lambda)|^2 / \pi\Gamma(2b). \tag{29}$$

If we choose (the normalizations are evidently determined only up to λ-dependent phase factors)

$$N(\lambda) = \sqrt{\rho_{1/4}(\lambda)}, \quad N'(\lambda) = \sqrt{\rho_{3/4}(\lambda)} \tag{30}$$

and define

$$|\lambda, e\rangle = \sum f_{2m}(\lambda)|2m\rangle, \quad f_{2m}(\lambda) = \sqrt{\rho_{1/4}(\lambda)}P_m(\lambda, 1/4), \tag{31}$$

$$|\lambda,o\rangle = \sum f_{2m+1}(\lambda)|2m+1\rangle, \quad f_{2m+1}(\lambda) = \sqrt{\rho_{3/4}(\lambda)}P_m(\lambda,3/4), \tag{32}$$

then the generalized orthonormality relations would be

$$\langle\lambda,e|\lambda',e\rangle = \langle\lambda,o|\lambda',o\rangle = \delta(\lambda-\lambda'), \quad \langle\lambda,e|\lambda',o\rangle = 0. \tag{33}$$

Whereas the last relation is obvious, the first two just express the completeness of the corresponding set of Pollaczek polynomials, and thus hold true.

In a similar way the orthonormality conditions for the Pollaczek polynomials imply the completeness relation for the generalized eigenfunctions, since this relation says that

$$\langle\psi|\psi\rangle = \int [|\langle\lambda,e|\psi\rangle|^2 + |\langle\lambda,o|\psi\rangle|^2]d\lambda \tag{34}$$

for an arbitrary state $|\psi\rangle = \sum c_n|n\rangle$, which can also be written in the form

$$\sum|c_n|^2 = \sum_{m,n}[c_{2m}{}^*c_{2n}\int f_{2m}(\lambda)^* f_{2n}(\lambda)d\lambda + c_{2m+1}{}^*c_{2n+1}\int f_{2m+1}(\lambda)^* f_{2n+1}(\lambda)d\lambda].. \tag{35}$$

As $P_0(\lambda,b)=1$ it follows from (31) and (32) that

$$\langle 0|\lambda,e\rangle = \sqrt{\rho_{1/4}(\lambda)}, \quad \langle 1|\lambda,o\rangle = \sqrt{\rho_{3/4}(\lambda)} \tag{36}$$

To derive the form of the eigenfunctions in the z-representation one could of course start from the corresponding eigenvalue equation

$$(-d^2/dz^2 + z^2)\phi_\lambda(z) = i4\lambda\phi_\lambda(z). \tag{37}$$

We shall instead use the result we have obtained in the n-representation and apply a standard integral representation for the F function to obtain from (26)

$$\phi_{\lambda e}(z) = \sum f_{2m}z^{2m}/\sqrt{(2m)!} = A(\lambda)\int_0^1 t^{-3/4-i\lambda}(1-t)^{-3/4+i\lambda}\left[\sum\frac{\sqrt{(1/2)_m}}{\sqrt{m!(2m)!}}[z(2t-1)]^m\right]$$

$$= A(\lambda)e^{-z^2/2}\int_0^1 t^{-3/4-i\lambda}(1-t)^{-3/4+i\lambda}e^{z^2t}dt, \quad \text{where } A(\lambda) = \sqrt{\rho_{1/4}(\lambda)}\frac{\Gamma(1/2)}{|\Gamma(1/4+i\lambda)|^2}. \tag{38}$$

Use of a standard integral representation for the confluent hypergeomertric function ${}_1F_1$ results in the final form

$$\phi_{\lambda,e}(z) = \sqrt{\rho_{1/4}(\lambda)}e^{-z^2/2}{}_1F_1(1/4-i\lambda,1/2;z^2). \tag{39}$$

In a similar way we obtain

$$\phi_{\lambda,o}(z) = \sqrt{\rho_{3/4}(\lambda)}e^{-z^2/2}z\,{}_1F_1(3/4-i\lambda,3/2;z^2). \tag{40}$$

As the second parameter of the ${}_1F_1$ function is 1/2 or 3/2, respectively, the functions can be expressed in terms of parabolic cylinder functions $D_\nu(\sqrt{2}\,z)$ and $D_\nu(-\sqrt{2}\,z)$ with $\nu = -1/2 + i2\lambda$ (see [8], p.117).

Finally we shall derive the form of the eigenfunctions in the q-representation. In this case we shall solve the corresponding eigenvalue equation, which can be written

$$qd\varphi_\lambda(q)/dq = -(1/2+i2\lambda)\varphi_\lambda(q). \tag{41}$$

We have a singularity at q=0 and can choose independent solutions on the positive and negative half-axes. These solutions can be put together to give solutions even and odd, respectively, in q, corresponding to even and odd number solutions. We then get

$$\varphi_{\lambda e}(q) = (2\pi)^{-1/2}\frac{1}{|q|^{1/2+i2\lambda}}, \quad \varphi_{\lambda o}(q) = (2\pi)^{-1/2}\frac{\mathrm{sgn}\,q}{|q|^{1/2+i2\lambda}}. \tag{42}$$

The factors $(2\pi)^{-1/2}$ are introduced to ensure the generalized orthonormality relation

$$\int \varphi_{\lambda'}(q)^* \varphi_\lambda(q)dq = (2\pi)^{-1}2\int_0^\infty |q|^{-1+i2(\lambda'-\lambda)}\, dq = (2\pi)^{-1}\int_{-\infty}^{\infty} \exp[i(\lambda'-\lambda)t]dt$$
$$= \delta(\lambda'-\lambda). \tag{43}$$

It is clear that $\varphi_{\lambda e}(q)$ and $\varphi_{\lambda o}(q)$ should correspond to the earlier determined eigenfunctions up to λ–dependent phase factors To determine these we calculate the scalar product with the vacuum and 1-photon state, respectively, and compare with (36):

$$\pi^{-1/4}\int e^{-q^2/2}\varphi_{\lambda e}(q)dq = \pi^{-1/4}\sqrt{2/\pi}\int_0^\infty q^{-1/2-i2\lambda}e^{-q^2/2}dq = (2\pi^3)^{-1/4}\frac{\Gamma(1/4-i\lambda)}{2^{i\lambda}}, \tag{44}$$

$$\pi^{-1/4}\sqrt{2}\int qe^{-q^2/2}\varphi_{\lambda o}(q)dq = 2(2\pi^3)^{-1/4}\frac{\Gamma(3/4-i\lambda)}{2^{i\lambda}}. \tag{45}$$

From this we finally get

$$\langle q|\lambda, e\rangle = \exp\{i[\arg\Gamma(1/4+i\lambda)+\lambda\ln 2]\}(2\pi)^{-1/2}\frac{1}{|q|^{1/2+i2\lambda}}, \tag{46}$$

$$\langle q|\lambda, o\rangle = \exp\{i[\arg\Gamma(3/4+i\lambda)+\lambda\ln 2]\}(2\pi)^{-1/2}\frac{\operatorname{sgn} q}{|q|^{1/2+i2\lambda}}. \tag{47}$$

It should be observed that although the eigenfunctions in the q-representation are ordinary functions (though with a singularity at the origin) they are not square integrable but should be considered as tempered distributions on the real line, similar to the eigenfunctions $e^{i\lambda q}$ of the momentum operator $p = -id/dq$. See e g [5] for more details on the mathematical nature of the generalized eigenfunctions of the su(1,1) generators.

3. Generalized Eigenfunctions of the Generator K_+

We have seen in (12) that the parabolic generator K_+ has the forms $(a^+ + a)^2/4$, $q^2/2$, and $(d^2/dz^2 + 2zd/dz - z^2 + 1)/4$ in the different representations. As in the previous case one could of course start from the eigenvalue equation in the n-representation and solve the corresponding difference equation, which again involves either even or odd number states, corresponding to a doubling of the spectrum. In this case it is simpler to use the form of the operator in the q-representation, where the eigenvalue equation

$$(q^2/2)\varphi_\eta(q) = \eta\varphi_\eta(q) \text{ or } (q^2 - 2\eta)\varphi_\eta(q) = 0 \tag{48}$$

shows that the spectrum of K_+ is the half axis $[0, \infty)$ and that we can choose as properly normalized even and odd solutions

$$\varphi_{\eta e}(q) = (1/2\sqrt{2\eta})[\delta(q-\sqrt{2\eta}) + \delta(q+\sqrt{2\eta})], \tag{49}$$
$$\varphi_{\eta o}(q) = (1/2\sqrt{2\eta})[\delta(q-\sqrt{2\eta}) - \delta(q+\sqrt{2\eta})]. \tag{50}$$

where the normalization is determined (up to a η–dependent phase factor) by demanding

$$\int \varphi_{\eta' e}(q)^* \varphi_{\eta e}(q)dq = \delta(\eta'-\eta) \text{ etc.} \tag{51}$$

Completeness of the set of eigenfunctions follows from the fact that together the two families give a family of δ-functions covering the whole q-axis.

The corresponding eigenfunctions in the n- and z-representations are obtained from (13) and (15) as

$$|\eta,e\rangle = \sum g_{2m}(\eta)|2m\rangle, \quad |\eta,o\rangle = \sum g_{2m+1}(\eta)|2m+1\rangle, \text{ where} \tag{52}$$

$$g_{2m}(\eta) = \int \varphi_{\eta e}(q) u_{2m}(q) dq = N_{2m}(1/\sqrt{2\eta}) e^{-\eta} H_{2m}(\sqrt{2\eta}), \tag{53}$$

$$g_{2m+1}(\eta) = \int \varphi_{\eta o}(q) u_{2m+1}(q) dq = N_{2m+1}(1/\sqrt{2\eta}) e^{-\eta} H_{2m+1}(\sqrt{2\eta}). \tag{54}$$

$$f_{\eta e}(z) = \int K(z,q)\varphi_{\eta e}(q) dq = \pi^{-1/4}(1/\sqrt{2\eta}) e^{-\eta} e^{-z^2/2} \cosh(2\sqrt{\eta} z), \tag{55}$$

$$f_{\eta o}(z) = \int K(z,q)\varphi_{\eta o}(q) dq = \pi^{-1/4}(1/\sqrt{2\eta}) e^{-\eta} e^{-z^2/2} \sinh(2\sqrt{\eta} z). \tag{56}$$

4. Spectrum of the General Second Degree Hamiltonian

A general hermitian second degree polynomial in $\{a,a^+\}$ (or in $\{q,p\}$) can be written

$$\begin{aligned} p_2(a,a^+) &= A(a^+a + aa^+) + Be^{i\Phi}a^2 + Be^{-i\Phi}a^{+2} + Ce^{i\Psi}a + Ce^{-i\Psi}a^+ + DI \\ &= 4[AJ_0 + B(\cos\Phi J_1 + \sin\Phi J_2)] + C\sqrt{2}(\cos\Psi\, q - \sin\Psi\, p) + DI. \end{aligned} \tag{57}$$

Since we have a real linear combination of a set of generators of the group HxM(2,**R**) we can conclude that if we have a Hamiltonian of type (57) with arbitrary time-dependent coefficients *A(t), B(t), C(t),* and *D(t),* the corresponding time development will be given by a unitary operator of the form (4), and hence a squeezed state will always (up to a phase factor) develop into a squeezed state. This clearly also holds for the set of displaced squeezed number states for a given n.

We shall now determine the various possible spectra for the operator (57). It is no essential restriction to assume A, B, and C non-negative, and D=0.

If $A \neq B$ we can transform away the linear terms (in C) and get the quadratic term to a form proportional to J_0 (if $A > B$) or to J_2 (if $B > A$) by a substitution $a \to \mu a + \nu a^+ + \alpha$, which according to (2) can be achieved by a unitary transformation We get the results

$$\text{If } A > B: \quad 4\sqrt{A^2 - B^2}J_0 - [C^2/2(A^2 - B^2)][A\cos(\Phi - \Psi) - B\cos\Psi]; \tag{58}$$

$$\text{If } A < B: \quad 4\sqrt{B^2 - A^2}J_2 - [C^2/2(A^2 - B^2)][A\cos(\Phi - \Psi) - B\cos\Psi]. \tag{59}$$

In both cases we should choose $\alpha = -C[Ae^{-i\Psi} - Be^{i(\Psi-\Phi)}]/[2(A^2 - B^2)]$.

If $A = B$ the phase rotation $a \to e^{-i\Phi/2}a$ results in the form

$$2Aq^2 + C\sqrt{2}\cos(\Psi - \Phi/2)q - C\sqrt{2}\sin(\Psi - \Phi/2)p. \tag{60}$$

Thus if $C = 0$, or $C \neq 0$ and $\Phi = 2\Psi (\bmod 2\pi)$ we get the form

$$2A(q \pm C/2\sqrt{2}A)^2 - C^2/4A \tag{61}$$

Otherwise, i e if $C \neq 0$ and $\Phi \neq 2\Psi (\bmod 2\pi)$ a translation in q leads to the form

$$2A[q^2 - (C/\sqrt{2}A)\sin(\Psi - \Phi/2)p] - (C^2/4A)\cos^2(\Psi - \Phi/2). \tag{62}$$

From the relation

$$\exp(iq^3/3b)(q^2 + bp)\exp(-iq^3/3b) = bp \tag{63}$$

it follows that in this case the operator can be transformed to the form

$$-C\sqrt{2}\sin(\Psi - \Phi/2)p \quad (\text{or } x\, q). \tag{64}$$

From (58), (59), (61), and (64) we can read off the different possible spectra of $p_2(a,a^+)$ in (57). $A = B$ is evidently a critical point. Here we can have as spectrum either a half-axis taken twice or the whole real axis. This situation is evidently not stable: a small perturbation can lead either to a discrete equidistant semi-bounded spectrum (if $A > B$) or the whole real line taken twice (if $A < B$).

5. The Two-Mode Case

In the two-mode case (two-dimensional harmonic oscillator) we get an infinite set of irreducible (true) representations of SU(1,1) by putting

$$J_0 = (a_1^+ a_1 + a_2^+ a_2 + 1)/2, \quad J_1 = (a_1^+ a_2^+ + a_1 a_2)/2, \quad J_2 = (a_1^+ a_2^+ - a_1 a_2)/2i. \tag{65}$$

For the Casimir operator we obtain

$$C_2 = J_0^2 - J_1^2 - J_2^2 = ([\Delta N]^2 - 1)/4, \quad \Delta N = N_1 - N_2 = a_1^+ a_1 - a_2^+ a_2. \tag{66}$$

Since the spectrum of ΔN is the set of all integers, we get for the spectrum of $C_2 = k(k-1)$ as posssible k values 1/2, 1, 3/2,.. i e all the irreducible SU(1,1) reps in the discrete class D_k^+. All except the first one (with $\Delta n = 0$) occur twice. It is evidently enough to study those with $\Delta n \geq 0$. We get as spectrum for J_0 $\{(\Delta n + 1)/2 + n_2; n_2 = 0,1,2,..\}$.
The derivations of the generalized eigenfunctions for J_2 and K_+ can be made in the n-representation (basis $\{|n_1 n_2\rangle\}$ and from there derived in the z- and q-representations (functions $\phi(z_1,z_2)$ and $\varphi(q_1,q_2)$) We shall only give the results for both operators in the first two representations. The integrals leading to the q-representation seem rather intractable in the general case; we only give the result for J_2 in the case $\Delta n = 0$.

The results for the two-mode squeezing generator J_2 ($\Delta n = 0,1,..$) are

$$|\lambda,\Delta n\rangle = \sum^{\infty} f_n(\lambda,\Delta n)|n+\Delta n, n\rangle, \quad f_n(\lambda,\Delta n) = \sqrt{\rho_c(\lambda)}\, P_n(\lambda,c) \tag{67}$$

$$\phi_\lambda(z_1,z_2;\Delta n) = \sqrt{\rho_c(\lambda)/\Gamma(2c)} \times z_1^{\Delta n} \exp(-z_1 z_2)\,{}_1F_1(c \quad i\lambda, 2c; 2z_1 z_2) \tag{68}$$

$P_n(\lambda,c)$ and $\rho_c(\lambda)$ are defined in (28) and (29). Here $c = (\Delta n + 1)/2$.
For $\Delta n < 0$ one should substitute $|\Delta n|$ for Δn in (67) and (68), interchange z_1 and z_2, and change $|n+\Delta n, n\rangle$ to $|n, n+|\Delta n|\rangle$.
The generalized orthonormality and completeness properties can be written

$$\langle\lambda',\Delta n'|\lambda,\Delta n\rangle = \delta_{\Delta n',\Delta n} \cdot \delta(\lambda' - \lambda), \tag{69}$$

$$\sum_{\Delta n} \int |\lambda,\Delta n\rangle\langle\lambda,\Delta n| d\lambda = I, \text{ identity operator.} \tag{70}$$

In the q-representation we only give the result for the case $\Delta n = 0$:

$$\varphi_\lambda(q_1,q_2;0) = \frac{1}{\pi|\Gamma(1/2 + i\lambda)|}\left|\frac{q_1+q_2}{q_1-q_2}\right|^{i\lambda} K_{i\lambda}(|q_1^2 - q_2^2|/2). \tag{71}$$

where K_ν is the modified Bessel function of the third kind.

For the two-mode generator $K_+ = J_0 + J_1$ we get

$$|\eta,\Delta n\rangle = \sum g_n(\eta,\Delta n)|n+\Delta n,n\rangle, \quad g_n = (-1)^n \sqrt{2}e^{-\eta}(2\eta)^{\Delta n/2}\sqrt{\frac{n!}{(n+\Delta n)!}}L_n^{\Delta n}(2\eta), \quad (72)$$

where L_n^{α} is the generalized Laguerre polynomial.

$$\phi_\eta(z_1,z_2:\Delta n) = \sqrt{2}e^{-\eta}(z_1/z_2)^{\Delta n/2}e^{-z_1 z_2}I_{\Delta n}(\sqrt{2\eta z_1 z_2}), \quad (73)$$

where I_ν is the modified Bessel function of the first kind.
Changes for $\Delta n < 0$, orthonormality and completeness relations are analogous to those in the previous case.

6. Dedication

This paper is a contribution to professor Guy Rideau on the occasion of his retirement. Rideau has been an important member of the group of scientists in Paris and Dijon that over the years has inspired my interest in group representation theory. May his retirement be as pleasant as mine!

7. References

1. See e g *J. Modern Optics* **34** (1987), No 6/7. Special issue on squeezed states, esp. article by R. Loudon, and P.L.Knight; Mandel, P. (Ed.), Quantum Optics, Solvay Conference Nov. 1991 Proc., *Physics Reports* **219** (1992), Nos 3-6; Walls, D.F. and Milburn, G.J.: *Quantum Optics*, Chapters 2 and 8, Springer-Verlag, Berlin, 1994.
2. We follow the convention used e.g. in the first and third references in [1]; sometimes the squeezing operator is defined with the opposite signs in the exponent.
3. We use the notation $\Delta(AB) = \langle (A'B' + B'A')/2\rangle$, where $A' = A - \langle A\rangle$ etc.; $\Delta(AA) = (\Delta A)^2$.
4. Lo, C.F.:Eigenfunctions and eigenvalues of squeeze operators, *Phys.Rev.* **A 42** (1990), 6752-6754, and references there.
5. Lindblad, G., and Nagel, B.: Continuous bases for unitary irreducible representations of SU(1,1), *Ann. Inst. Henri Poincaré* XIII (1970), Section A, 27-56.
6. See e g Perelomov, A.: *Generalized Coherent States and Their Applications,* Springer-Verlag, Berlin, 1986; Garcia-Bullé, M., Lassner, W., and Wolf, K.B.: The metaplectic group within the Heisenberg-Weyl ring, *J. Math. Phys.* **27** (1986), 29- 36.
7. Erdelyi, A, (Ed.) *Higher Transcendental Functions,* Vol. 1, p. 103, 2.8. (31), McGraw-Hill , New York, 1953.
8. Erdelyi, A, (Ed.) *Higher Transcendental Functions,* Vol. 2, p. 220, McGraw-Hill , New York, 1953, and references there to the original articles. Our normalization of the Pollaczek polynomials is different from the one given in Erdelyi, but is the correct one resulting in an orthonormal set of polynomials.

A HIGGS-FREE MODEL FOR FUNDAMENTAL INTERACTIONS

MAREK PAWŁOWSKI
Soltan Institute for Nuclear Studies, Warsaw, Poland

AND

RYSZARD RĄCZKA
Soltan Institute for Nuclear Studies, Warsaw, Poland
and
Interdisciplinary Laboratory for Natural and Humanistic Sciences
International School for Advanced Studies (SISSA), Trieste, Italy

Abstract.

A model for strong, electroweak and gravitational interactions based on the local symmetry group $G = SU(3) \times SU(2)_L \times U(1) \times C$ where C is the local conformal symmetry group is proposed. The natural minimal G-invariant form of total lagrangian is postulated. It contains all Standard Model fields and the gravitational interaction, however the Higgs mass term $\mu^2\Phi^\dagger\Phi$ is forbidden. Using the unitary gauge and the conformal scale fixing conditions we can eliminate all four real components of the Higgs field in this model. In spite of that the tree level masses of vector mesons, leptons and quarks are automatically generated and are given by the same formulas as in the conventional Standard Model. In this manner one gets the mass generation without the mechanism of spontaneous symmetry breaking. The gravitational sector of the model is also analyzed and it is shown that the model admits in the classical limit the Einsteinian form of gravitational interactions.

J. Bertrand et al. (eds.), Modern Group Theoretical Methods in Physics, 221–232.

1. Introduction

The recent evidence for top quark production with the top mass estimated as $m_t = 174 \pm 10^{+13}_{-12} GeV$ [1] implies that the Higgs particle – if exists – may have the mass of the order of many hundred of GeV: in fact the central value of m_H read off from the present central value of m_t and electroweak (EW) data is $m_H \approx 300 GeV$ [2],[3]. It should be stressed however that some observables give very high central value for m_H: for instance using the value of m_W as the input information one obtains that $m_H(m_W) \approx 500 GeV$ and taking forward-backward asymmetry of $b\bar{b}$ pair production $A^{0,b}_{FB}$ as the input information one obtains that $m_H(A^{0,b}_{FB}) \approx 750 GeV$ with an enormous error however. Some authors obtained even higher values of Higgs mass [4]. It is noteworthy that before the publication of the work [1] in most of electroweak calculations one assumed $m_H \approx 100 GeV$ whereas in the most recent works one uses in calculations $m_H \approx 300 GeV$ [2],[3],[5],[6]. Since in the lowest order $\lambda = \frac{1}{2}(\frac{m_H}{v})^2$ one can afraid that the Higgs self-coupling λ would be also very large ($\lambda \approx 0.75$ for $m_H = 300 GeV$ and even $\lambda \approx 4.6$ for $m_H = 750 GeV$). Such strong Higgs self-interaction would mean that the loops with Higgs particles would dominate all other contributions. Therefore the perturbative predictions in SM for many quantities would become unreliable. Consequently the predictive power of the Standard Model (SM) and its consistency may be questionable.

The Higgs particle with such a large mass becomes suspicious. It is natural therefore to search for a modification of SM in which all confirmed by experiment particles would exist but the Higgs particle as the observed object would be absent.

We show in this work that such a modification of SM is possible under the condition that one joints to strong and electroweak interactions also the gravitational interaction. This extension of the class of SM interactions is in fact very natural. Indeed whenever we have the strong and electroweak interactions of elementary particles, nuclei, atoms or other objects we have also at the same time the gravitational interactions. It seems natural therefore to consider a unified model for strong, electroweak and gravitational interactions which would describe simultaneously all four fundamental interactions. It is well known that gravitational interactions give a negligible effect to most of strong or electroweak elementary particle processes. We show however that they may play the crucial role in a determination of the physical fields and their masses in the unified model and that their presence allows to eliminate all Higgs fields from the final lagrangian.

In turn we recall that in the conventional Standard Model the Higgs mechanism of spontaneous symmetry breaking (SSB) provides a simple and effective instrument for mass generation of weak gauge bosons, quarks and leptons. However, despite of many efforts of several groups of experimentalists [7] the postulated Higgs particle of the SM was not observed. Hence one might expect that the model for strong and electroweak interactions supplemented by the gravitational interaction in which all dynamical Higgs fields may be eliminated can provide a natural frame–work for a description of elementary particle fundamental interactions.

In order to construct a unique form of the theory of strong and electroweak interactions extended by the gravitational interactions we observe that the gauge symmetry $SU(3) \times SU(2)_L \times U(1)$ of the fundamental interactions may be naturally extended by the local conformal symmetry. The choice of the unitary gauge condition for $SU(2)_L$ gauge group allows to eliminate the three out of four real Higgs fields from the complex Higgs doublet. In turn the choice of the scale fixing condition connected with the local conformal symmetry allows to eliminate the last Higgs field. In that manner all four Higgs fields can be gauged away completely! It is remarkable that in spite of the elimination of all Higgs fields in our model the vector meson, lepton and quark masses are generated and at the tree level they are given by the same analytical formulas as in the conventional SM.

Thus it may be that the dynamical real Higgs field and the associated Higgs particles are in fact absent and it is therefore not surprising that they could not be detected in various experiments [7].

The Higgs sector and the Higgs mechanism of mass generation looked suspicious to many physicists since the beginning of its introduction. In fact Kuminasa and Goto already in 1967 have proposed a Higgs-free model of gauge field theory for massive vector mesons interacting with fermions [8]. Next the Higgs-free models for electroweak interactions were considered from various points of view [9]-[12]. It should be also noted that Flato and Guenin have proposed a natural alternative mechanism of mass generation in gauge field theories entirely different from the Higgs mechanism [13].

It seems to us that our Higgs-free model based on the extension of electroweak and strong interactions by gravitational interactions, which leads to the extension of gauge symmetry by the local conformal symmetry, presents a most natural frame–work for a description of fundamental interactions.

The present work is the extension of our two previous papers and contains the answer to several questions raised by theirs readers [14], [15]. The detailed presentation of our Higgs-Free Model (HFM) including the proofs of some results presented in this work is given in [15] and [16]. The predictions obtained from the proposed HFM for leptonic and hadronic electroweak observables are presented in our work [17].

2. A Higgs-Free Model for Strong, Electroweak and Gravitational Interactions

Let $M^{3,1}$ be the pseudo–Riemannian space time with the metric $g_{\alpha\beta}$ with the signature $(+,-,-,-)$. Let $\Omega(x)$ be a strictly positive function on $M^{3,1}$ which has the inverse $\Omega^{-1}(x)$. Then the local conformal transformation in $M^{3,1}$ is defined as the transformation which changes the metric by the formula

$$g_{\mu\nu}(x) \rightarrow \tilde{g}_{\mu\nu}(x) = \Omega^2(x) g_{\mu\nu}(x). \tag{2.1}$$

The set of all local conformal transformations forms the multiplicative abelian infinite–dimensional group C with the obvious group multiplication law.

It should be stressed that a conformal transformation is not a diffeomorphism of space time. The physical meaning of the conformal transformations follows from the transformation law of the length element

$$dl(x) = \sqrt{-g_{ij}dx^i dx^j} \qquad \rightarrow \qquad d\tilde{l}(x) = \Omega(x) dl(x). \tag{2.2}$$

Hence a local conformal transformation changes locally the length scale.

The conformal group C has the representation in the field space given by the formula

$$\tilde{\Psi}(x) = U(\Omega)\Psi(x) = \Omega^s(x)\Psi(x), \qquad s \in R \tag{2.3}$$

The number s is determined by the condition of conformal invariance of a field equation for Ψ. One finds for instance that the Maxwell field $F_{\mu\nu}$ and the Yang-Mills field $F^a_{\mu\nu}$ has the conformal weight $s = 0$ whereas the scalar massless field Φ has the conformal weight $s = -1$ [18], [19].

We postulate that the searched unified theory of strong, electroweak and gravitational interactions will be determined by the condition of invariance with respect to the group G

$$G = SU(3) \times SU(2)_L \times U(1) \times C \tag{2.4}$$

where C is the local conformal group defined by (2.1). Let Ψ be the collection of vector meson, fermion and scalar fields which appear in the conventional minimal SM for electroweak and strong interactions. Then the minimal natural conformal and $SU(3) \times SU(2)_L \times U(1)$ -gauge invariant total lagrangian $L(\Psi)$ may be postulated in the form:

$$L = [L_G + L_F + L_Y + L_\Phi + L_{grav}]\sqrt{-g} \tag{2.5}$$

Here L_G is the total lagrangian for the gauge fields A^a_μ, W^b_μ and B_μ, $a = 1, ..., 8$, $b = 1, 2, 3$ associated with $SU(3) \times SU(2)_L \times U(1)$ gauge group

$$L_G = -\frac{1}{4} F^a{}_{\mu\nu} F^{a\mu\nu} - \frac{1}{4} W^b{}_{\mu\nu} W^{b\mu\nu} - \frac{1}{4} B_{\mu\nu} B^{\mu\nu}, \tag{2.6}$$

and $F^a{}_{\mu\nu}$, $W^b{}_{\mu\nu}$ and $B_{\mu\nu}$ are the conventional field strengths of gauge fields; L_F is the lagrangian for fermion field interacting with the gauge fields; L_Y represents the Yukawa interactions of fermion and scalar fields; L_Φ is the G-invariant lagrangian for the scalar fields, which may be written in the form:

$$L_\Phi = (D\Phi)^\dagger (D\Phi) - \lambda(\Phi^\dagger\Phi)^2 + \beta\partial_\mu|\Phi|\partial^\mu|\Phi| - \frac{1}{6}(1+\beta)R\Phi^\dagger\Phi, \tag{2.7}$$

where D denotes the covariant derivative with connections of all symmetry groups. Notice that the condition of conformal invariance does not admit the Higgs mass term $\mu^2\Phi^\dagger\Phi$ which assures the mechanism of spontaneous symmetry breaking and mass generation in the conventional formulation. Instead we have two additional terms: the Penrose term [20]

$$-\frac{1}{6}(1+\beta)R\Phi^\dagger\Phi \tag{2.8}$$

which assures that the lagrangian (2.7) is conformal invariant, and the term

$$\beta\partial_\mu|\Phi|\partial^\mu|\Phi| \tag{2.9}$$

which together with the term$-\frac{1}{6}\beta R\Phi^\dagger\Phi$ is conformal and gauge invariant.

The last term in (2.5) is the Weyl term

$$L_{grav} = -\rho C^2, \qquad \rho > 0, \tag{2.10}$$

where $C^\delta_{\alpha\beta\gamma}$ is the Weyl tensor which is conformally invariant.

We see that the condition of conformal invariance does not admit in (2.5) the conventional gravitational Einstein lagrangian

$$L = \kappa^{-2}R\sqrt{-g}, \qquad \kappa^2 = 16\pi G. \tag{2.11}$$

We show however in Section 4. that Einstein Lagrangian is reproduced from (2.5) when the scale fixing condition is chosen properly.

The theory given by (2.5) is our conformally invariant proposition alternative to the standard Higgs–like theory with SSB. Its new, most important feature is the local conformal invariance. It means that simultaneous rescaling of all fields (including the field of metric tensor) with a common, arbitrary, space–time dependent factor $\Omega(x)$ taken with a proper power for each field (the conformal weight) will leave the Lagrangian (2.5) unaffected. The symmetry has a clear and obvious physical meaning [21], [18]. It changes in every point of the space–time all dimensional quantities (lengths, masses, energy levels, etc) leaving theirs ratios unchanged. It reflexes the deep truth of the nature that nothing except the numbers has an independent physical meaning.

The freedom of choice of the length scale is nothing but the scale fixing freedom connected with the conformal symmetry group. In the conventional approach we define the length scale in such a way that elementary particle masses are the same for all times and in all places. This will be the case when we rescale all fields with the x–dependent conformal factor $\Omega(x)$ in such a manner that the length of the rescaled scalar field doublet is fixed i.e.

$$\tilde{\Phi}^{\dagger}\tilde{\Phi} = \frac{v^2}{2} = const. \tag{2.12}$$

Obviously we can choose other scale. We have proven however in [15] the following theorem:

Theorem.

Let $C(\Psi)$ be the conformal invariant function of field operators. Then the vacuum expectation value $< C(\Psi) >_0$ is independent on the scale fixing condition.

The above theorem implies that we can calculate the vacuum expectation values of conformal invariant function of field operators using the most convenient scale fixing condition. Since the condition (2.12) together with the unitary gauge fixing condition for $SU(2)_L$ group eliminates all four Higgs fields from the action integral $S_T(\Psi)$ we shall use it exclusively in all following calculations. We note that the scattering operator $\hat{S}$ is dimensionless and therefore conformal invariant. Consequently if we use the normalization of asymptotic states such that they are also dimensionless we can use the scale fixing condition (2.12) for calculation of probability amplitudes of all physical processes.

3. Generation of lepton, quark and vector boson masses

We demonstrate now that using the conformal group scale fixing condition (2.12) we can generate the same lepton, quark and vector meson masses as in the conventional SM without however use of any kind of Higgs mechanism and SSB.

In fact inserting the scale fixing condition (2.12) into the Lagrangian (2.5) we obtain

$$\tilde{L} = L^{scaled} = [L_G + L_F + L_Y^{scaled} + L_{\Phi}^{scaled} + L_{grav}]\sqrt{-g}, \tag{3.1}$$

in which the condition (2.12) was inserted into L_Φ and L_Y. We should use the symbol $\tilde{\Phi}$, $\tilde{\Psi}$ etc. for the rescaled fields in (3.1), however for the sake of simplicity we shall omit "~" sign over fields in the following considerations.

The condition (2.12) together with the unitary gauge fixing of $SU(2)_L\times U(1)$ gauge group, reduce by (2.12) the Higgs doublet to the form

$$\Phi^{scaled} = \frac{1}{\sqrt{2}}\begin{pmatrix}0\\ v\end{pmatrix}, \qquad v > 0 \tag{3.2}$$

and produce the tree level mass terms for leptons, quarks and vector bosons associated with $SU(2)_L$ gauge group. For instance the Φ–lepton Yukawa interaction L_Y^l reads

$$L_Y^l = -\sum_{i=e,\mu,\tau} G_i \bar{l}_{iR}(\Phi^\dagger l_{iL}) + h.c.$$

where

$$l_{eL} = \begin{pmatrix}\nu_e\\ e_L\end{pmatrix} \quad etc.$$

It passes into

$$L_Y^{l\ scaled} = -\frac{1}{\sqrt{2}}v(G_e\bar{e}e + G_\mu\bar{\mu}\mu + G_\tau\bar{\tau}\tau) \tag{3.3}$$

giving the conventional, space–time independent lepton masses

$$m_e = \frac{1}{\sqrt{2}}G_e v, \qquad m_\mu = \frac{1}{\sqrt{2}}G_\mu v, \qquad m_\tau = \frac{1}{\sqrt{2}}G_\tau v. \tag{3.4}$$

Similarly one generates from Φ–quark Yukawa interaction L_Y^q the corresponding quark masses. In turn from L_Φ-lagrangian (2.7) using the scaled scalar field (3.2) one obtains

$$(D_\mu\Phi)^\dagger D^\mu\Phi = \frac{g_2^2v^2}{4}W_\mu^+W^{\mu-} + \frac{g_1^2+g_2^2}{8}v^2Z^2$$

where

$$Z_\mu = -\sin\theta_W B_\mu + \cos\theta_W {W^3}_\mu, \qquad \cos\theta_W = \frac{g_2}{\sqrt{g_1^2+g_2^2}}.$$

Hence one obtains the following vector mesons masses

$$m_W = \frac{v}{2}g_2, \qquad m_Z = \frac{m_W}{\cos\theta_W}. \tag{3.5}$$

It is remarkable that the analytical form for tree level fermion and vector meson masses in terms of coupling constants and the parameter v is the same as in the conventional SM. We see therefore that the Higgs mechanism and SSB is not indispensable for the fermion and vector mesons mass generation!

We note that the fermion–vector boson interactions in our model are the same as in SM. Hence analogously as in the case of conventional formulation of SM one can deduce the tree level relation between v and G_F – the four–fermion coupling constant of β–decay:

$$v^2 = (2G_F)^{-1} \to v = 246 GeV. \tag{3.6}$$

Here we have used the standard decomposition $g^{\mu\nu}\sqrt{-g} = \eta^{\mu\nu} + \sqrt{2}\kappa h^{\mu\nu}$ (see e.g. [22]) which reduces the tree level problem for the matter fields to the ordinary flat case task.

We see therefore that the resulting expressions for masses of physical particles are identical as in the conventional SM.

Let us stress that the scale fixing condition like (2.12) does not break $SU(2)_L \times U(1)$ gauge symmetry. The symmetry is broken (or rather one of gauge equivalent description is fixed) when (2.12) is combined with unitary gauge condition of electroweak group leading to (3.2). However, also after imposing of a gauge condition like (3.2) we have a remnant of both the conformal and $SU(3) \times SU(2)_L \times U(1)$ initial gauge symmetries: this is reflected in the special, unique relations between couplings and masses in our model

4. Gravity Sector

Let us impose the scale fixing condition (2.12) on the lagrangian (2.5) and collect all gravitational terms. The lagrangian reads:

$$L^{scaled} = [L^{scaled}_{matter} - \frac{1}{12}(1+\beta)v^2 R - 2\rho(R^{\mu\nu}R_{\mu\nu} -$$

$$\frac{1}{3}R^2) - \frac{\lambda}{4}v^4]\sqrt{-g} \tag{4.1}$$

where we have selected the part L^{scaled}_{matter} (describing the matter interacting with gravity) from the remaining purely gravitational terms.

The variation of (4.1) with respect to the metric $g^{\mu\nu}$ leads to the following classical equation of motion:

$$\rho[-\frac{2}{3}R_{;\mu;\nu} + 2R_{\mu\nu}{}^{;\eta}{}_{;\eta} - \frac{2}{3}g_{\mu\nu}R^{;\eta}{}_{;\eta} -$$

$$4R^{\eta\lambda}R_{\mu\eta\nu\lambda} + \frac{4}{3}RR_{\mu\nu} + g_{\mu\nu}(R^{\eta\lambda}R_{\eta\lambda} - \frac{1}{3}R^2)]+$$

$$\frac{1}{12}(1+\beta)v^2(R_{\mu\nu} - \frac{1}{2}g_{\mu\nu}R) + \frac{\lambda}{8}v^4 g_{\mu\nu} = \frac{1}{2}T_{\mu\nu}. \tag{4.2}$$

In the empty case $T_{\mu\nu} = 0$ this equation is satisfied by all solutions of an empty space Einstein equation with a properly chosen cosmological constant Λ:

$$R_{\mu\nu} - \frac{1}{2}g_{\mu\nu}R + \Lambda g_{\mu\nu} = 0. \tag{4.3}$$

In fact (4.3) implies that

$$R_{\mu\nu} \sim g_{\mu\nu} \qquad \Rightarrow \qquad R_{\mu\nu} = \frac{1}{4}Rg_{\mu\nu} \tag{4.4}$$

and then

$$R_{\mu\nu} = \Lambda g_{\mu\nu}. \tag{4.5}$$

Inserting (4.4) into (4.2) we find that the part proportional to ρ vanishes. The remnant can be collected leading to the relation

$$\frac{1}{8}v^2 g_{\mu\nu}(\frac{2}{3}(1+\beta)\Lambda - \lambda v^2) = 0 \tag{4.6}$$

where the empty space condition $T_{\mu\nu} = 0$ were used for the right hand side of (4.6).

Equation (4.6) implies

$$\Lambda = \frac{3}{2(1+\beta)}\lambda v^2. \tag{4.7}$$

Equation (4.7) relates the undetermined so far coupling constant λ with a potentially observable cosmological constant Λ.

Let us go back to the case with the matter. Observe that the term linear in the curvature appears in (4.1) with the coefficient $-\frac{1}{12}(1+\beta)v^2$. If we want to reproduce the correct gravitational sector already at the classical level we have to admit for nonzero β coupling. This would lead us to a model which is equivalent to the nonrenormalizable gauged nonlinear sigma model in the material sector. Accepting this price we can put

$$-\frac{1}{12}(1+\beta)v^2 = \kappa^{-2} \tag{4.8}$$

reproducing the Newtonian coupling in front of curvature R in (4.1). This would mean that $\beta \approx -10^{38}$! Notice however that taking the scale fixing condition (2.12) the term $\beta\partial_\mu|\Phi|\partial^\mu|\Phi|$ vanishes. Hence it looks like that the only role of this term is to generate the proper value of Newton constant in the Einstein–Hilbert tree level lagrangian resulting from the Penrose term. (For further discussion see [14].)

The cosmological constant Λ given by (4.7) was obtained from the analysis of gravitational interactions in the empty space-time. In reality the matter is always present and modifies the formula for Λ. In this case the most natural definition of effective cosmological constant was given by Zel'dovich [23] and by Adler [24] by means of the partition function determined by the lagrangian (3.1). (See also the excellent analysis of this problem in [25].)

5. Discussion

The elementary particle physics is at present at a crossroad. We have in fact three drastically different alternatives:

I^o The Higgs particle exists, its mass will be experimentally determined and will have the value predicted by the radiative corrections of SM. This will confirm the SSB mechanism for mass generation, the validity of SM framework and it will represent an extraordinary success of quantum gauge field theory.

II° The Higgs particle exists but its mass is considerable different from that predicted by the radiative corrections of SM. This would signal some kind of "New Physics" which will imply a reformulation of the present version of SM.

III° The Higgs particle does not exists. This will lead to a rejection of SM with Higgs sector and it will give preference to Higgs-free models for fundamental interactions. Presumably the obtained physical Higgs-free models will be nonrenormalizable [26].

It may be that the renormalizability of Quantum Gravity determined by Einstein–Hilbert action integral coupled with matter fields is not an "accident at work in quantum field theory" but it represents a universal feature that physical fundamental interactions considered simultaneously are nonrenormalizable. In this situation we are compulsed to use the nonrenormalizable models of quantum field theory for a description of fundamental interactions and we have to learn how to deduce predictions for experiments from such models. We have shown in [17] how to deduce the predictions for observables in our nonrenormalizable model. We have demonstrated that our predictions are in the surprising agreement with the experimental data. In fact the direct calculations of electroweak universal parameters ε_1, ε_2 and ε_3 introduced by Altarelli *et. al.* [27] demonstrate that the Standard Model and the present model results differ effectively by the term proportional to $\log \frac{\Lambda^2}{m_H^2}$: thus it looks like that the very high Higgs mass m_H plays in SM the role of the UV cutoff which in the present model may be replaced by parameter Λ. Hence the predictive power of our model may be comparable with that of the conventional SM.

In view of the possibility that nonrenormalizable nonabelian massive gauge field theories have to be used for a description of fundamental interactions it seems necessary to develop perturbative and nonperturbative methods for extracting predictions for scattering amplitudes and observables from such models [26]. In particular one should develop the corresponding Generalized Equivalence Theorems and determine explicitly the high energy behavior of cross sections in such models. The comparison of the obtained results with analytic formulas coming from Lipatov calculations [30] would be very inspiring. It would be also useful to develop systematic two–loop calculus with UV cutoff Λ for electroweak processes. We plan in a near future to present several examples of such calculations.

The present model allows to obtain the Einsteinian form of gravitational interactions in the classical limit. It can be also analyzed by means of effective action for induced gravity [25].

ACKNOWLEDGMENTS

The authors are grateful to Prof. Iwo Białynicki Birula, Prof. M.F. Flato, Dr. B. Grządkowski, Prof. Z. Haba, Dr. M. Kalinowski and Prof. J. Werle for interesting discussions and Dr. S.D. Odintsov for sending to us the results of his group. They are especially grateful to Dr. S. Dittmaier for sending them his computer code and to Prof. D. Schildknecht for the extensive discussion of properties of his model.

References

1. CDF Collaboration, FERMILAB-PUB-94/116-E.
2. P. Langacker, Tests of the Standard Model and searches for new physics, *Precision tests of the Standard Electroweak Model,* ed. P. Langacker (World Scientific, Singapore, 1994); J. Erler and P. Langacker, Implications of high precision experiments and the CDF top quark candidates; preprint UPR-0632T, *Bull. Board: hep-ph 9403303.*
3. Review of Particle Data, Part I, Phys. Rev. **D50** (1994) No. 3, §26.
4. Z. Hioki and R. Najima, Is the Standard Electroweak Theory Happy with $m_t \sim 174GeV$?, preprint TOCUSHIMA 94-02, YCCP-9404; Z. Hioki, Talk presented at INS Workshop "*Physics of e^+e^-, $e^-\gamma$ and $\gamma\gamma$ collisions at linear accelerators*", Inst. for Nucl. Study (INS), Univ. of Tokyo, Japan, December 20-22, 1994.
5. G. Montagna *et al.*, The Top Quark and the Higgs Boson Mass from LEP SLC and CDF Data. *Bull. Board: hep-ph 9411203.*
6. A. Blondel, Precision electroweak physics at LEP; preprint CERN-PPE/94-133
7. For recent review see e.g ICHEP94, Glasgow.
8. T. Kuminasa and T. Goto, Progr. Theor. Phys. **37** (1967) 452.
9. P.Q. Hung and J.J. Sakurai, Nucl. Phys. **B143** (1978) 81; J.D. Bjorken, Phys. Rev. **D19** (1979) 335; A.C. Longitano, Nucl. Phys **B188** (1981) 118; T. Sonoda and S.Y. Tsai, Prog. Theor. Phys. **71** (1984) 878.

10. S. Dittmaier, C. Grosse-Knetter and D. Schildknecht, On the Role of the Higgs Mechanism in Present Electroweak Precision Tests *Bull. Board: hep-th 9406378* (to appear in Zeitschrift für Physik C).
11. D. Caerupeel and M. Leblanc, Mass Generation for Gauge Fields without Scalars *Bull. Board: hep-ph 9407029.*
12. i) R. Mańka and J. Syska, Boson Condensation in the GSW Electroweak Theory, preprint University of Silesian 1993 (to be published in Phys. Rev. D), ii) G. Pocsik, E. Lendvai and G. Cynolter, Acta Phys. Pol. **B24** (1993) 1495; see also G. Cynolter, E. Lendvai and G. Pocsik, *ρ-parameter in the vector condensate model of electroweak interactions*, preprint Inst. of Theor. Physics Eotvos Lorad Univ. Budapest.
13. M.Flato and M. Guenin, Czechoslovak J. of Physics **B32** (1982) 490; *erratum* p. 1056.
14. M. Pawłowski and R. Rączka, Found. of Phys. **24** (1994) 1305.
15. M. Pawłowski and R. Rączka, A Higgs-Free model for Fundamental Interactions and its Predictions, preprint ILAS/EP-1-1995, *Bull. Board: hep-ph 9501370.*
16. M. Pawłowski and R. Rączka, A Higgs-Free model for Fundamental Interactions. Part I: Theoretical analysis, preprint ILAS/EP-2-1995.
17. M. Pawłowski and R. Rączka, A Higgs-Free model for Fundamental Interactions. Part II: Predictions for electroweak LEP observables, preprint ILAS/EP-3-1995.
18. R.M. Wald, *General Relativity* (The University of Chicago Press,Chicago 1984).
19. M.A. Castagnino and J.B. Sztrajman, J. Mat. Phys. **27** (1986) 1037.
20. R. Penrose, Ann. of Phys. **10** (1960) 171.
21. J.V. Narlikar, *Introduction to Cosmology* (Jones and Bartlet Pub. 1983).
22. D.M. Capper, G. Leibbrandt and M.R. Medrano, Phys. Rev. **D8** (1973) 4320.
23. Ya.B. Zel'dovich, Zh. Exp. Teor. Fiz. Pis'ma Red. **6** (1967) 883 [JETP Lett. **6** (1967) 316].
24. S.L. Adler, Rev. Mod. Phys. **54** (1982) 729.
25. S.D. Buchbinder, S.D. Odintsov and I.L. Shapiro, *Effective Action in Quantum Gravity* (IOP, Bristol, 1992).
26. A. Burnel, Phys. Rev. **D33** (1986) 2981; *ibid.* **D33** (1986) 2985; R. Delbourgo, S. Twisk and G. Thompson, Int. Jour. Mod. Phys. *A3* (1988) 435.
27. G. Altarelli, R. Barbieri and F. Caravaglios, Nucl. Phys. **B405** (1993) 3.
28. See e.g. M.J. Herrero and E.R. Morales, Nucl. Phys. *B418* (1994) 341; M.J. Herrero and E.R. Morales, Non-decoupling effects of the SM Higgs boson to one loop, preprint FTUAM 94/11, *Bull. Board: hep-ph 9411207* and references contained therein.
29. C. Grosse–Knetter, The Equivalence Theorem for Heavy-Higgs Standard Model and the Gauged Nonlinear σ–Model *Bull. Board: hep-ph 9405412.*
30. See e.g. L.N. Lipatov, JETF **90** (1986) 1536; R. Kirschner, L.N. Lipatov, L. Szymanowski, Nucl. Phys. B (Proc. Suppl.) **29A** (1992) p. 84; L.N. Lipatov, ibid p. 89; V.S.Fadin and L.N. Lipatov, ibid p. 93 and references contained therein.

COUPLED HODGKIN HUXLEY NEURONS WITH STOCHASTIC SYNAPTIC INPUTS

A Guy Rideau

ROGER RODRIGUEZ
Centre de Physique Théorique
CNRS-Luminy, case 907,F13288 Marseille CEDEX 9,France.

Abstract. The fluctuations of synaptic transmissions in neuronal systems are described, at the scale of refraction time, as inhomogeneous compound jump processes. Active properties of membrane potentials are also introduced and analytical estimates of useful probabilistic quantities can be deduced for biologically plausible dimensional reductions of Hodgkin Huxley type neuron models.

1. Introduction

The biologically plausible modeling of neurons populations dynamics with emphasis on some analytical control still remains an important task. A realistic approach has to take into account the stochastic character of behaviour as it clearly appears in all experimental results. Actually, the use of probabilistic methods in Neurosciences has a rather long history. Tuckwell [Tu] gives an interesting review on the subject with some focus on development of one cell theory in fluctuating environment.

A second aspect which has to be considered for the modeling of realistic neurons is the highly non linear dynamics of individual cells. Since the seminal work of Hodgkin and Huxley [H.H] on an experimentally confirmed mathematical model of spiking activity of some neural cells, there has been developed many models with less or more possible analytical control ([K.K], [M.L], [R.E], [Tu]). In the following (section [2.1]), we propose a formulation of the synaptic interactions between cells with some emphasis on their

J. Bertrand et al. (eds.), Modern Group Theoretical Methods in Physics, 233–242.

random nature. This fact being analyzed since many years [K.M] at an experimental level, there has been many works taking into account such behaviour ([Bl], [St], [Ka], [In], [Pe], [Am], [S.V], [L.S]).

Our model is built on an inhomogeneous compound Poisson process and may give, as in [Bl] and [In] , some biological insight on the nature of temperature parameter which is usually introduced in the statistical dynamics of the network (see e.g [Li]). Moreover, we introduce in it the active properties of membranes, at least along some low dimensional phase space reduction method. In section [2.2], we give analytical formulas for firing probabilities of cells whose dynamics is of Hodgkin Huxley type and synaptic inputs from the network are stochastic. In this way, one has a generalization of usual, discrete time, discrete states, theory.

2. The Hodgkin Huxley type neural network.

2.1. INPUTS AS INHOMOGENEOUS COMPOUND POISSON PROCESSES

In this section, we introduce and describe our model of synaptic transmission which extends previous work ([Bl]). We shall call N the number of cells in the network and we shall consider chemical synapses $\{j \to i\}$ between neurons j and i, which act mainly as current sources modulated by neurotransmitters activity on postsynaptic receptors. For a $\{j \to i\}$ synapse, the number of activated vesicles is taken as an inhomogeneous Poisson process $N_{ij}(t)$. This latter is such that: $N_{ij}(0) = 0$ and $N_{ij}(t)$ has independent increments. Moreover, for some $\lambda_{ij}(t)$, a positive real function of time, one has :

$$\mathrm{Prob}\Big(N_{ij}(t+h) - N_{ij}(t) = 1 \mid N_{ij}(t) = k\Big) = \lambda_{ij}(t)h + o(h) \; k = 1, 2, \dots \tag{2.1}$$

$$\mathrm{Prob}\Big(N_{ij}(t+h) - N_{ij}(t) \geq 2 \mid N_{ij}(t) = k\Big) = o(h) \quad k = 1, 2, \dots \tag{2.2}$$

This process may be viewed as an homogeneous one by the well known change of time scale $\tau_{ij} = \int_0^t \lambda_{ij}(s)ds = \Lambda_{ij}(t)$. One has $E(N_{ij}(t)) = \Lambda_{ij}(t)$. At this stage, we define $\lambda_{ij}(t)$ as a factor depending on presynaptic activity $a_j(t)$ and on values of synaptic efficacies $J_{ij}(t)$ of the synapse. This positive latter variable, which takes its full meaning from synaptic plasticity mechanism, has a dynamical behaviour with a time scale much bigger than the one occurring in (2.1) (2.2).

The activity $a_j(t)$ is postulated in a rather natural way as $a_j(t) = \sigma(V_j(t - \delta_{ij}))$ where σ is some sigmoïdal function with value in [0,1], $V_j(t)$ is the presynaptic membrane potential, δ_{ij} is some delay for transmission in the $\{j \to i\}$ connection. Naturally, evaluation of these delays is generally a very complex task and we shall not take up this question. Even if they can be introduced here as phenomenological parameters, they are not essential for our purpose and are neglected.

Finally, $\lambda_{ij}(t)$ are taken of the form $\lambda_{ij}(t) = J_{ij}(t)a_j(t)$. Now, to each activated vesicle is associated a (random) number of released transmitters in the synaptic $\{j \to i\}$ cleft. We postulate the produced synaptic current $I_i(t)$ resulting from all synapses on a given cell i to be the (formal) derivative $I_i(t) = \frac{d\xi_i(t)}{dt}$ of the following inhomogeneous compound Poisson process

$$\xi_i(t) = \sum_{\alpha,\{j\to i\}_\alpha} \sum_{k=1}^{N_{ij}^\alpha(t)} \zeta_k^\alpha \tag{2.3}$$

where $\{\zeta_k^\alpha\}$, $k = 1, 2, \ldots$, $\alpha = \mp$ are i.i.d random variables, with some density probability law $f_\alpha(\zeta)$. According to the excitatory $(+)$ or inhibitory $(-)$ character of the synaptic junction, the support of $f_+(\zeta)$ (resp. $f_-(\zeta)$) is mainly contained in R^+ (resp. R^-) in such a way that $\bar{\zeta}_+ > 0$ (resp. $\bar{\zeta}_- < 0$). The first sum in (2.3) has the obvious meaning of a sum on the two kinds of synapses with a dependence of the counting processes $N_{ij}^\alpha(t)$ on α. Thus we introduce two kinds of synaptic efficacy $J_{ij}^\alpha(t), \alpha = +$ or $-$ and $\Lambda_{ij}^\alpha(t) = \int_0^t J_{ij}^\alpha(s)a_j(s)ds$ is the parameter of $N_{ij}^\alpha(t)$. These latter processes are all supposed independent between them and independent of the $\{\zeta_k^\alpha\}_k$. This model can be easily extended to many synaptic buttons for a given connection.

As a random sum of random variables, the mean of $\xi_i(t)$ is easily found to be

$$E(\xi_i(t)) = \sum_{\alpha,\{j\to i\}_\alpha} \bar{\zeta}_\alpha \Lambda_{ij}^\alpha(t) \tag{2.4}$$

The time derivative of this term has the usual meaning of a weighted sum of inputs where the functions $a_j(t)$ are mainly concentrated on the time occurrences $(t_j^k)_k$ of spikes emitted by the presynaptic neuron j. As in [Bl], it is useful to look for the process $\xi_i(t)$ as a martingale, that is an Ito integral ([G.S]) with respect to some random Poisson measure. By the time scale change $t \to \tau_{ij}$, the processes in (2.3) take the form

$$\xi_i(t) = \sum_{\alpha,\{j\to i\}_\alpha} \sum_{k=1}^{\widetilde{N}_{ij}^\alpha(t)} \zeta_k^\alpha \tag{2.5}$$

where $\widetilde{N}_{ij}^\alpha(t)$ are homogeneous Poisson processes. Let us define the Poisson random measure $\nu_{ij}^\alpha(d\zeta, ds)$ associated to $\widetilde{N}_{ij}^\alpha(t)$, then

$$\xi_i(t) = \sum_{\alpha,\{j\to i\}_\alpha} \int_0^{\Lambda_{ij}^\alpha(t)} \int_R \zeta \nu_{ij}^\alpha(d\zeta, ds) \tag{2.6}$$

The mean of $\nu_{ij}^{\alpha}(d\zeta, ds)$ is $E(\nu_{ij}^{\alpha}(d\zeta, ds)) = ds f_{\alpha}(\zeta) d\zeta$. With the centered random measure $\tilde{\nu}_{ij}^{\alpha}$ associated to ν_{ij}^{α}, $\xi_i(t)$ takes a form well adapted for characterization of fluctuations

$$\xi_i(t) = \sum_{\alpha,\{j \to i\}_\alpha} \zeta_\alpha^- \Lambda_{ij}^{\alpha}(t) + \sum_{\alpha,\{j \to i\}_\alpha} \int_0^{\Lambda_{ij}^{\alpha}(t)} \int_R \zeta \tilde{\nu}_{ij}^{\alpha}(d\zeta, ds) \qquad (2.7)$$

One has , with $\zeta_\alpha^{-2} = \int_R \zeta^2 f_\alpha(\zeta) d\zeta$, $\quad \mathrm{var}(\xi_i(t)) = \sum_{\alpha,\{j \to i\}_\alpha} \Lambda_{ij}^{\alpha}(t) \zeta_\alpha^{-2}$.

It may be interesting to look for so called diffusion approximations (see e.g [Tu]) of processes like $\xi_i(t)$. Such processes described in terms of Brownian motion are built in such a way that first and second differential moments of the original process and the approximated one coincide. This method may be useful for technical reasons related to the structure of Kolmogorov equation which is simpler to solve in the diffusive case. In our case, one has

$$\xi_i(t) \approx X_i(t) = \sum_{\alpha,\{j \to i\}_\alpha} \zeta_\alpha^- \Lambda_{ij}^{\alpha}(t) + \int_0^t \left(\sum_{\alpha,\{j \to i\}_\alpha} \zeta_\alpha^{-2} \lambda_{ij}^{\alpha}(s) \right)^{1/2} dW_{i,s} \qquad (2.8)$$

2.2. HODGKIN HUXLEY TYPE MODELS OF MEMBRANE EXCITABILITY

We shall now look for the behaviour of membrane potential of neurons models under synaptic inputs of the previously described form. We shall focus on [H.H] type systems which have shown, since forty years, their ability to reproduce correctly the most important properties of nervous cells.

Here, we are interested in modeling approaches which can lead to analytical results. Thus, we shall concentrate on some low dimensional models corresponding to reduction in the number of dynamical variables representing the activity of the numerous channels types present in the cell. These models ([K.K], [M.L]) may exhibit most interesting properties like spike and spike train generation with and without latency, bistability, bursting and others.

Generally, in an [H.H] system, the ionic currents $(I_j)_{j=1,2,\ldots}$ through various channels types are viewed as ohmic ones of the form $I_j = \overline{g}_j \gamma_j(V, z_1, z_2, \ldots, z_M)(V - V_j)$ where $\overline{g}_j$ is the maximum conductance with all type j open channels, γ_j is the proportion of open channels and V_j is the associated Nernst inversion potential. $(z_i(t))_{i=1,2,\ldots M}$ are so called time and potential dependent (in)activation variables with prescribed kinetics. These latter will depend on some equilibrium distribution $\overline{z}_i(V)$ and time constant $k_i^{-1}(V)$ which are generally non linear functions of V. These kinetics are of the form $\frac{dz_i}{dt} = k_i(V)(\overline{z}_i(V) - z_i) i = 1, 2, \ldots M$.

Many models have been considered since the work of Hodgkin and Huxley ([H.H]) with different kinds of currents, different (in)activation kinetics. Technical difficulties related to nonlinearities associated to high number of variables have involved the development of low dimensional approximations.

In the following, we shall take advantage of the reduction approach of Morris and Lecar ([M.L] see also [R.E]) .We shall look for their two dimensional dynamical system describing neural dynamics under the stochastic influence of synaptic inputs described in the previous section. In the model, two kinds of channels are present. Voltage gated Ca^{++} channels and voltage gated delayed rectifier K^+ channels. The calcium current plays here the role of Na current in the original [H.H] system. The dynamical system with external input $I_{\text{ext}}(t)$, in presence of noise $\xi(t)$, as considered in section [2.2], has the following form :

$$(\Sigma)\qquad \begin{aligned} CV_t &= CV_0 + \int_0^t f(V_s, X_s)ds + \int_0^t I_{\text{ext}}(s)ds + \xi(t) \\ \frac{dX}{dt} &= k_X(V)(\overline{X}(V) - X) \end{aligned}$$

where

$$f(V, X) = -\overline{g}_{Ca}\overline{m}(V)(V - V_{Ca}) - \overline{g}_K X(V - V_K) - \overline{g}_L(V - V_L) \tag{2.9}$$

The functions $\overline{m}(V)$ and $\overline{X}(V)$ are of the form

$$\overline{m}(V) = \frac{1}{2}\left(1 + \text{th}\left(\frac{V - V_1}{V_2}\right)\right) \tag{2.10}$$

$$\overline{X}(V) = \frac{1}{2}\left(1 + \cosh\left(\frac{V - V_3}{V_4}\right)\right) \tag{2.11}$$

In these equations, X has the meaning of fraction of K^+ open channels, $\overline{m}(V)$ acts as an instantaneous activation of Ca^{++} channels, $\overline{g}_{Ca}$, $\overline{g}_K$ and $\overline{g}_L$ are constant with conductance dimension, V_{Ca} and V_K are rest potentials for the two different kinds of ions, C is the membrane capacity and $(V_i)_{i=1,\dots,4}$ are constants.

Without the noise term, this system has proved to be a very interesting simple model for biologically plausible dynamical behaviour. In particular, it is able to reproduce quite correctly all or none signal emission, emergence of oscillations associated to Hopf bifurcations leading to spike trains, bistability when a stable steady state and a stable oscillation coexist, burst generation with slowly varying injected current.

There exists up to now other possible dimensional reduction well adapted for recovering of experimental evidence ([Ke]). Nevertheless, for our purpose, the nice structure of Morris and Lecar model may be helpful for a probabilistic study when noisy input is taken on.

2.3. FIRING PROBABILITY ESTIMATES.

One of the possible ways to get informations from the non linear stochastic process (Σ) is to look for the dynamics in the vicinity of well identified singular points of the unnoisy system. Actually, there are at most three such singular points for that latter system, for all possible realistic values of the parameters $(V_i)_{i=1...4}$, V_{Ca}, V_K, $\overline{g}_{Ca}$, $\overline{g}_K$, $\overline{g}_L$ and C. Let us now introduce the synaptic input current (2.8) for the membrane potential of neuron i. Let us take the external source $I_{\text{ext}}(t) = 0$ for simplicity. One has, with function f_i in (2.9) depending on cell i:

$$CV_{i,t} = CV_{i,0} + \int_0^t f_i(V_{i,s}, X_{i,s})ds + \sum_{\alpha,\{j\to i\}_\alpha} \zeta_\alpha^- \int_0^t \lambda_{ij}^\alpha(s)ds \quad (2.12)$$

$$+ \int_0^t \left(\sum_{\alpha,\{j\to i\}_\alpha} \zeta_\alpha^{-2} \lambda_{ij}^\alpha(s) \right)^{1/2} dW_{i,s}$$

$$\frac{dX_i}{dt} = k_X(V_i)\left(\overline{X}(V_i) - X_i\right) \quad (2.13)$$

The third and last term in (2.12) introduce a time dependent perturbation of the unnoisy system and fluctuations with time dependent diffusive character. If one wants to evaluate probabilistic estimates for crossing threshold of the random variable $V_{i,t}$, it may be uneasy to control the action of these time dependent terms. However, by looking at a sufficiently small time scale where $\lambda_{ij}^\alpha(s)$ may be considered as nearly constant quantities, one can get such estimates. This may correspond to the characterization of membrane potentials just after the occurrence of spikes from presynaptic neurons in a synchronized activity at this time scale.

Thus, let us call t_0^i the time occurrence of a presynaptic spike on neuron i. The term $\lambda_{ij}^\alpha(s)$ in (2.8), will be well approximated by

$$\lambda_{ij}^\alpha(s) = J_{ij}^\alpha(t_0^i)S_j(t_0^i) = \widetilde{\lambda}_{ij}^\alpha(t_0^i) \quad \text{for } s \text{ in } \left[t_0^i, t_0^i + \tau_r\right] \quad (2.14)$$

where τ_r is of the order of the refractory period, $S_j(t_0^i) \cong 1$ if neuron j has emitted a spike at the time t_0^i and $S_j(t_0^i) \cong 0$ if it is not the case (that is the role of the sigmoïdal function in the previous defined activity $a_j(t)$).

On the time interval $[t_0^i, t_0^i + \tau_r]$, the stochastic system (Σ) is of the form, for $\tau \leq \tau_r$,

$$dV_i = a_i(V_i, X_i)d\tau + \sigma_i dW_{i,\tau} \tag{2.15}$$

$$dX_i = k_X(V_i)\Big(\overline{X}(V_i) - X_i\Big)d\tau \quad i = 1, 2, \; N \tag{2.16}$$

with obvious definitions of the drift terms a_i and coefficients σ_i from (2.12). Now, for each i, let us denote as (V_i^S, X_i^S) the components of a singular point S for the system (Σ) without random term. Namely, (V_i^S, X_i^S) verifies the equations

$$f_i(V_i^S, X_i^S) + \sum_{\alpha,\{j\to i\}_\alpha} \zeta_\alpha^- J_{ij}^\alpha(t_0^i) S_j(t_0^i) = 0 \tag{2.17}$$

$$X_i^S = \overline{X}_i(V_i^S) \tag{2.18}$$

The structure of the isoclines for that system is of N shaded and parabolic type (see [R.E]) in the two dimensional phase space (V_i, X_i) for each i and there exists at most 3 singular points. The tangent mapping A_i around such a point S has the following components

$$A_{11} = -\overline{g}_{Ca}\left(\frac{d\overline{m}}{dV}(V^S, X^S)(V^S - V_{Ca}) + \overline{m}(V^S, X^S)\right) - \overline{g}_K X^S - \overline{g}_L \tag{2.19}$$

$$A_{12} = -\overline{g}_K(V^S - V_K) \tag{2.20}$$

$$A_{21} = \frac{dk_X}{dV}(V^S, X^S)\Big(\overline{X}(V^S) - X^S\Big) + k_X(V^S)\frac{d\overline{X}}{dV}(V^S, X^S) \tag{2.21}$$

$$A_{22} = -k_X(V^S) \qquad \text{(the index } i \text{ has been omitted)} \tag{2.22}$$

Thus, letting $v_i = (V_i', X_i') = (V_i - V_i^S, X_i - X_i^S)$, $(\lambda_i^j)_{j=1,2}$ the eigenvalues of A_i and U_i the transformation which diagonalizes A_i (which can be characterized), one has in the basis of eigenvectors of A_i the following system of stochastic equations describing the dynamics around S

$$\chi_{i,\tau} = \chi_{i,0} + \int_0^\tau L_i \chi_{i,\tau} ds + K_i z_{i,\tau} \quad i = 1, 2, \ldots N \tag{2.23}$$

where $\chi_i = U_i v_i$, L_i is the diagonal matrix $(L_i)_{j,k=1,2} = \lambda_i^j \delta_{j,k}$, K_i is the vector with components U_i^{11} and U_i^{12} and $z_{i,\tau} = \sigma_i W_{i,\tau}$ where $W_{i,\tau}$ is the standard Brownian motion. Thus equation (2.23) describes two independent Ornstein Uhlenbeck processes with drifts λ_i^1 and λ_i^2 and diffusion coefficients

$D_i^{1m} = (U_i^{1m})^2(\sigma_i)^2, m = 1,2 \quad i = 1,2,\dots N$. The well known transition probability density function for them is given, for each i, by

$$\rho(\chi_i^1, \chi_i^2, \tau \mid \chi_{i0}^1, \chi_{i0}^2, 0) = \tag{2.24}$$

$$\prod_{m=1,2} \left\{ \sqrt{\frac{\lambda_i^m}{\pi \mu_i^m(\tau)}} \exp \left\{ \frac{-\lambda_i^m \left(\chi_i^m - \chi_{i0}^m \exp(\lambda_i^m \tau)\right)^2}{\mu_i^m(\tau)} \right\} \right\}$$

$$\text{with } \mu_i^m(\tau) = D_i^{1m}\left(-1 + \exp(2\lambda_i^m \tau)\right).$$

Now, the behaviour of the neuron i under stochastic synaptic inputs will crucially depend on the nature of its rest state S as it has been obtained in (2.17) (2.18). Let us examine two possible simple cases: the first will correspond to S as a unique stable rest state. The second generic one will be associated to the existence of the three singular points with one as a saddle point as it has been shown in [R.E]. In that case, there exists a separatrix with distinct behaviours: spike emission or not according to the position of initial conditions with respect to the separatrix.

In the first case, the two eigenvalues $(\lambda_i^m)_{m=1,2}$ are negative and the firing probability may be the probability that $\chi_i^1 > \theta_i^1$ and $\chi_i^2 > \theta_i^2$ at the time τ_r for some thresholds θ_i^1 for the potential and θ_i^2 for the activation variable X_i. This can be simply deduced from (2.24) but it is not truly useful without precise knowledge of θ_i^1 and θ_i^2.

The second case is more interesting since the firing question is related to the crossing of the separatrix. Thus, we calculate the probability for that crossing (see also [L.N]). If λ_i^1 denotes the eigenvalue of the tangent mapping for the unstable manifold around the saddle point S, the process will cross the separatrix if the first component $\chi_{i\tau}^1$ of $\chi_{i\tau}$ is positive for $\tau \approx \tau_r$. Let us recall that we are looking for the system in the time interval $[t_0^i, t_0^i + \tau_r]$ where at time t_0^i, $\left(S_j(t_0^i)\right)_j$ are known. Thus $\lambda_i^1 = \lambda_i^1(t_0^i)$, $D_i^{11} = D_i^{11}(t_0^i)$ and the firing probability is

$$P_{t_0^i} \text{ firing } = \text{Prob}\left(\chi_i^1(\tau_r) > 0 \mid \chi_i^1(0) = \chi_{i0}^1\right) \tag{2.25}$$

$$= \frac{1}{2} \left\{ \text{erf} \left(\sqrt{\frac{\lambda_i^1(t_0^i)}{D_i^{11}(t_0^i)}} G\left(2\lambda_i^1(t_0^i)\tau_r\right) \chi_{i0}^1 \right) + 1 \right\}$$

where $G(u) = \left(1 - \exp(-u)\right)^{-1}$ for $u > 0$ and erf is the error function. This expression may be related to the one obtained in [L.N] with different model of noise, without the G term.The above derivation was done with the implicit assumption that individual cells have rather well defined behaviour, namely they can send one spike or not. Interactions of other cells are, in some sense, weak enough to not fundamentally modify these regimes but act as random perturbations with random spike emission as a result.

However, we know that these cells have much richer individual dynamics like spike train emission or bursting activity. In a forthcoming paper, we develop the formulation of the role of stochastic interactions when non perturbed cells behave as oscillators and when the coupling is sufficiently weak to enable the use of perturbation theory on the oscillator cycles along a random phase approximation.

References

[Tu] Tuckwell H.C., *Introduction to Theoretical Neurobiology*, Cambridge University Press, Cambridge (1988) ; *Stochastic processes in the Neurosciences*, S.I.A.M. Philadelphia, Pennsylvania (1989).

[H.H] Hodgkin A.L., Huxley A.F., J. Physiol. (London) **117** (1952) 500 .

[K.K] Krinskii V.I., Kokoz Y.M., Analysis of equations of excitable membranes, *Biofizika*, **18**, (1973), p. 506.

[M.L] Morris C., Lecar H., Voltage oscillations in the barnacle giant muscle fiber, *Biophy. Journ.*, **35**, (1981), p. 195.

[R.E] Rinzel J., Ermontrout G.B., Analysis of neural excitability and oscillations, in *Methods of neural neuronal modeling. From synapses to networks.*, Koch C., Seguev I., Eds M.I.T Press Cambridge Mass.

[K.M] Katz B., Miledi R., A study of synaptic transmission in the absence of nerve impulses, *J. Physiol.*, **192** (1967), p. 407.

[Bl] Blanchard P., Combe P., Nencka H., Rodriguez R., Stochastic dynamical aspect of neuronal activity, *Jour. of Math. Biology*, **31**, (1993), p. 189.

[St] Stein R.B., A theoretical analysis of neuronal variability,*Biophys. Journ.*, **5**, (1965), p. 173 ; Some models of neuronal variability, *Biophys. Journ.*, **7**, (1967), p. 37.

[Ka] Kallianpur G., Wolpert R.L., Weak convergence of stochastic neuronal models, In *Stochastic methods in Biology*, Proceedings Nagoya Japan (1985), Kimura M., Kallianpur G., Hida T. (Eds.), *Lecture Notes in Biomathematics*, **70**, Springer Verlag (1987).

[In] Ingber L., Stastistical mechanics of neocortical interactions, *Physica*, **5D**, (1982), p. 83.

[Pe] Peretto P., *An introduction to the Modeling of Neural Networks*, Cambridge University Press (1992).

[Am] Amit D.J., *Modeling brain function*, Cambridge University Press (1989).

[S.V] Shaw G.L., Vasudevan R., Persistent states of neural networks and the random nature of synaptic transmission, *Math. Biosciences*, **21**, (1974), p. 207.

[L.S] Little W.A., Shaw G.L., Analytic study of the memory storage capacity of a neural network, *Math. Biosciences*, **39**, (1978), p. 281.

[Li] Little W.A., The existence of persistent states in the brain, *Math. Biosciences*, **19**, (1974), p. 101.

[G.S] Gihman I.I., Skorohod A.V., *Stochastic differential equations*, Springer Verlag (1972).

[L.N] Lecar H., Nossal R., Theory of threshold fluctuations in nerves I, II, *Biophysical Journal*, **11**, (1971), p. 1048.

[Ke] Kepler T.B, Abbott L.F., Marder E., Reduction of conductance-based neuron models, *Biol. Cyber.*, **66**, (1992), p. 381.

INITIAL DATA FOR NON-LINEAR EVOLUTION EQUATIONS AND DIFFERENTIABLE VECTORS OF GROUP REPRESENTATIONS

JACQUES C.H. SIMON
Département de Mathématiques,
Université de Bourgogne,
B.P. 138, 21004 Dijon Cedex, France

AND

ERIK TAFLIN
Direction Scientifique,
Union des Assurances de Paris,
9, Place Vendôme, 75052 Paris Cedex 01, France

A Guy Rideau, en hommage amical

Abstract. Regularity properties of non-linear Lie algebra representations are defined. These properties are satisfied in examples given by evolution equations. We prove that this regularity implies that the set of C^ω vectors for the non-linear group representation obtained by integration of the Lie algebra representation coincide with the set of C^∞ vectors of the linear part (the order one term) of this group representation.

1. Introduction

A relativistic covariant wave equation defines a non-linear representation (for this notion see [1]) T of the Poincaré Lie algebra $\mathfrak{p} = \mathbb{R}^4 \oplus \mathfrak{sl}(2,\mathbb{C})$ on a Fréchet space E^0 of initial data; $T_X, X \in \mathfrak{p}$, is a C^∞ mapping from a neighbourhood of zero in E^0 to E^0. We define $T^1_X(u) = (D_0 T_X)(u), u \in E^0, D_0$ being the differential at 0. T^1 is a linear representation of $\mathfrak{p}$ on E^0 which is integrable to a group representation U^1 of the Poincaré group $\mathcal{P}_0 = \mathbb{R}^4 \otimes SL(2,\mathbb{C})$ on a Banach space $E(E^0 \subset E)$. We denote by E_∞ the Fréchet space of C^∞ vectors of U^1; $E^0 \subset E_\infty$.

J. Bertrand et al. (eds.), Modern Group Theoretical Methods in Physics, 243–253.

We proved for the Maxwell-Dirac equations [2], [3] and the relativistic non-linear Klein-Gordon equation [6] that the representation T is integrable to a non-linear representation (for this notion see [1]) U of the Poincaré group on E_∞ which is *global* in a sense which we precise now:

There exists a neighbourhood $\mathcal{O}$ of zero in E such that, if $\mathcal{O}_\infty = \mathcal{O} \cap E_\infty$, the map $(g, u) \mapsto U_g(u)$ maps $\mathcal{P}_0 \times \mathcal{O}_\infty$ to $\mathcal{O}_\infty$ and, if $u \in \mathcal{O}_\infty$, $t \mapsto U_{\exp tX}(u), X \in \mathfrak{p}$ is C^∞ and $\frac{d}{dt} U_{\exp tX}(u) \mid_{t=0} = T_X(u)$ if $u \in \mathcal{O} \cap E^0$. We keep the same notation $T_X(u)$ for the left-hand side of the previous equality when $u \in \mathcal{O}_\infty$.

An important point in the existence proof of the invariant open set $\mathcal{O}$ is the fact that the set of $u \in \mathcal{O}'$ (an open neighbourhood of 0 in E) such that $g \mapsto U_g(u)$ is C^∞ around the identity in $\mathcal{P}_0$ coincides with $\mathcal{O}' \cap E_\infty$.

The purpose of this note is to specify a large enough class of non-linear Lie algebra representations (for general finite-dimensional Lie algebras) called x-regular, $x \in]0, 1]$, for which the above property is satisfied.

2. The mathematical framework

Let G be a connected real Lie group. We denote by $\mathfrak{g}$ its Lie algebra, $\dim \mathfrak{g} = d$. Let U^1 be a linear continuous representation of G in a Banach space E. We denote by T^1 the differential of U^1 which is a linear representation of $\mathfrak{g}$ on the Fréchet space E_∞ of C^∞ vectors to U^1.

For $i \in \mathbb{N} \cup \{\infty\}$, let E_i be the space of C^i vectors of the representation U^1, i.e. the subset of all $u \in E$ such that $g \mapsto U^1_g u$ is C^i from G to E.

Let Π be a basis of $\mathfrak{g}$ and suppose given an ordering on Π. Let Π' be the corresponding basis of the universal enveloping algebra $U(\mathfrak{g})$ of $\mathfrak{g}$, as defined by the Poincaré-Birkhoff-Witt Theorem. If $Y \in \Pi'$, let $|Y|$ be the degree of Y in the natural graduation of $U(\mathfrak{g})$. If $i \in \mathbb{N}, E_i$ coincides with the Banach space obtained by the completion of E_∞ with respect to the norm

$$\|u\|_{E_i} = \Big(\sum_{\substack{Y \in \Pi' \\ |Y| \leq i}} \|T^1_Y u\|^2_E \Big)^{1/2}, \tag{2.1}$$

where $T^1_Y, Y \in U(\mathfrak{g})$ is the canonical extension of T^1 to $U(\mathfrak{g})$. We have $E_\infty \subset E_j \subset E_i \subset E_0 = E$ for $i \leq j$.

Given two Banach spaces X and Y, and U(resp.V) an open subset of X(resp.Y) we denote by $\mathcal{H}(U, V)$ the space of analytic functions from U to V. Then $f \in \mathcal{H}(U, V)$ if and only if given $x \in U$, for any $y \in U$, we have

$$f(y) - f(x) = \sum_{n \geq 1} F_n(y - x), \tag{2.2}$$

where $F_n \in \mathcal{L}_n(X, Y)$, the Banach space of n-linear symmetric continuous maps from X to Y, the power series being absolutely convergent in the ball $\|y - x\|_X < r(x),\ r(x) > 0$.

The nth derivative of a C^n function $f : U \longrightarrow V$ at the point $x \in U$ in the directions $x_1, \ldots, x_n \in E$ will be denoted by $(D^n f)(x; x_1, \ldots, x_n)$. The bracket $[A, B]$ of C^1 functions is defined by

$$[A, B](x) = (DA)(x; B(x)) - (DB)(x; A(x)) \tag{2.3}$$

We suppose given a continuous linear representation U^1 of G in a Banach space E.

DEFINITION 1. *Let $\mathcal{O}$ be an open neighbourhood of zero in E and $\mathcal{O}_i = E_i \cap \mathcal{O}$ for $i \in \mathbb{N} \cup \{\infty\}$. A regular analytic representation T of $\mathfrak{g}$ on $\mathcal{O}_\infty$ is a linear map $X \mapsto T_X$ form $\mathfrak{g}$ to $\mathcal{H}(\mathcal{O}_1, E_0)$ such that*

i) $T_X(0) = 0, X \in \mathfrak{g}$, and $T^1_X(u) = \frac{d}{dt} U^1_{\exp tX}(u) \mid_{t=0}, u \in \mathcal{O}_1$;

ii) $[T_X, T_Y] = T_{[X,Y]}, X, Y \in \mathfrak{g}$;

iii) $T_X \in \mathcal{H}(\mathcal{O}_{n+1}, E_n), X \in \mathfrak{g}, n \in \mathbb{N}$.

We shall now introduce different spaces of analytic maps from $\mathcal{O}_\infty$ to E_∞ which will be of interest in the following. If $-s \in \mathbb{N}$, we define

$$\mathcal{A}_s(\mathcal{O}_\infty, E_\infty) = \bigcap_{i \geq 0} \mathcal{H}(\mathcal{O}_{i-s}, E_i). \tag{2.4}$$

We note that $\mathcal{A}_{s_1}(\mathcal{O}_\infty, E_\infty) \subset \mathcal{A}_{s_2}(\mathcal{O}_\infty, E_\infty)$ if $s_1 \geq s_2$.

Now, if $n, i, -s \in \mathbb{N}$, and if B is a closed bounded subset of $\mathcal{O}_{i-s}$ (for the Banach space topology of E_{i-s}), we define on $\mathcal{A}_s(\mathcal{O}_\infty, E_\infty)$ the seminorm

$$|||f|||_{n,i,s,B} = \sup_{u \in B \cap E_\infty, \|v_k\|_{E_{i-s}} = 1} \Big((1 + \|u\|_{E_{i-s}})^{-1} \times$$

$$\times \| (D^n f)(u, v_1, \ldots, v_n) \|_{E_i} \Big). \tag{2.5a}$$

If $-s \in]0, +\infty[- \mathbb{N}$, let $-l \in \mathbb{N}$ be such that $l - 1 < s < l$. If in addition $n, i \in \mathbb{N}$ and B is a closed bounded subset of $\mathcal{O}_{i-l}$, we define, for $f \in \mathcal{A}_{[s]}(\mathcal{O}_\infty, E_\infty)$, where $[s] \in \mathbb{Z}$ is such that $[s] \leq s < [s] + 1$,

$$|||f|||_{n,i,s,B} = \sup_{\substack{u \in B \cap E_\infty \\ v_k \in E_\infty, \|v_k\|_{E_{i-l}} = 1}} (\lambda(u; v_1, \ldots, v_n))^{-1} \times$$

$$\times \|(D^n f)(u; v_1, \ldots, v_n)\|_{E_i}, \tag{2.5b}$$

where

$$\lambda(u; v_1, \ldots, v_n) = 1 \quad + \|u\|_{E_{i-l}}^{1-l+s} \quad \|u\|_{E_{i-l+1}}^{l-s} \\ + \|v_1\|_{E_{i-l+1}}^{l-s} + \ldots + \|v_n\|_{E_{i-l+1}}^{l-s}.$$

Let $\mathcal{A}_s(\mathcal{O}_\infty, E_\infty)$ be the subset of $f \in \mathcal{A}_{[s]}(\mathcal{O}_\infty, E_\infty)$ such that $|||f|||_{n,i,s,B} < +\infty$. It follows directly from the definition that

$$\mathcal{A}_{s_1}(\mathcal{O}_\infty, E_\infty) \subset \mathcal{A}_{s_2}(\mathcal{O}_\infty, E_\infty) \text{ for } s_2 \leq s_1 \leq 0. \tag{2.6}$$

We shall say that $\mathcal{A}_s(\mathcal{O}_\infty, E_\infty)$ is the space of s-smooth maps forms $\mathcal{O}_\infty$ to E_∞. Since $s \leq 0$ this means that we loose $|s|$ derivatives in the scale of Banach spaces $(E_i)_{i \geq 0}$.

DEFINITION 2. *Let $r \in [0,1]$, let T be a regular analytic representation of $\mathfrak{g}$ on $\mathcal{O}_\infty$, and let $T_X = T_X^1 + \widetilde{T}_X$. We say that T is r-regular if we have $\widetilde{T}_X \in \mathcal{A}_{r-1}(\mathcal{O}_\infty, E_\infty), X \in \mathfrak{g}$.*

EXAMPLE 1: The non linear representation of the Poincaré Lie algebra associated with the non-linear Klein Gordon equation with a non-linearity which is polynomial, in a space-time of dimension $n+1, n \geq 2$, is 1 regular [6].

EXAMPLE 2: The non linear representation of the Poincaré Lie algebra associated to the Maxwell-Dirac equations in $E^\rho, \frac{1}{2} < \rho < 1$,(see [2] or [3] for the definition) is $\frac{3}{2} - \rho$ regular.

Let T be regular as in definition 1. We temporarily suppose in addition that there exists a non-linear representation U of the group G in $E_\infty, g \mapsto U_g(u), u \in \mathcal{O}_\infty$, which is C^∞ on a neighbourhood of the identity in G and such that $\frac{d}{dt} U_{\exp tX}(u) \mid_{t=0} = T_X(u), u \in \mathcal{O}_\infty (\mathcal{O}_\infty$ is as in definition 1).
We define, if $\mathbb{I}$ is the identity in $U(\mathfrak{g}), T_{\mathbb{I}} u = u, u \in E_\infty$, and if $Y = X_1^{\alpha 1}, \ldots, X_d^{\alpha d} \in \Pi'$,

$$T_Y(u) = \frac{\partial^{\alpha_1 + \ldots + \alpha_d}}{\partial s_1^{\alpha_1} \ldots \partial s_d^{\alpha_d}} U_{\exp(s_1 X_1)} \cdots U_{\exp(s_d X_d)}(u) \mid_{s_1 = \ldots = s_d = 0} \tag{2.7}$$

if $u \in \mathcal{O}_\infty$. We then define $T_Y u$ for $Y \in U(\mathfrak{g})$ by linearity if $u \in \mathcal{O}_\infty$. One can verify that this definition of $T_Y, Y \in U(\mathfrak{g})$ coincides with the one introduced in [3] and [6]. Namely $T_{\mathbb{I}}(u) = u$ and

$$T_{YX}(u) = (DT_Y)(u; T_X u), \quad Y \in U(\mathfrak{g}), \quad X \in \mathfrak{g}. \tag{2.8}$$

This is the non linear analogue of the extension of T^1 to a linear representation of $U(\mathfrak{g})$.

THEOREM 1. *If T is a regular analytic representation of $\mathfrak{g}$ and $Y \in \Pi'$ then $T_Y \in \mathcal{H}(\mathcal{O}_{n+|Y|}, E_n)$ for $n \in \mathbb{N}$. Moreover, if T is r-regular, $r \in [0,1]$, and $|Y| \geq 1$, then $\widetilde{T}_Y \in \mathcal{A}_{r-|Y|}(\mathcal{O}_\infty, E_\infty)$.*

PROOF: The proof of the first statement is based on the fact that if F and G are Banach spaces, if U(resp. V) is an open subset of F(resp. G) then $f \in \mathcal{H}(U,V)$ implies that $df \in \mathcal{H}(U,\mathcal{L}(F,G))$, where $\mathcal{L}(F,G)$ is the Banach space of continuous linear maps from F to G. The statement follows then from definition 1 and by induction in $|Y|$ for $Y \in \Pi'$, using (2.8).

The second statement for $r = 0$ follows from the first statement since $T_Y^1 \in \mathcal{L}(E_{n+|Y|}, E_n)$. If $r = 1$, then $\widetilde{T}_X \in \mathcal{H}(\mathcal{O}_n, E_n)$ for $n \in \mathbb{N}$ by definition 2 and by (2.4) with $s = 0$. Since $\widetilde{T}_{YX} = T_Y^1 \widetilde{T}_X + D\widetilde{T}_Y.T_X, Y \in \Pi', X \in \Pi$, it then follows from the first statement that $\widetilde{T}_{YX} \in \mathcal{H}(\mathcal{O}_{n+|Y|}, E_n)$ for $|Y| \geq 1$, if $\widetilde{T}_Y \in \mathcal{H}(\mathcal{O}_{n+|Y|-1}, E_n)$. The result follows by induction. Let $r \in]0,1[$, then $T_X \in \mathcal{A}_{-1}(\mathcal{O}_\infty, E_\infty)$ according to (2.6). Therefore $X \mapsto T_X$ is 0-regular, so $\widetilde{T}_Y \in \mathcal{A}_{-|Y|-1}(\mathcal{O}_\infty, E_\infty)$ for $Y \in \Pi', X \in \mathfrak{g}$. Suppose that $\widetilde{T}_Y \in \mathcal{A}_{r-|Y|}(\mathcal{O}_\infty, E_\infty)$, which is true for $|Y| = 1$. According to (2.5b) it follows that

$$\begin{aligned} &\|(D^n\widetilde{T}_X)(u; v_1, \ldots, v_n)\|_{E_{i+|Y|}} \leq |||\widetilde{T}_X|||_{n,i+|Y|,r-1,B_{i+|Y|}} \times \\ &\times (1 + \|u\|^r_{E_{i+|Y|}} \|u\|^{1-r}_{E_{i+|Y|+1}} + \textstyle\sum_{k=1}^n \|v_k\|^{1-r}_{E_{i+|Y|+1}}), \end{aligned} \tag{2.9}$$

when $u \in B_{i+|Y|} \cap E_{i+|Y|+1}, v_k \in E_{i+|Y|+1}, \|v_k\| = 1$ for $1 \leq k \leq n, where i \in \mathbb{N}, Y \in \Pi'$, and $B_{i+|Y|}$ is a closed bounded subset of $\mathcal{O}_{i+|Y|}$. Since $\widetilde{T}_Y \in \mathcal{A}_{-|Y|}(\mathcal{O}_\infty, E_\infty)$ it follows from (2.5a) that

$$\begin{aligned} &\|(D^n\widetilde{T}_Y)(u; v_1, \ldots, v_n)\|_{E_i} \leq |||\widetilde{T}_Y|||_{n,i,-|Y|,B_{i+|Y|}} \times \\ &\times (1 + \|u\|_{E_{i+|Y|}}) \Pi_{k=1}^n \|v_k\|_{E_{i+|Y|}}, \end{aligned} \tag{2.10}$$

when $u \in B_{i+|Y|}, v_k \in E_{i+|Y|}$, for $1 \leq k \leq n, i \in \mathbb{N}$, and $B_{i+|Y|}$ is a closed bounded subset of $\mathcal{O}_{i+|Y|}$. If $n_1 + n_2 = n, n_1, n_2 \in \mathbb{N}$, it follows from inequalities (2.9) and (2.10) that

$$\begin{aligned} &\|(D^{1+n_1}\widetilde{T}_Y)(u; (D^{n_2}\widetilde{T}_X)(u; v_1, \ldots, v_{n_2}), v_{n_2+1}, \ldots, v_n)\|_{E_i} \\ &\leq |||\widetilde{T}_Y|||_{n_1+1,i,-|Y|,B_{i+|Y|}} (1 + \|u\|_{E_{i+|Y|}}) |||\widetilde{T}_X|||_{n_2,i+|Y|,r-1,B_{i+|Y|}} \times \\ &\times (1 + \|u\|^r_{E_{i+|Y|}} \|u\|^{1-r}_{E_{i+|Y|+1}} + \textstyle\sum_{k=1}^{n_2} \|v_k\|^{1-r}_{E_{i+|Y|+1}}), \end{aligned} \tag{2.11}$$

when $u \in B_{i+|Y|} \cap E_{i+|Y|+1}, v_k \in E_{i+|Y|+1}, \|v_k\|_{E_{i+|Y|}} = 1, i \in \mathbb{N}$.

Leibniz rule and inequality (2.11) show $D\widetilde{T}_Y.\widetilde{T}_X \in \mathcal{A}_{r-|Y|-1}(\mathcal{O}_\infty, E_\infty)$. The fact that $T_Y^1 \widetilde{T}_X \in \mathcal{A}_{r-|Y|-1}(\mathcal{O}_\infty, E_\infty)$ follows from inequality (2.9). This shows that $\widetilde{T}_{YX} = T_Y^1 \widetilde{T}_X + D\widetilde{T}_Y.T_X$ belongs to $\mathcal{A}_{r-|Y|-1}(\mathcal{O}_\infty, E_\infty)$, and proves the theorem by induction.

If $a : \Pi' \longrightarrow E$ we define

$$\mathcal{P}_n(a) = \Big(\sum_{\substack{Y \in \Pi' \\ |Y| \le n}} \|a_Y\|_E^2 \Big)^{1/2}, n \in \mathbb{N}. \tag{2.12}$$

THEOREM 2. *Let $r \in]0,1]$, and let T be a r-regular analytic representation of $\mathfrak{g}$, F is a closed bounded subset of $\mathcal{O}$ and $F_\infty = F \cap E_\infty$, there exist increasing functions $C_N : \mathbb{R}^+ \longrightarrow \mathbb{R}^+, N \in \mathbb{N}$ such that*

i) $\mathcal{P}_N(T(u)) \le C_N(\|u\|_{E_N})\|u\|_{E_N}$ and

$\|u\|_{E_N} \le C_N(\mathcal{P}_N(T(u))\mathcal{P}_N(T(u))$ *for* $u \in \mathbb{N}, u \in F_\infty$.

ii) $\mathcal{P}_N(T(u) - T(v)) \le C_N(\|u\|_{E_N} + \|v\|_{E_N})\|u - v\|_{E_N}$ and

$\|u - v\|_{E_N} \le C_N(\mathcal{P}_N(T(u)) + \mathcal{P}_N(T(v))\mathcal{P}_N(T(u) - T(v))$ *for* $n \in \mathbb{N}$ *and* $u, v \in F_\infty$ *such that* $\{w \in E \mid \ \|u - w\|_E \le \|u - v\|_E\} \subset F$.

PROOF: We outline the proof of statement i). According to theorem 1, $\tilde{T}_Y \in \mathcal{A}_{r-|Y|}(\mathcal{O}_\infty, E_\infty)$ for $Y \in \Pi', |Y| \ge 1$. Since $\tilde{T}_X(u)$ has a zero of order two at $u = 0$ the definitions (2.5a) and (2.5b) of the seminorms in $\mathcal{A}_{r-1}(\mathcal{O}_\infty, E_\infty)$, for $r \in]0,1]$, show that there exists $K_F \in \mathbb{R}^+$ such that

$$\|\tilde{T}_X(u)\|_{E_0} \le K_F \|u\|_{E_0}^{r+1} \|u\|_{E_1}^{1-r}, u \in F \cap E_1, \ X \in \mathfrak{g}. \tag{2.13a}$$

For $n \in \mathbb{N}$ and $R \in \mathbb{R}^+$, let $B_n(R) = F \cap \{u \in E_n \mid \ \|u\|_{E_n} \le R\}$. Using definitions (2.5a) and (2.5b) for $|||\tilde{T}_Y|||_{0,0,r-|Y|,B_N(R)}$ it follows that there exists a function $H_{F,N} : \mathbb{R}^+ \longrightarrow \mathbb{R}^+$ such that

$$\|\tilde{T}_Y(u)\|_{E_0} \le H_{F,|Y|}(\|u\|_{E_{|Y|-1}})\|u\|_{E_{|Y|-1}}^{1+r} \|u\|_{E_{|Y|}}^{1-r}, \tag{2.13b}$$

for $u \in F \cap E_{|Y|}, Y \in \Pi', |Y| \ge 1$. The first part of statement i) follows from (2.13b). According to (2.12) we have for some $k_N > 0$,

$$\|u\|_{E_N} \le k_N \mathcal{P}_N(T^1(u)) \le k_N(\mathcal{P}_N(T(u)) + \mathcal{P}_N(\tilde{T}(u))).$$

Inequality (2.13b) then gives for $N \ge 1$ and $u \in F \cap E_N$:

$$\|u\|_{E_N} \le k_N \mathcal{P}_N(T(u)) + H_{F,N}(\|u\|_{E_{N-1}})\|u\|_{E_{N-1}}^{1+r} \|u\|_{E_N}^{1-r},$$

where we have redefined the functions $H_{F,N} : \mathbb{R}^+ \longrightarrow \mathbb{R}^+$.

Since $\|u\|_{E_0} = \mathcal{P}_0(T(u))$ and $0 \le 1 - r < 1$, the second part of statement i) of the theorem follows by induction in N. The second statement is proved in a similar way.

In order to state next theorem we shall introduce a local group action U of G on a dense (in the topology of E_0) subset $\mathcal{O}^0$ of $\mathcal{O}$. Let $\Omega \subset G \times \mathcal{O}^0$ be given by

$$\Omega = \cup_{\Omega \in \mathcal{O}^0} V_u \times \{u\},$$

where V_u is an open neighbourhood of the identity e of G and let the function $U : \Omega \longrightarrow \mathcal{O}^0((g,u) \mapsto U_g(u))$ satisfy $U_g(0) = 0$ and

a) $g \in V_u, h \in V_{U_g(u)}, hg \in V_u \Longrightarrow U_h(U_g(u)) = U_{hg}(u)$

b) $V_u \ni g \mapsto U_g(u) \in E_0$ is continuous for $u \in \mathcal{O}^0$

We denote by $(Y_1, \ldots, Y_{c(k)})$ the lexicographic ordering of the elements in the set $\{Y \in \Pi' \mid 1 \leq |Y| \leq k\}$, with cardinality $c(k)$.

Let $\mathcal{O}^n$ be the subset of $u \in \mathcal{O}^0$ for which $V_u \ni g \mapsto U_g(u) \in E_0$ is C^n. For $u \in \mathcal{O}^n$, define $S_{\mathbb{I}}(u) = u$,

$$S_Y(u) = \frac{\partial^{|Y|}}{\partial \epsilon_1^{\alpha_1} \ldots \partial \epsilon_d^{\alpha_d}} U_{\exp(\epsilon_1 X_1)} \cdots U_{\exp(\epsilon_d X_d)}(u) \mid_{\epsilon_1 = \ldots = \epsilon_d = 0}$$

for $Y \in \Pi', 1 \leq |Y| \leq n$, where $Y = X_1^{\alpha_1} \ldots X_d^{\alpha_d}, X_k \in \Pi$.

We introduce, for $n \geq 1$, the condition

c_n) The set $\mathcal{O}^n$ is dense in $\mathcal{O}^0$ and the subset $\{(u, S_{Y_1}(u), \ldots, S_{Y_{c(k)}}(u)) \mid u \in \mathcal{O}^k\}$ of $\mathcal{O}^k \times E_0^{c(k)}$ is closed (with its inherited topology from $E_0 \times E_0^{c(k)}$) for $1 \leq k \leq n$.

THEOREM 3. *Let $n \geq 1, r \in]0,1]$, let T be a r-regular analytic representation of $\mathfrak{g}$ on $\mathcal{O}_\infty$ and let $U : \Omega \longrightarrow \mathcal{O}^0$ be a local group action of G on a dense subset $\mathcal{O}^0$ of $\mathcal{O}$, satisfying conditions (a), (b) and (c_n). If $\mathcal{O}^n \cap \mathcal{O}_n$ is dense in $\mathcal{O}$ and if $S_X(u) = T_X(u)$ for every $X \in \mathfrak{g}$ and $u \in \mathcal{O}^n \cap \mathcal{O}_n$ then $\mathcal{O}^k \subset \mathcal{O}_k$ for $0 \leq k \leq n$.*

We give an outline of the proof. $\mathcal{O}^0 \subset \mathcal{O}_0 = \mathcal{O}$ by definition. Let $n = 1$. For $v \in \mathcal{O}^1$, we can choose $\{u_l\}_{l \geq 0} \subset \mathcal{O}^1 \cap \mathcal{O}_1$ converging in E_0 to v. According to condition (c_1), $\{T_X(u_l)\}_{l \geq 0}$ converges in E_0 to $S_X(v)$.

The second part of statement (ii) of Theorem 2 shows that $\{u_l\}_{l \geq 0}$ is a Cauchy sequence in E_1. Therefore $v \in \mathcal{O}^1 \cap E_1 \subset \mathcal{O} \cap E_1 = \mathcal{O}_1$.

Suppose that the theorem is true for $n = N \geq 1$ and that the hypotheses of the theorem are satisfied for $n = N + 1$. Then according to hypotheses $S_X(u) = T_X(u)$ for $X \in \mathfrak{g}$ and $u \in \mathcal{O}^{N+1} \cap \mathcal{O}_{N+1}$. Since $\mathcal{O}^{N+1} \cap \mathcal{O}_{N+1}$ is dense in $\mathcal{O}$ it follows from condition (c_{N+1}) that $S_X(u) = T_X(u)$ for $X \in \mathfrak{g}$ and $u \in \mathcal{O}^N \cap \mathcal{O}_N$. Therefore, the hypotheses of Theorem 3 with $n = N$ are satisfied, so $\mathcal{O}^k \subset \mathcal{O}_k$ for $0 \leq k \leq N$. Let $v \in \mathcal{O}^{N+1}$. Then we can choose $\{u_l\}_{l \geq 0} \subset \mathcal{O}^{N+1} \cap \mathcal{O}_{N+1}$ converging in E_0 to v. According to condition (c_{N+1}), $\{T_Y(u_l)\}_{l \geq 0}$ converges in E_0 to $S_Y(v), Y \in \Pi', |Y| \leq N + 1$. The second part of statement (ii) of Theorem 2 shows that $\{u_l\}_{l \geq 0}$ is a Cauchy sequence in E_{N+1}. This shows that $v \in \mathcal{O}^{N+1} \cap E_{N+1} \subset \mathcal{O} \cap E_{N+1} = \mathcal{O}_{N+1}$, which proves the theorem.

COROLLARY 4. *Let $r \in]0,1]$, let T be a r-regular analytic representation of $\mathfrak{g}$ on $\mathcal{O}_\infty$ and let $U : \Omega \longrightarrow \mathcal{O}^0$ be a local group action of G on a dense subset $\mathcal{O}^0$ of $\mathcal{O}$ satisfying conditions (a), (b) and (c_n) for $n \geq 1$. If $\mathcal{O}^\infty \cap \mathcal{O}_\infty$ is dense in $\mathcal{O}$, where $\mathcal{O}^\infty = \bigcap_{n\geq 0} \mathcal{O}^n$, and if $S_X(u) = T_X(u)$ for $X \in \mathfrak{g}$ and $u \in \mathcal{O}^\infty \cap \mathcal{O}_\infty$, then $\mathcal{O}^i \subset \mathcal{O}_i$ for $i \in \mathbb{N} \cup \{\infty\}$.*

If moreover $v \in \mathcal{O}^i, i \in \mathbb{N} \cup \{\infty\}$, then there exists a neighbourhood V^v of the identity in G such that $U_g(v) \in \mathcal{O}^i$ for $g \in V^v$.

The linear map $Y \mapsto T_Y, Y \in U(\mathfrak{g})$ has a precise transformation law, when in definition (2.7) and (2.8) we consider the conjugated representation $g \mapsto U_h \circ U_g \circ U_{h^{-1}}$ instead of $g \mapsto U_g$, namely

$$Y \mapsto T_{Ad_h(Y)}, Y \in U(\mathfrak{g}), \tag{2.14}$$

where Ad_h is the adjoint representation of G on $U(\mathfrak{g})$, i.e.

$$Ad_h(\Pi_{j=1}^k X_j) = \Pi_{j=1}^k Ad_h(X_j), \quad X_j \in \mathfrak{g} \tag{2.15a}$$

and

$$Ad_h X = \frac{d}{dt} h \exp(tX) h^{-1} \mid_{t=0}, \quad X \in \mathfrak{g}. \tag{2.15b}$$

We note that if $ad_X Y = [X, Y]$ for $X \in \mathfrak{g}$ and $Y \in U(\mathfrak{g})$ then

$$ad_X Y = \frac{d}{dt} Ad_{\exp(tX)}(Y) \mid_{t=0} . \tag{2.15c}$$

Let $h(t) = \exp(tX), X \in \mathfrak{g}$, be a one-parameter subgroup of G. Then

$$\frac{d}{dt} Ad_{h(t)} Y = ad_X(Ad_{h(t)} Y), \quad Y \in U(\mathfrak{g}). \tag{2.16}$$

COROLLARY 5. *In the situation of Corollary 4, let $X \in \mathfrak{g}$ and $h(t) = exp(tX)$. If $u \in \mathcal{O}^\infty$, then there exists $\epsilon > 0$ such that*

$$\frac{d}{dt} T_{Ad_{h(t)}(Y)}(U_{h(t)}(u)) = T_{X\,Ad_{h(t)}(Y)}(U_{h(t)}(u)), \tag{2.17}$$

for $|t| < \epsilon$ and $Y \in U(\mathfrak{g})$.

To prove this we use (2.8) and (2.16), which give that

$$\tfrac{d}{dt}T_{Ad_{h(t)}(Y)}(U_{h(t)}(u)) = T_{\frac{d}{dt}(Ad_{h(t)}(Y))}(U_{h(t)}(u))$$

$$+(DT_{Ad_{h(t)}(Y)})(U_{h(t)}(u); \tfrac{d}{dt}U_{h(t)}(u))$$

$$= T_{ad_X(Ad_{h(t)}(y))}(U_{h(t)}(u)) + (DT_{Ad_{h(t)}(Y)})(U_{h(t)}(u); T_X(U_{h(t)}(u))$$

$$= T_{ad_X(Ad_{h(t)}(Y))}(U_{h(t)}(u)) + T_{(Ad_{h(t)}(Y))X}(U_{h(t)}(u))$$

$$= T_{X\,Ad_{h(t)}(Y)}(U_{h(t)}(u)).$$

In the situation of Corollary 4 let us consider the question of the existence of a solution $v \in C^\infty([0,\tau[, \mathcal{O}_\infty)$ of the evolution equation

$$\frac{d}{dt}v(t) = T_{P_0}(v(t)), v(0) = u \in \mathcal{O}_\infty, \tag{2.18}$$

where $P_0 \in \mathfrak{g}$ and for simplicity we suppose that $\mathcal{O}^i = \mathcal{O}_i$ for $i \in \mathbb{N}$ in Corollary 4 and that $\widetilde{T}_{P_o} \in \mathcal{A}_0(\mathcal{O}_\infty, E_\infty)$. Then $\mathcal{O}^\infty = \mathcal{O}_\infty$. Here $\tau \in]0,\infty]$. For $Y \in \Pi'$, we introduce

$$a_Y(t) = T_{Ad_{h(t)}(Y)}(v(t)), \tag{2.19}$$

where $h(t) = \exp(tP_0)$. Let $\mathbb{I} < Y_1 < \ldots < Y_{c(k)}$ be the lexicographic ordering of the set of $Y \in \Pi'$ such that $|Y| \leq k$, and let

$$b_n(t) = (a_{Y_0}(t), a_{Y_1}(t), \ldots, a_{Y_{c(n)}}(t)), n \geq 0. \tag{2.20}$$

According to Corollary 5 and equation (2.18) we have

$$\frac{d}{dt}T_{Ad_{h(t)}(Y)}(v(t)) = T_{P_0 Ad_{h(t)}(Y)}(v(t)). \tag{2.21}$$

Definitions (2.19) and (2.20), equation (2.21) and Corollary 2.5 of [6] give that there exist analytic functions $F_n \in \mathcal{H}(\mathcal{O} \times (E^{c(n)}), E)$ such that

$$\frac{d}{dt}b_n(t) = T^1_{P_o}b_n(t) + F_n(b_n(t)), \tag{2.22}$$

which after integration gives

$$b_n(t) = U^1_{\exp(tP_0)}b_n(0) + \int_0^t U^1_{\exp((t-s)P_0)}F_n(b_n(s))ds, \tag{2.23a}$$

where $t \in [0,\tau[$ and

$$b_n(0) = (u, T_{Y_1}(u), \ldots, T_{Y_{c(n)}}(u)), u \in \mathcal{O}_\infty. \tag{2.23b}$$

The following theorem reduces the study of solutions of equation (2.18) in E_∞ to a study of a system of equations in E.

THEOREM 6. *Let $\tau \in]0,\infty]$. If $u \in \mathcal{O}_\infty$ is such that equation (2.23a) has a solution $b_n \in C^0([0,\tau[,\mathcal{O} \times E^{c(n)})$ for each $n \in \mathbb{N}$, then $v = b_0 \in C^\infty([0,\tau[,E_\infty)$ is a solution of equation (2.18).*

This result follows from Theorem 2 and Corollary 5.

The hypotheses of Theorem 6 are satisfied in the case of Examples 1 and 2. In the case of the non-linear Klein-Gordon equation in Example 1, we shall make the connection between the concept of non-linear representations of Lie groups and Lie algebras as introduced in [1] and the use of the quasi-regular representations of the Poincaré Lie algebra as utilized in [4] for non-linear wave equations and in [5] for non-linear Klein-Gordon equation. Let

$$\begin{aligned}
\xi_{P_0} &= \tfrac{\partial}{\partial t}, \xi_{P_i} = \partial_i && \text{for } 1 \le i \le n, \\
\xi_{M_{ij}} &= -x_i \tfrac{\partial}{\partial x_j} + x_j \tfrac{\partial}{\partial x_i} && \text{for } 1 \le i < j \le n, \\
\xi_{M_{0i}} &= x_i \tfrac{\partial}{\partial t} + t \tfrac{\partial}{\partial x_i} && \text{for } 1 \le i \le n.
\end{aligned}$$

and let the representation $X \mapsto \xi_X$ of $\mathfrak{g} = \mathbb{R}^n \oplus \mathfrak{so}(n,1)$ on $C^\infty(\mathbb{R}^n)$ be defined by linearity. Let $Y \longrightarrow \xi_Y, Y \in u(\mathfrak{g})$ be the canonical extension to $U(\mathfrak{g})$. It then follows that

$$((\xi_Y\phi)(t), (\xi_{P_0 Y}\phi)(t)) = T_{Ad_{\exp(tP_0)(Y)}}(U_{\exp(tP_0)}(u))$$

where

$$(\phi(t), \dot{\phi}(t)) = (\phi(t), (\xi_{P_0}\phi)(t)) = U_{\exp(tP_0)}(u), u = (\phi(0), (\xi_{P_0}\phi)(0)),$$

$$\begin{aligned}
&(\Box + m^2)\phi = P(\phi, \dot{\phi}, \nabla\phi), \\
&T_{P_0}(f, \dot{f}) = (\dot{f}, (\Delta - m^2)f) + (0, P(f, \dot{f}, \nabla f)), \\
&T_{P_i}(f, \dot{f}) = (\partial_i f, \partial_i \dot{f}), 1 \le i \le n, \\
&T_{M_{ij}}(f, \dot{f}) = (\xi_{M_{ij}} f, \xi_{M_{ij}} \dot{f}), 1 \le i < j \le n, \\
&(T_{M_{0i}}(f, \dot{f}))(x) = x_i((T_{P_0})(f, \dot{f}))(x) + (0, (\partial_i f)(x)), 1 \le i \le n.
\end{aligned}$$

Acknowledgements. We wish to thank M. Flato for stimulating discussions and D. Sternheimer for a careful reading of the manuscript.

References

1. Flato M., Pinczon G., Simon J.C.H.: "Non linear representations of Lie groups". *Ann. Sci. Ec. Norm. Super.* **10**, 405-418 (1977).

2. Flato M., Simon J.C.H., Taflin E.: "The Maxwell-Dirac equations: Asymptotic completeness and infrared problem." *Reviews in Math. Phys.*, **6**(5a), 1071-1083 (1994).

3. Flato M., Simon J.C.H., Taflin E.: "Asymptotic completeness, global existence and the infrared problem for the Maxwell-Dirac equations". Preprint (Feb. 1995) available by anonymous ftp from xxx.lanl. gov as hep-th/9502061.

4. Klainerman S. :"Global existence for non linear wave equations". *Commun. Pure Appl. Math.* **33**, 43-101 (1980).

5. Klainerman S.:"Global existence of small amplitude solutions to non linear Klein-Gordon equations in four space-time dimensions". *Commun. Pure Appl. Math.* **38**, 631-641 (1985).

6. Simon J.C.H., Taflin E.: "The Cauchy problem for non linear Klein-Gordon equations". *Commun. Math. Phys.* **152**, 433-478 (1993).

STAR PRODUCTS: THEIR UBIQUITY AND UNICITY

DANIEL STERNHEIMER

Laboratoire de Physique Théorique Mathématique,
Université Denis Diderot - Paris VII,
Tour Centrale 3e étage, 2 place Jussieu,
F-75251 Paris Cedex 05, France.
(Tél. 44275741; fax 44277793; E-mail: daste@ccr.jussieu.fr)

Dedicated to my long-time friend and colleague in CNRS, Guy Rideau.

Abstract. The "headlines" of the theory of star products and their ramifications are briefly presented, focusing on uniqueness of star products and their presence in many domains.

FOREWORD. Since there was little time for the presentation and the space in these Proceedings is limited, I chose to present the "headlines" of this vast subject and comment them in the talk. These headlines will be reproduced here, with a few comments (in smaller print) after each item. A list of references follows. Though extensive, it has no claim to be exhaustive except perhaps for our own works (a still incomplete bibliography of 102 pages on part of the subject covered here can be found in [86]); only a fraction will be quoted in the text, mainly as a support to the text and a starting point for further reading that should also include references quoted in the references. The aim of this contribution is to show that, like Mr. Jourdain in Molière was speaking prose without knowing it, many mathematicians and theoretical physicists are (or were) in fact dealing with star products even if they are not aware of it. Star products can thus be found (almost) everywhere, and yet there are very few of them (up to equivalence).

J. Bertrand et al. (eds.), Modern Group Theoretical Methods in Physics, 255–265.

1. Presentation of the question.

1.1. PRELIMINARIES: DEFORMATION VIEW OF PHYSICAL THEORIES

(geometrical symmetries, mechanics, nonlinearities).
The passage from one level of physical theory to a next one occurs as a deformation of some algebraic structure [42]. This interpretation turns out to be extremely fruitful in many domains and can be seen e.g. on the geometric level by deformation from Galilei to Poincaré to De Sitter invariance (with far-reaching applications of the latter [43]), in mechanics by passage from Poisson to Moyal brackets and star products (the basis of the present review) and when interactions bring in nonlinearities by the occurrence of nonlinear representations of invariance groups [48] [49].

1.2. PHASE SPACE, POISSON BRACKETS AND QUANTIZATION: PREHISTORY

(Weyl, Wigner, Moyal; semi-classical approximations; constraints; geometric quantization; the 1-differentiable case).
Physicists had introduced Weyl quantization [96] in the beginning of quantum theory, in 1927, and its inverse a few years later [98]; the inverse images of products and brackets were then obtained [59] [76], but with the impossible aim of realizing quantum mechanics as a statistical theory over classical observables. Many developments followed, e.g. consideration of various orderings [1] and associated pseudodifferential operators [60], semi-classical approximations [94], etc.. All this was essentially based on $\mathbb{R}^{2n}$ as phase space, the "constrained mechanics" of Dirac [34] permitting some treatment of the nonflat case. In the early seventies appeared geometric quantization [67], originating from the theory of group representations, aiming at a general approach to quantization but that ran into serious problems in practical situations. At the same time we looked at deformations of Poisson brackets, but first within the frame of bidifferentiable operators of order at most one, also for general symplectic manifolds [45].

1.3. STAR PRODUCTS: THE BEGINNING

(Vey [93], BaFFLiS [12]). This is an *autonomous* theory.
Inspired by our works on 1-differentiable deformations of Poisson brackets Jacques Vey managed to develop a complete theory of differentiable deformations on symplectic manifolds with vanishing third De Rham cohomology, of which the model was the Moyal bracket (that he then ignored but rederived). Taking notice of this fact we pushed further the theory, developing systematically the notion of star products (already touched by Vey) and making full use of the interplay between the Hochschild cohomology of the associative algebra of classical observables and the Chevalley-Eilenberg Poisson bracket Lie algebra cohomology. Our main point

was to show, by concrete and typical examples (such as the harmonic oscillator and the hydrogen atom), that quantum mechanics was in fact, and could be developed in an autonomous manner as, a deformation of classical mechanics – with no need of the usual Hilbert space operatorial formulation. A parallel formulation in the complex domain, corresponding to normal ordering, was around the same time developed independently by Berezin [14], but not in terms of deformations.

1.4. PSEUDODIFFERENTIAL OPERATORS (AND INDEX THEOREMS)

Weyl quantization is the Fourier transform of the inverse Fourier transform of classical observables where in the last Fourier transform the usual exponential kernel is replaced by the von Neumann representation of the Heisenberg group. Mathematicians had in the fifties introduced procedures of this kind, singular integral operators, to define (partial) differential operators of noninteger order, which gave birth to the theory of pseudodifferential operators [27]. This amounted to nothing else than what physicists had called "standard ordering" [1], an avatar of the Weyl quantization that was giving [60] [94] very similar operators. The theory of pseudodifferential operators became most popular in mathematics when, in addition to its specialized use in partial differential equations, it became linked by the index theorems [80] to many frontier domains of mathematics – a link that grew immensely since its beginning 30 years ago. Not surprisingly it turns out that star products are very present in this context [23] [28] [29] [41].

1.5. QUANTUM GROUPS (ALGEBRAIC AND TOPOLOGICAL THEORIES)

Quantum groups were discovered "experimentally" fairly recently [69] and shortly afterwards systematized using Hopf algebras [38], though some aspects of them can be traced back to the 19th century. Two "dual" aspects were considered, that amount to the deformation of a product or of a coproduct [38] [65] [40] and numerous variations on that theme were developed, including notably connection with noncommutative geometry [28] [99]. As was noticed already by Drinfeld, the deformation of the product is in fact an example of star product. The duality however poses problem in the algebraic context and requires some adequate topologizations of the algebras (with locally convex vector space topologies) that were developed very recently [21] and show more clearly the underlying star product structure.

2. Star Products: Existence and Equivalence.

2.1. DEFORMATIONS AND COHOMOLOGIES

(differentiable, differentiable null on constants (d.n.c.), closed star products; bialgebras; Leibniz pairs and their bicomplex with Hochschild and Chevalley-Eilenberg cohomologies; rigidity).

In accordance with the general theory of deformations of algebras [57], existence and equivalence are governed by third and second cohomologies for the adjoint action. Here the algebras are those with underlying vector space N, the C^∞ functions on a symplectic or Poisson manifold, and various classes of cochains and algebras can be considered: Lie, associative, Hopf and bialgebras, with differentiable, d.n.c. or closed cochains. A crucial remark is that star products are associative deformations for which the corresponding commutator starts with the Poisson bracket; this permits an interplay between the (infinite) Hochschild and (finite) Chevalley cohomologies that was put into a very interesting and fruitful geometrical context (based on symplectic connections) by Fedosov [41] and systematized recently [44] with the concept of Leibniz pairs. The notion of closed star product was introduced [29] in order to have a character (a trace) on the star product algebra and thus to be able to "plug in" the algebraic formalism of noncommutative geometry [28].

2.2. EXISTENCE OF STAR PRODUCTS

(Case $b_3 = 0$; cotangent; general case: De Wilde - Lecomte, Fedosov and Omori - Maeda - Yoshioka, Boutet de Monvel - Guillemin).
The existence of star products on general symplectic manifolds can be shown by gluing together Moyal products taken on Darboux charts. At first this required $b_3 = 0$ to take care of multiple intersections [77] [93] but was soon extended to cotangent bundles and in an abstract manner to the general case [32]; the notion of Weyl bundle introduced by Fedosov permitted a more geometrical and even algorithmic construction [41] [79] [39]. As can be expected, it turned out that the composition of symbols of pseudodifferential operators provides proof of existence in a quite general context [23].

2.3. EQUIVALENCE

(classification by b_2; closed $\sim$ d.n.c $\sim$ diff.; noninnocent question!).
The second Chevalley differentiable cohomology of N has dimension $1 + b_2$, where b_n denotes the nth Betti number of the manifold. Here 1 "stands for Moyal" and b_2 for the 1-differentiable case (amounting to deformations of the Poisson 2-tensor). Star products are thus classified, up to equivalence, by b_2 [95] [61]. Since, from standard deformation theory, the deformation of a unital algebra is unital and equivalent to a deformation with the same unit, this result proved in the d.n.c case holds for general differential star products; the existence of a trace shows [78] it holds also for closed star products. As shown by the problem of ordering of physical observables and by the question of spectrality [25], mathematical equivalence and "physical" equivalence are very different issues.

3. Star Products and:

3.1. NONCOMMUTATIVE GEOMETRY (AND ALGEBRAIC INDEX THEOREMS)

In all of section 3 we indicate the main domains of direct application of star products. Noncommutative geometry developed out of the theory of C^* algebras into a very ramified field of mathematics with physical roots and applications; it has a lot in common with the theory of (closed) star products; for all this we refer to [28].

3.2. PSEUDODIFFERENTIAL CALCULUS

As indicated above, there is a deep interplay between the mathematical theories of pseudodifferential operators and star products that starts only now to be exploited.

3.3. QUANTUM GROUPS

Quantum groups, in their various flavors, are in essence an example of star products; the topological theory of quantum groups presented at this Conference [20] [15] is especially relevant in this connection. However "quantum groups" – which are not groups – are not "quantum" in the same sense as star products; the latter are deformations with $\hbar$, the Planck constant, as parameter [47] [50] and directly linked with quantization, while the quantum character of the former is related to their appearance in the quantization of 2-dimensional field theoretical models.

3.4. STATISTICAL MECHANICS, NAMBU MECHANICS

Star products can be modified so as to make appear another constant, the Boltzmann constant β, as a second parameter [9]. The Poisson bracket is then replaced by a Jacobi bracket related to a conformal symplectic structure (possibly on an infinite-dimensional manifold) and the Kubo-Martin-Schwinger (KMS) condition is then expressed similarly to the vanishing of the trace of a commutator. This is only a simple example of what can develop into a much more involved formalism. Another generalization deals with n-ary composition laws such as those appearing in the so-called Nambu mechanics [90], the quantization of which is still to be found, and in the symplectic operads of O. Mathieu [72].

3.5. QUANTUM FIELD THEORY

The uniqueness (for $\mathbb{R}^{2n}$) of star products is (as exemplified also below) an avatar of the von Neumann uniqueness theorem, that is completely untrue in the infinite-dimensional case of field theory. Poisson structures are however known in infinite-dimensional cases, e.g. on spaces of initial data for classical field equations [85]. As can be expected (from the $\frac{n}{2}$ ground state eigenvalue of the harmonic oscillator in Moyal ordering) the Moyal bracket cannot be defined here but normal orderings can; moreover orderings "close to", equivalent to but different of, normal ordering produce "more finite" results, exhibiting a very promising "cohomological renormalization" [35] [36] [37].

3.6. LIE GROUP (STAR) REPRESENTATIONS

(nilpotent and solvable; compact; special cases incl. Poincaré; wavelets; some series of representations in semi-simple case, unipotent orbits).

From the beginning group theory and quantum mechanics were interrelated [96]. Obviously deformation quantization has permitted the development of a parallel theory of group representations by star products on coadjoint orbits (or the whole dual of the Lie algebra), including nilpotent and solvable groups [5] where the Moyal product gives essentially everything, compact groups and some series of representations in the semi-simple case [4], and other classes of representations. The theory of wavelets is also related to this aspect (star representations of the affine group of $\mathbb{R}$ and of other groups).

References

1. G.S. Agarwal and E. Wolf. Calculus for functions of noncommuting operators and general phase-space methods in quantum mechanics I, II, III, Phys. Rev. D(3) **2** (1970), 2161-2186, 2187-2205, 2206-2225.
2. C. Alcalde and D. Sternheimer. Analytic vectors, anomalies and star representations. Lett. Math. Phys. **17**, 117-127 (1989).
3. A.Yu. Alekseev and A.Z. Malkin. Symplectic structures associated to Poisson group, Commun. Math. Phys. **162**, 147-173 (1994). See also:
J.H. Lu and A. Weinstein. Poisson-Lie groups, dressing transformations and Bruhat decompositions, J. Diff. Geom. **31**, 501-526 (1990).
Y. Kosmann-Schwarzbach and F. Magri. Poisson-Lie groups and complete integrability, Ann. Inst. H. Poincaré Phys. Théor. **49**, 433-460 (1988); and: Poisson Nijenhuis structures, ibid. **53**, 35-81 (1990).
4. D. Arnal, M. Cahen and S. Gutt. Representations of compact Lie groups and quantization by deformation, Bull. Acad. Royale Belg. **74** (1988), 123-141; Star exponential and holomorphic discrete series, Bull. Soc. Math. Belg. **41** (1989), 207-227. See also: C. Moreno. Invariant star products and representations of compact semisimple Lie groups, Lett. Math. Phys. **12**, 217-229 (1986).
5. D. Arnal and J.C. Cortet. Nilpotent Fourier transform and applications, Lett. Math. Phys. **9** (1985), 25-34; Star-products in the method of orbits for nilpotent Lie groups, J. Geom. Phys. **2** (1985), 83-116; Représentations star des groupes exponentiels, J. Funct. Anal. **92** (1990), 103-135.
D. Arnal, J.C. Cortet and J. Ludwig. Moyal product and representations of solvable Lie groups, J. Funct. Anal. (to be published).
6. D. Arnal, J.C. Cortet, M. Flato and D. Sternheimer. Star products: quantization and representations without operators, pp. 85-111 in: *Field theory, quantization and statistical physics* (E. Tirapegui, ed.), D. Reidel, Dordrecht (1981).

7. D. Arnal, J.C. Cortet and P. Molin. Star-produit et représentation de masse nulle du groupe de Poincaré, C.R. Acad. Sci. Paris A **293**, 309-312 (1981).
8. D. Arnal, J.C. Cortet, P. Molin and G. Pinczon. Covariance and geometrical invariance in star-quantization, J. Math. Phys. **24**, 276-283 (1983).
9. H. Basart, M. Flato, A. Lichnerowicz and D. Sternheimer. Deformation theory applied to quantization and statistical mechanics, Lett. Math. Phys. **8**, 483-494 (1984); Mécanique statistique et déformations, C.R. Acad. Sci. Paris **299** I, 405-410 (1984).
10. H. Basart and A. Lichnerowicz. Déformations d'un star-produit sur une variété symplectique. C. R. Acad. Sci. Paris **293** I, 347-350 (1981).
11. F. Bayen, M. Flato, C. Fronsdal, A. Lichnerowicz and D. Sternheimer. Quantum mechanics as a deformation of classical mechanics. Lett. Math. Phys. **1**, 521-530 (1977).
12. F. Bayen, M. Flato, C. Fronsdal, A. Lichnerowicz and D. Sternheimer. Deformation theory and quantization, Ann. Phys. (NY) (1978) **111**,
I - Deformation of symplectic structures. pp. 61-110
II - Physical applications. pp. 111-151.
13. F. Bayen and C. Fronsdal. Quantization on the sphere, J. Math. Phys. **22**, 1345-1349 (1982).
14. F.A. Berezin. Quantization, Math. USSR Izvestija **8**, 1109-1165 (1974); Quantization in complex symmetric spaces, ibid. **9**, 341-379 (1975); General concept of quantization, Commun. Math. Phys. **40**, 153-174 (1975). See also:
F.A. Berezin and M.A. Shubin. Symbols of operators and quantization, pp. 21-52 in: Coll. Math. Soc. Janos Bolyai 5, "Hilbert Space Operators", Tihany (1970).
15. F. Bidegain. Noncompact topological quantum groups, these proceedings.
16. F. Bidegain. A candidate for a noncompact quantum group, Lett. Math. Phys. (to be published).
17. F. Bidegain and G. Pinczon. A star product approach to noncompact quantum groups, Lett. Math. Phys. **33**, 231-240 (1995).
18. P. Bonneau. Cohomology and associated deformations for not necessarily coassociative bialgebras, Lett. Math. Phys. **26** (1992), 277-283.
19. P. Bonneau. Topological quantum double, Reviews in Math. Phys. **6**, 305-318 (1994).
20. P. Bonneau. Compact topological quantum groups, these proceedings.
21. P. Bonneau, M. Flato, M. Gerstenhaber and G. Pinczon. The hidden group structure of quantum groups: strong duality, rigidity and preferred deformations. Commun. Math. Phys. **161**, 125-156 (1994).
22. P. Bonneau, M. Flato and G. Pinczon. A natural and rigid model of quantum groups. Lett. Math. Phys. **25**, 75-84 (1992).
23. L. Boutet de Monvel and V. Guillemin. The spectral theory of Toeplitz operators, Ann. of Math. Studies no. 99, Princeton (1981). See also:
L. Boutet de Monvel. On the index of Toeplitz operators of several complex variables, Invent. Math. **50**, 249-272 (1979); V. Guillemin. Star products on compact pre-quantizable manifolds, Lett. Math. Phys. (1995); L. Boutet de Monvel. Star products on conic Poisson manifolds of constant rank (preprint 1994, lecture at M. Vergne's seminar in ENS).
24. C. Cadavid and M. Nakashima. The star-exponential and path integrals on compact groups, Lett. Math. Phys. **23** (1991), 111-115.
25. M. Cahen, M. Flato, S. Gutt and D. Sternheimer. Do different deformations lead to the same spectrum? J. Geom. Phys. **2**, 35-48 (1985).
26. M. Cahen and S. Gutt. Discrete spectrum of the hydrogen atom: an illustration of deformation theory methods and problems, J. Geom. Phys. **1**, 65-83 (1984); Regular $*$-representations of Lie algebras, Lett. Math. Phys. **6**, 395-404 (1982).
M. Cahen, S. Gutt and J. Rawnsley. Quantization of Kähler manifolds, I: geometric interpretation of Berezin's quantization, J. Geom. Phys. **7**, 45-65 (1990); II, Trans. Amer. Math. Soc. **337**, 73-98 (1993); III, Lett. Math. Phys. **30**, 291-305 (1994).

27. A.P. Calderon and A. Zygmund. Singular integral operators and differential equations, Amer. J. Math. **77**, 901-921 (1957);
L. Hörmander. Pseudo-differential operators and non-elliptic boundary problems, Ann. Math. **83**, 129-209 (1966); Fourier Integral Operators I, Acta Math. **127**, 79-183 (1971).
28. A. Connes. Spectral sequences and homology of currents for operator algebras, in: Math. Forschunginstitut Oberwolfach Tatungsbericht 41/81, Funktionalanalysis und C*-Algebren, 27-9/3-10 (1981); Noncommutative differential geometry, Publ. Math. IHES **62**, 41-144 (1986); *Géométrie Non-Commutative*, Interéditions, Paris (1990) [expanded English edition, preprint IHES/M/93/12, March 1993, published by Academic Press].
29. A. Connes, M. Flato and D. Sternheimer. Closed star-products and cyclic cohomology, Lett. Math. Phys. **24**, 1-12 (1992).
30. A.Connes, M. Gromov and H. Moscovici. Conjecture de Novikov et fibrés presque plats, C.R. Acad. Sci. Paris **I 310**, 273-277 (1990).
31. A. Connes and H. Moscovici. Cyclic cohomology, the Novikov conjecture and hyperbolic groups, Topology **20**, 345-388 (1990).
32. M. De Wilde and P. Lecomte. Existence of star-products revisited, Note di Mat. vol. X (1992); Existence of star-products and of formal deformations of the Poisson Lie algebra of arbitrary symplectic manifolds, Lett. Math. Phys. **7**, 487-496 (1983). See also S. Gutt. An explicit $*$-product on the cotangent bundle to a Lie group, Lett. Math. Phys. **7**, 249-258 (1983).
33. M. De Wilde, S. Gutt and P. Lecomte. Ann. Inst. Henri Poincaré **A 40**, 77-89 (1984).
34. P.A.M. Dirac, *Lectures on Quantum Mechanics*, Belfer Graduate School of Sciences Monograph Series No. 2 (Yeshiva University, New York, 1964).
35. J. Dito. Star-product approach to quantum field theory : the free scalar field, Lett. Math. Phys. **20**, 125-134 (1992).
36. J. Dito. Star-products and nonstandard quantization for Klein-Gordon equation, J. Math. Phys. **33**, 791-801 (1992).
37. J. Dito. An example of cancellation of infinities in star-quantization of fields, Lett. Math. Phys. **27**, 73-80 (1993).
38. V.G. Drinfeld. Quantum Groups, in: *Proc. ICM86, Berkeley*, **1**, 101-110, Amer. Math. Soc., Providence (1987); Quasi-Hopf algebras, Leningrad Math. J. **1**, 1419-1457 (1990); On almost co-commutative Hopf algebras, ibid. pp. 321-431.
39. C. Emmrich and A. Weinstein. The differential geometry of Fedosov's quantization, in: *Lie Theory and Geometry: In Honor of B. Kostant*, J.L. Brylinski et al., eds., Progress in Mathematics, Birkhäuser, Boston (1994). See also:
S. Gutt. Lectures at ICTP Workshop (Trieste, March 1993); P. Bieliavsky, M. Cahen and S. Gutt. Symmetric symplectic manifolds and deformation quantization, these proceedings.
40. L.D. Faddeev, N.Yu. Reshetikhin and L.A. Takhtajan. Quantization of Lie groups and Lie algebras, Leningrad Math. J. **1**, 193-225 (1990).
41. B.V. Fedosov. A simple geometrical construction of deformation quantization, J. Diff. Geom. **40**, 213-238 (1994) and "Formal quantization" in *Some topics of modern mathematics and their applications to problems of mathematical physics* (in Russian), 129-139 (Moscow, 1985); see also: Sov. Phys. Dokl. **34**, 319-321 (1989); Proof of the Index Theorem for Deformation Quantization, Advances in partial differential equations (1994); A trace density in deformation quantization (preprint Potsdam Max-Planck-Arbeitsgruppe, 1994); Analytical formulas for the index of elliptic operators, Trans. Moscow Math. Soc. **30**, 159-240 (1974).
42. M. Flato. Deformation view of physical theories, Czechoslovak J. Phys. **B32** (1982), 472-475.

43. M. Flato and C. Fronsdal. Fundamental Gauge Theory, pp. 273-290 in: "Topological and Geometrical Methods in Field Theory" (J. Hietarinta and J. Westerholm, eds.), World Scientific (1986); Composite Electrodynamics, J. Geom. Phys. **5**, 37-61 (1988);
C. Frønsdal. A model for QCD in three dimensions, these proceedings.
44. M. Flato, M. Gerstenhaber and A. Voronov. Cohomology and deformation of Leibniz pairs, Lett. Math. Phys. (1995).
45. M. Flato, A. Lichnerowicz and D. Sternheimer. Déformations 1-différentiables d'algèbres de Lie attachées à une variété symplectique ou de contact, C.R. Acad. Sci. Paris, **279**, A877-881 (1974) and Compositio Mathematica, **31**, 47-82 (1975); Algèbres de Lie attachées à une variété canonique, J. Math. Pures Appl. **54**, 445-480 (1975).
46. M. Flato, A. Lichnerowicz and D. Sternheimer. Deformations of Poisson brackets, Dirac brackets and applications. J. Math. Phys. **17**, 1754-1762, (1976).
47. M. Flato and Z.C. Lu. Remarks on quantum groups. Lett. Math. Phys. **21**, 85-88 (1991); Z.C. Lu. Quantum algebras on phase-space, J. Math. Phys. **33**, 446-453 (1992); M. Flato, Z.C. Lu and D. Sternheimer. From where do quantum groups come? Foundations of Physics **23**, 587-598 (1993).
48. M. Flato, G. Pinczon and J. Simon. Non Linear Representations of Lie Groups, Ann. Sc. Ec. Norm. Sup. 4e série, **10**, 405-418 (1977).
49. M. Flato and D. Sternheimer. Deformations of Poisson brackets, separate and joint analyticity in group representations, nonlinear group representations and physical applications, in "Harmonic Analysis and Representations of Semi-Simple Lie Groups", J. Wolf et al. ed., pp. 385-448, MPAM, Vol.5, D. Reidel (1980).
50. M. Flato and D. Sternheimer. On a possible origin of quantum groups, Lett. Math. Phys. **22**, 155-160 (1991).
51. M. Flato and D. Sternheimer. Quantum groups, star-products and cyclic cohomology, in: "Quantum and Non-Commutative Analysis", Math. Phys. Studies vol. 16, pp. 221-237 (H. Araki, K. Ito, A. Kishimoto and I. Ojima, eds.), Kluwer Acad. Publ. (1993).
52. M. Flato and D. Sternheimer. Star-products, quantum groups, cyclic cohomology and pseudo-differential calculus. American Mathematical Society Contemporary Mathematics vol. 175 (P. Sally et al. eds.), pp. 53-72 (1994).
53. M. Flato and D. Sternheimer. Topological Quantum Groups, star products and their relations, St. Petersburg Mathematical Journal (Algebra and Analysis) **6**(3), 242-251 (1994).
54. M. Flato and D. Sternheimer. Closedness of star products and cohomologies, pp. 241-259 in: *Lie Theory and Geometry: In Honor of B. Kostant*, J.L. Brylinski et al., eds., Progress in Mathematics, Birkhäuser, Boston (1994).
55. C. Fronsdal. Some ideas about quantization, Reports On Math. Phys. **15**, 111-145 (1978).
56. C. Fronsdal. Invariant $*$-product quantization and the 1-dimensional Kepler problem, J. Math. Phys. **20**, 2226-2232 (1979).
57. M. Gerstenhaber. On the deformation of rings and algebras, Ann. Math. **79**, 59-103 (1964); and (IV), *ibid.* **99**, 257-276 (1974).
58. M. Gerstenhaber and S.D. Schack. Bialgebra cohomology, deformations and quantum groups, Proc. Nat. Acad. Sci. USA **87** (1990), 478-481.
59. A. Groenewold. On the principles of elementary quantum mechanics, Physica **12**, 405-460 (1946).
60. A. Grossman, G. Loupias and E.M. Stein. An algebra of pseudodifferential operators and quantum mechanics in phase space, Ann. Inst. Fourier **18**, 343-368 (1968).
61. S. Gutt. *Déformations formelles de l'algèbre des fonctions différentiables sur une variété symplectique*, Thesis, Bruxelles (1980); Cohomology associated with the Poisson algebra, Lett. Math. Phys. **3**, 297-309 (1979); Ann. Inst. Henri Poincaré **A 32**, 1-31 (1980).

62. F. Hansen. The Moyal product and spectral theory for a class of infinite-dimensional matrices, Publ. RIMS, Kyoto Univ., **26**, 885-933 (1990); Quantum Mechanics in Phase Space, Reports On Math. Phys. **19**, 361-381 (1984).
63. G. Hochschild, B. Kostant and A. Rosenberg. Differential forms on regular affine algebras, Trans. Am. Math. Soc. **102**, 383-406 (1962).
64. L. Hörmander. The Weyl calculus of pseudodifferential operators, Commun. Pure Appl. Math. **32**, 353-443 (1979).
65. M. Jimbo. A q-difference algebra of $U(g)$ and the Yang-Baxter equation, Lett. Math. Phys. **10**, 63-69 (1985).
66. J.M. Kammerer. Analysis of the Moyal product in a flat space, J. Math. Phys. **27**, 529-535 (1986).
67. B. Kostant. *Quantization and unitary representations*, in: Lecture Notes in Math. **170**, 87-208, Springer Verlag, Berlin (1970). See also:
L. Auslander and B. Kostant. Polarization and unitary representations of solvable groups, Inv. Math. **14**, 255-354 (1971);
J.M. Souriau. *Structure des systèmes dynamiques*, Dunod, Paris (1970);
A.A. Kirillov. *Elements of the theory of representations*, Springer, Berlin (1976).
68. B. Kostant. Symplectic spinors, Symposia Mathematica **14** (1974), 139-152.
69. P.P. Kulish and N. Yu. Reshetikhin. Quantum linear problem for the sine-Gordon equation and higher representations. Zap. Nauch. Sem. LOMI **101** (1981), 101-110 (in Russian; English translation in Jour. Sov. Math. **23** (1983), 24-35).
70. A. Lichnerowicz. Déformations d'algèbres associées à une variété symplectique (les $*_\nu$-produits), Ann. Inst. Fourier, Grenoble, **32** (1982), 157-209.
71. J.M. Maillard. On the twisted convolution product and the Weyl transform of tempered distributions, J. Geom. Phys. **3**, 231-261 (1986).
72. O. Mathieu. The symplectic operad, preprint (1994, dedicated to I.M. Gelfand).
73. P. Molin. Existence de produits star fortement invariants sur les orbites massives de la coadjointe du groupe de Poincaré, C.R. Acad. Sci. Paris A **293**, 309-312 (1981); Application of the star product method to the angular momentum quantization, Lett. Math. Phys. **25**, 213-225 (1992).
74. C. Moreno and F. Galvez. The basic lemma in the theory of formal groups, these proceedings.
75. C. Moreno and L. Valero. Star-products and quantization of Poisson-Lie groups, J. Geom. Phys. **9**, 369-402 (1992).
76. J.E. Moyal. Quantum mechanics as a statistical theory, Proc. Cambridge Phil. Soc. **45**, 99-124 (1949).
77. O.M. Neroslavsky et A.T. Vlasov. Sur les déformations de l'algèbre des fonctions d'une variété symplectique, C.R. Acad. Sc. Paris **292** I, 71-76 (1981).
78. R. Nest and B. Tsygan. Algebraic Index Theorem, to appear in Commun. Math. Phys.; Algebraic Index Theorem For Families, to appear in Advances in Mathematics.
79. H. Omori, Y. Maeda and A. Yoshioka. Weyl manifolds and deformation quantization, Adv. in Math. **85**, 225-255 (1991); Existence of a closed star product, Lett. Math. Phys. **26**, 285-294 (1992).
80. R. Palais. Seminar on the Atiyah-Singer index theorem, Annals of Mathematics Studies, Vol. 57, Princeton (1965). See also:
Séminaire Cartan-Schwartz 1963/64, "Théorème d'Atiyah-Singer sur l'indice d'un opérateur différentiel elliptique".
M.F. Atiyah and I.M. Singer. The index of elliptic operators I and III, Ann. Math. **87**, 484-530 and 546-604 (1968).
P. Gilkey. *The index theorem and the heat equation*, Mathematics Lecture Series vol.4 (1974); *Invariance theory, the heat equation and the Atiyah-Singer index theorem*; ibid. vol.11 (1984) [Publish or Perish, Boston].
81. Pankaj Sharan. Star-product representation of path integrals, Phys. Rev. D **20**, 414-418 (1979).

82. M. Rieffel. Deformation quantization of Heisenberg manifolds, Commun. Math. Phys. **122**, 531-562 (1989).
83. M. Rosso. Représentations des groupes quantiques, Séminaire Bourbaki, exposé 744 (Juin 1991).
84. J. Schwinger. The special canonical group, Proc. Natl. Acad. Sci. **46**, 1401-1415 (1960).
85. I.E. Segal. Symplectic structures and the quantization problem for wave equations, Symposia Mathematica **14** (1974), 79-117.
86. S. Shnider and S. Sternberg. *Quantum Groups*, Graduate Texts in Mathematical Physics vol. II, International Press (Boston and Hong-Kong, 1993). [With an extended reference list of 1264 items!].
87. D. Sternheimer. *Basic Notions in Topological Vector Spaces*, in: Proc. Adv. Summer Inst. in Math. Phys. (Istanbul 1970, A. Barut ed.), 1-51, Studies in Math. Phys., D. Reidel, Dordrecht (1973);
F. Trèves. *Topological Vector Spaces, Distributions and Kernels*, Academic Press (1967).
88. D. Sternheimer. Deformation theory applied to quantization and group representations, Proceedings of the 1981 IAMP Conference on Mathematical Physics, Lecture Notes in Physics, Vol. 153, pp. 314-318, Springer Verlag (1982); Phase-space representations, pp. 255-267 in: "Applications of Group Theory in Physics and Mathematical Physics" (M.Flato, P.Sally and G.Zuckerman, eds.), Lectures in Applied Math. vol. 21, Amer. Math. Soc. (1985); Approche invariante de l'analyse nonlinéaire et de la quantification, pp. 260-293 in: "Systèmes dynamiques nonlinéaires: intégrabilité et comportement qualitatif" (Séminaire de Mathématiques Supérieures n° 102, P. Winternitz ed.), Presses de l'Université de Montréal (1986).
89. L.A. Takhtajan. *Introduction to quantum groups*, in: Springer Lecture Notes in Physics, **370**, 3-28 (1990); *Lectures on quantum groups*, pp. 69-197 in: Nankai Lectures on Math. Phys. (M. Ge and B. Zhao eds.), World Scientific, Singapore (1990).
90. L.A. Takhtajan. On foundation of the generalized Nambu mechanics, Commun. Math. Phys. **160**, 295-315 (1994);
Y. Nambu. Generalized Hamilton dynamics. Phys. Rev **D7**, 2405-2412 (1973);
F. Bayen and M. Flato. Remarks concerning Nambu's generalized mechanics, Phys. Rev. **D11**, 3049-3053 (1975).
91. B. Tsygan. Homology of matrix algebras over rings and Hochschild homology, Uspekhi Mat. Nauk (= Russ. Math. Surveys) **38**, 217-218 (= 198-199) (1983).
92. A. & J. Unterberger. Quantification et analyse pseudodifférentielle, Ann. Sc. Ec. Norm. Sup. 4e série, **21**, 133-158 (1988).
93. J. Vey. Déformation du crochet de Poisson sur une variété symplectique, Comment. Math. Helv. **50** (1975), 421-454.
94. A. Voros. *Développements semi-classiques*, Thèse (Orsay et Saclay), mai 1977.
95. A. Weinstein. Deformation Quantization, Séminaire Bourbaki, exposé 789 (Juin 1994).
96. H. Weyl. *The theory of groups and quantum mechanics* Dover, New-York (1931), translated from *Gruppentheorie und Quantenmechanik*, Hirzel Verlag, Leipzig (1928); see also Quantenmechanik und Gruppentheorie, Z. Physik **46**, 1-46 (1927).
97. H. Widom. A complete symbolic calculus for pseudodifferential operators, Bull. Sci. Math. (2), **104**, 19-63 (1980).
98. E. Wigner. Quantum corrections for thermodynamic equilibrium, Phys. Rev. **40**, 749-759 (1932).
99. S.L. Woronowicz. Compact matrix pseudogroups, Commun. Math. Phys. **111**, 613-665 (1987); Twisted SU(2) group. An example of noncommutative differential calculus, Publ. RIMS, Kyoto Univ. **23**, 117-181 (1987); Differential calculus on compact matrix pseudogroups (quantum groups), Commun. Math. Phys. **160**, 125-170 (1989); Quantum E(2) group and its Pontryagin dual, Lett. Math. Phys. **23**, 251-263 (1991).

FINITE BRAID GROUPS FOR THE $SU(2)$ KNIZHNIK ZAMOLODCHIKOV EQUATION

IVAN T. TODOROV
Physique Mathématique, Université de Bourgogne
B.P.138, F-21004 Dijon Cedex, France
and
Institute for Nuclear Research, Bulgarian Academy of Sciences
Tsarigradsko Chaussée 72, BG-1784 Sofia, Bulgaria

A Guy RIDEAU, avec appréciation et amitié.

We consider the monodromy representations of the mapping class group $\mathcal{B}_4$ of the 2-sphere with 4 punctures acting in the solutions space of the $\mathfrak{su}(2)$ Knizhnik-Zamolodchikov equation [3] (note that the monodromy representations of the braid group have a more general geometric definition [4]).

The 4 point correlation function

$$(1) \qquad G(\bar{\mathcal{Z}}, \mathcal{Z}; k, I) = \overline{W_\lambda(\mathcal{Z})} A^{\lambda\mu} W_\mu(\mathcal{Z}), \mathcal{Z} = (z_1, z_2, z_3, z_4)$$

is invariant under the $\mathcal{B}_4$ action: $B^* A B = A$ for $A = (A^{\lambda\mu}) = A^*$, $B \in \mathcal{B}_4$. Here k is the Kac Moody level, I is the isospin, $2I \leq k$, $\{W_\mu\}$ span a basis of SU_2 invariant tensor valued solutions of the KZ equation [6]

$$(2) W_\mu(\mathcal{Z}) \in \mathrm{inv} V_I^{\otimes 4} = \mathcal{H}_4^I \; (\dim V_I = 2IH = \dim \mathcal{H}_4^I) \; \mu = 0, 1, \ldots, 2I$$

The braid invariant form A is positive semi definite. For $4I > k$ it has a $4I - k$ dimensional kernel ker A. Only the factor representation $\mathcal{B}^{(k,I)}$ of $\mathcal{B}_4$ acting in $\mathcal{H}^{kI} = \mathcal{H}_4^I/\mathrm{ker}A$ is unitarizable. We ask the question: for which pairs (k, I) is $\mathcal{B}^{kI}$ a finite (matrix) group, so that the KZ equation has an algebraic solution? There is a classical solution (by H.A. Schwarz, 1873) to a problem of this type for the Gauss hypergeometric equation; more than 100 years later the result was extended to higher order hypergeometric equations [1]. Our solution [5] relies on the observation that in the regular basis introduced in [6] the elements of $\mathcal{B}_4$ have entries in a cyclotomic field and so does the matrix A. To formulate this property more precisely, we

J. Bertrand et al. (eds.), Modern Group Theoretical Methods in Physics, 267–269.

first notice that if $\mathcal{C}^{kI}$ is the commutator subgroup of $\mathcal{B}^{kI}$ then the factor group $\mathcal{B}^{kI}/\mathcal{C}^{kI}$ is a finite cyclic group, so that we can replace $\mathcal{B}^{kI}$ by $\mathcal{C}^{kI}$ in our study. We find that

$$\mathcal{C}^{kI} \subset SL(m+1, \mathbb{Q}(q)), m = \frac{1}{2}(k - \mid k - 4I \mid) \tag{3}$$

where $\mathbb{Q}(q)$ is the field of polynomials with rational coefficients of the undeterminate q that is a primitive hth root of -1 for $h = k+2$. Define the *Galois group* Gal_h of the field $\mathbb{Q}(q)$ as the group of all substitutions of the form

$$q \longrightarrow q^l, (l, 2h) = 1, \;\; l \in \mathbb{Z}/2h\mathbb{Z} \tag{4}$$

(in other words, l runs through the coprime with $2h$ integers in the interval $1 \leq l \leq 2h-1$. The number of elements of Gal_h (which, clearly, depends on the divisibility of h) is equal to the degree of the irreducible polynomial P_h such that $P_h(q) = 0$ (for example when $h = 6$ we have $P_6 = q^4 - q^2 + 1$ and $\mathrm{Card}(\mathrm{Gal}_6) = 4$). Then the entries $A^{\lambda\mu}$ of the braid invariant hermitean from A (in our regular basis) belong to the real subfield $\mathbb{Q}(q + \bar{q})$ of $\mathbb{Q}(q)$ where the complex conjugate $\bar{q}$ of q is defined algebraically by $\bar{q} = q^{2h-1}$.

A powerful sufficient condition ensuring the finiteness of $\mathcal{B}^{kI}$ is given by the *total positivity* of A (A is totally positive if it is positive together with all its Galois images). Using the explicit knowledge of A ([8] [7]), we find that the total positivity of A is equivalent to the total positivity of the *quantum dimensions*,

$$[2\lambda + 1] = \frac{q^{2\lambda+1} - \bar{q}^{2\lambda+1}}{q - \bar{q}} > 0 \text{ for } 0 \leq \lambda \leq m. \tag{5}$$

On top of the cyclic groups $\mathbb{Z}_4, \mathbb{Z}_2$ and the trivial braid group (consisting of the identity), all three encountered for $m = 0$, we are left with just four possibilities for $k : k = 1, 2, 4, 8$ (corresponding to the four integral quadratic algebras).

For $I = \frac{1}{2}$ they correspond to the commutator groups $\{1\}, \tilde{\mathfrak{A}}_4$ (the binary tetrahedral group), H_8 (the group of quaternonic units) and $\tilde{\mathfrak{A}}_5$ (the binary icosahedral group). The latter three also appear as the commutator subgroups of the (double coverings) of the symmetry groups of Platonic solids (cf. [2]). We also find a single finite group of 3×3 matrices (corresponding to $I = 1$). In this case $\mathcal{C}^{4,1}$ is isomorphic to the 27 elements Heisenberg group of genereators $b, \bar{b}$ and ω satisfying the relations

$$b^3 = \bar{b}^3 = \omega^3 = 1, b\bar{b} = \omega\bar{b}b \tag{6}$$

with ω a central element.

This presentation is essentially a summary of the results that are being published in [5].

It is a pleasure to thank the friends of Guy Rideau (who organized the meeting) for including me in their company.

References

1. F. Beukers, G. Heckman: "Monodromy for the hypergeometric function nF_{n-1}" *Invent. Math.* **95** (1989), 325-354.
2. V.F.R. Jones: "Braid groups, Hecke algebras and type II_1 factors", in *Geometric Methods in Operator Algebras*, Pitman Research Notes in Math. **123** (1986) 242-273.
3. V.G. Knizhnik, A.B. Zamolodchikov: "Current algebra and Wess-Zumino model in two dimensions", *Nucl. Phys.* **B247** (1984), 83-103.
4. R.J. Lawrence: "Homological representations of the Hecke algebra", *Commun. Math. Phys.*.**135** (1990), 141-191.
5. Ya. S. Stanev, I.T. Todorov: "On Schwarz problem for the $\widehat{\mathfrak{su}}_2$ Knizhnik-Zamolodchikov equation", *Vienna preprint*, ESI **121** (1994), *Lett. Math. Phys.* (to appear).
6. Ya. S. Stanev, I.T. Todorov, L.K. Hadjiivanov:"Braid invariant rational conformal models with a quantum group symmetry, Phys. Lett. B276 (1992), 87-94.
7. Ya. S. Stanev, I.T. Todorov, L. K. Hadjiivanov: "Braid invariant chical conformal models with a quantum group symmetry", in *Quantum Symmetries*, H-D. Doebner, V.K. Dobrev eds. (World Scientific, Singapore 1993) pp. 24-40.
8. I.T. Todorov, Ya. S.Stanev: "Chiral current algebras and 2-dimensional conformal models", Troisième cycle de la physique en Suisse Romande, Lausanne lectures, 1992.

CONTINUOUS SYMMETRIES OF DIFFERENTIAL-DIFFERENCE EQUATIONS

P. WINTERNITZ AND D. LEVI
Centre de recherches mathématiques, Université de Montréal, C.P. 6128, succursale Centre-ville, Montréal, Québec H3C 3J7, Canada.

Abstract. It is shown that the methods of Lie group theory, developed for analyzing differential equations, provide equally powerful tools for solving differential-difference equations.

1. Introduction

The purpose of this presentation is to show how the theory of Lie groups can be used to study and solve differential-difference equations (DDEs). The power of Lie group theory in the treatment of differential equations is well known [1, 2]. Group theory is used to obtain new solutions from old ones, to reduce ordinary differential equations to lower order ones, to reduce the number of independent variables in partial different ial equations, to classify equations into equivalence classes and for many other purposes. Here we shall show that the theory of Lie groups plays the same role for equations that involve discrete variables, in addition to continuous ones [3, 4, 5, 6, 7]. Applications to DDEs involve the following aspects:

1. An efficient algorithm for finding the symmetry group of a DDEs is needed (and presented). Infinitesimal methods are applied for calculating the connected subgroup of the symmetry group.

2. Symmetry reduction (to equations with fewer variables) is performed, using both continuous and discrete subgroups of the symmetry group.

3. The concept of "conditional symmetries", i.e. group transformations that leave only a subclass of solutions invariant [8, 9, 10] can be extended

[1]Permanent address of D. Levi: Dipartimento di Fisica, Terza Universitá di Roma, Via C. Segre, 2–00136 Roma, Italia.

J. Bertrand et al. (eds.), Modern Group Theoretical Methods in Physics, 271–281.

to DDEs and applied to obtain reductions of equations and also Bäcklund transformations.

4. As for differential equations, intriguing relations between symmetries and integrability exist.

5. It is possible to construct DDEs with prechosen symmetry groups, just as one can for differential equations [11, 12, 13].

6. The group classification of DDEs according to their symmetries can be performed [14].

We shall discuss some of these aspects below.

2. Algorithms for constructing the connected part of the symmetry group

For brevity and notational simplicity we restrict ourselves to equations of the type

$$(1)\quad \Delta\Big(x, n, u(n+k, x) \,\Big|_{k=-a}^{b}, u_x(n+k, x) \,\Big|_{k=-a}^{b}, u_{xx}(n+k, x) \,\Big|_{k=-a}^{b}\Big) = 0,$$

where $x \in \mathbb{R}^n$, $n \in \mathbb{Z}$, $a, b \in \mathbb{Z}^>$, $u \in \mathbb{R}$, u_x, u_{xx} denote all first and second order partial derivatives, and Δ is some given function.

We shall look for Lie point symmetries only, i.e. local transformations of the form

$$(2)\qquad \widetilde{u}(n, \widetilde{x}) = \Omega_g(n, x, u(n, x)), \quad \widetilde{x} = \Lambda_g(x, u(n, x)),$$

such that $\widetilde{u}(n, \widetilde{x})$ is a solution of eq. (1), whenever $u(n, x)$ is a solution (g denotes group parameters).

The Lie algebra of the symmetry group is realized in terms of vector fields of the form

$$(3)\qquad \widehat{v} = \sum_{i=1}^{p} \xi_i(x, u(n))\partial_{x_i} + \phi(n, x, u(n))\partial_{u(n)}.$$

In order to generate symmetries, the prolongations of the vector fields must annihilate the equation on its solution surface:

$$(4)\qquad \mathrm{pr}\, \widehat{v} \cdot \Delta \,\Big|_{\Delta=0} = 0.$$

The prolongation $\mathrm{pr}\, \widehat{v}$ acts on functions of x, $u(n, x)$ and of derivatives and differences of $u(n, x)$. From the requirement that eq. (1) should hold for the transformed quantities (2) we obtain

$$
\begin{aligned}
\operatorname{pr}\widehat{v} &= \sum_{i=1}^{p} \xi_i(x, u(n))\partial_{x_i} + \sum_{k=-a}^{b} \Big\{ \phi(k, x, u(n+k))\partial_{u(n+k)} \\
&\quad + \sum_{i=1}^{p} \phi^{x_i}(k, x, \dots)\partial_{u_{x_i}(n+k)} + \sum_{i,j=1}^{p} \phi^{x_i x_j}(k, x, \dots)\partial_{u_{x_i x_j}(n+k)} \Big\} \\
\phi^{x_i}(k) &= D_{x_i}\phi(k) - \sum_{j=1}^{p} (D_{x_i}\xi_j) u_{x_j}(k) \\
\phi^{x_i x_j}(k) &= D_{x_j}\phi^{x_i}(k) - \sum_{\ell=1}^{p} (D_{x_j}\xi_\ell) u_{x_i x_\ell}
\end{aligned} \tag{5}
$$

Here D_{x_i} denotes the total derivate and eq. (5) differs from the prolongation for a purely differential equation [1] precisely by the summation over k, taking into account differences (in addition to derivatives).

Eq. (4) provides a system of linear differential equations for the coefficients ξ_i and ϕ of the vector field $\widehat{v}$. These are the so called determining equations that can be read off as the coefficients of the different linearly independent expressions in $u(n)_x$, $u(n+k)_x$, $u(n+k)_{xx}$, $u(n+k)$, ..., etc, occurring in eq. (4). As for differential equations, the determining system is usually overdetermined and relatively easy to solve. The solution may depend on a finite number of significant integration constants, or on a finite number of arbitrary functions of the independent, or dependent variables. In the first case we obtain a finite dimensional symmetry algebra (and group), in the second case an infinite one.

In Ref. 4 we called the above algorithm the "intrinsic method". An alternative algorithm was also proposed [4], in which the system (1) is treated as an infinite system of differential equations for the separate functions $\dots u_{-1}(x), u_0(x), u_1(x) \dots$, i.e. n is not treated as an independent variable, but simply as a label, numbering different functions. This "differential equation method" is, in general, much more difficult to apply. It may however provide more general results, since the coefficients ξ_i and ϕ_n can now depend on all functions $u(k) \equiv u_k$ rather than on the current one, $u(n)$, only. As far as the discrete label is concerned, the additional transformations are no longer point ones, but rather contact symmetries, or generalized (higher) symmetries.

3. Symmetry group and symmetry reduction for the Toda lattice

Let us apply the algorithm described above to the two-dimensional Toda system (the Mikhailov-Fordy-Gibbons system) [15, 16]

$$u_{xt}(n) = e^{u(n-1)-u(n)} - e^{u(n)-u(n+1)}. \tag{6}$$

The result is an infinite dimensional symmetry algebra [3, 4] the basis of which can be written as

$$\begin{aligned} T(f) &= f(t)\partial_t + f'(t)n\partial_{u_n}, \qquad U(k) = k(t)\partial_{u_n} \\ X(g) &= g(x)\partial_x + g'(x)n\partial_{u_n}, \quad W(h) = h(x)\partial_{u_n} \end{aligned} \tag{7}$$

where $f(t)$, $k(t)$, $g(x)$ and $h(x)$ ($h'(x) \neq 0$) are arbitrary functions of t or x, respectively and the primes denote derivatives with respect to the argument. The commutation relations are

$$[T(f_1), T(f_2)] = T(f_1 f_2' - f_1' f_2), \quad [T(f), U(k)] = U(fk') \tag{8}$$

and similarly for $X(g)$ and $W(h)$. Thus $\{T(f), U(k)\}$ and $\{X(g), W(h)\}$ are two $\widehat{u}(1)$ Kac-Moody-Virasoro algebras.

Thus, eq. (6), involving two continuous and one discrete variable, shares one of the distinctive properties of integrable equations in three (continuous) dimensions, like the equations of the Kadomtsev-Petviashvili hierarchy, the Davey-Stewartson equation and many others. Namely, their Lie point symmetry algebra is not only infinite dimensional, but has a distinctive Kac-Moody-Virasoro structure [17, 18, 19, 20].

The local Lie point transformations generated by the algebra (7) and, by construction, transforming solutions of eq. (6) into solutions, are

$$\begin{aligned} t &= F_{\lambda_1}(\tilde{t}), \quad x = G_{\lambda_2}(\tilde{x}), \\ \tilde{u}_n(\tilde{x}, \tilde{t}) &= u_n(x,t) + n \ln[F'_{\lambda_1}(\tilde{t}) G'_{\lambda_2}(\tilde{x})] + \lambda_3 h(\tilde{x}) + \lambda_4 k(\tilde{t}) \end{aligned} \tag{9}$$

where the functions F and G satisfy the group properties

$$F_0(\tilde{t}) = t, \quad F_{-\lambda_1}(t) = \tilde{t}, \quad G_0(\tilde{x}) = x, \quad G_{-\lambda_2}(x) = (\tilde{x}). \tag{10}$$

Eq. (6) is also invariant under certain discrete transformations, in particular integer translations of n:

$$n' = n + N, \quad N \in \mathbb{Z} \tag{11}$$

as well as simultaneous reflections of x and t and their permutation.

Continuous and discrete subgroups of the symmetry group can be used to perform symmetry reduction. It is convenient to extend the symmetry algebra (7) by a "generator" $Z = \partial/\partial_n$ of the discrete translations (11). Let us consider examples of reductions of eq. (6).

3.1. REDUCTION BY A *CONTINUOUS SUBGROUP*

Consider the group generated by the translation operator $T(1) - vX(1)$. An invariant solution has the form

$$u(n,x,t) = u(n,\xi), \quad \xi = x + \lambda t \tag{12}$$

where $u(n,\xi)$ is a solution of the one-dimensional Toda equation

$$\lambda u_{\xi\xi}(n,\xi) = e^{u(n-1,\xi)-u(n,\xi)} - e^{u(n,\xi)-u(n+1,\xi)}. \tag{13}$$

Eq. (13) is invariant under a four-dimensional symmetry algebra, generated by

$$D = \xi\partial_\xi + 2n\partial_{u(n)}, \quad T = \partial_\xi, \quad W = \xi\partial_{u_n}, \quad U = \partial_{u_n}, \tag{14}$$

to which we add the "discrete" generator $Z = \partial/\partial_n$.

3.2. REDUCTION BY A *DISCRETE SUBGROUP*

The subgroup (11) of translations of N can be used to impose periodicity in n, i.e. put $u(n+N) = u(n)$ for some fixed value of N. For instance, setting $N = 2$ or $N = 3$ we obtain the sine-Gordon, or the Mikhailov-Dodd-Bullough equations, respectively

$$\begin{aligned} \phi_{xt} &= -4\sin\phi, \quad u(2k+1) = -u(2k) \equiv i\phi(x,t) \\ \phi_{xt} &= e^{\,2\phi} - e^{\phi}, \quad u(3) = -u(1) \equiv \phi(x,t), \quad u(2) = 0. \end{aligned} \tag{15}$$

3.3. REDUCTION BY A NON-SPLITTING *MIXED SUBGROUP*

Let us now consider a further reduction of the Toda lattice equation (13). The subgroup we consider is "generated" by the element $Z + aD + bU$, $a \neq 0$, i.e.:

$$\begin{aligned} n' &= n + N, \quad \xi' = e^{aN}\xi, \\ u'(n',\xi') &= u(n,\xi) + (2an+b)N + aN^2 \end{aligned} \tag{16}$$

where $N \in \mathbb{Z}$ is the group parameter. The invariants of this transformation are

$$I_1 = \eta = \xi e^{-an}, \quad I_2 = u - n^2 a - bn. \tag{17}$$

We can hence put:

$$u(n,\xi) = n^2 a + nb + F(\eta) \tag{18}$$

and reduce eq. (13) to a differential-dilation equation in one variable η.

$$\lambda F''(\eta) = e^{-b}\left[e^{F(\eta e^{a})-F(\eta)+a} - e^{F(\eta)-F(\eta e^{-a})-a}\right] \tag{19}$$

We mention that differential dilation equations typically occur in the theory of q-functions [21] and representations of quantum algebras [22, 23, 24].

4. Symmetry classification of discrete dynamical systems

Let us consider a discrete dynamical system with nearest neighbour interactions described by the differential difference equation:

$$\ddot{u}_n(t) = F_n(t, u_{n+1}(t), u_n(t), u_{n-1}(t)). \tag{20}$$

We wish to classify these equations into equivalence classes according to their Lie point symmetries. To do this, we follow the procedure adopted in Ref. 14 to classify variable coefficient KdV equations. We first introduce a classification group, or "group of allowed transformations" that will take equation (20) into a similar equation, in general with a different function $\widetilde{F}_n(\tilde{t}, \widetilde{u}_{n+1}, \widetilde{u}_n, \widetilde{u}_{n-1})$.

We restrict the allowed transformations to the form

$$u_n(t) = \Omega_n(\widetilde{u}_n(\tilde{t}), t), \quad \tilde{t} = \tilde{t}(t), \quad \widetilde{n} = n. \tag{21}$$

Substituting into eq. (20) and requiring that the terms $(\widetilde{u}_{n,t})^p$, $p = 1, 2$ be absent, we find that for any F_n the allowed transformations will be

$$\begin{gathered} u_n(t) = \frac{A_n}{\sqrt{\tilde{t}_t}} \widetilde{u}_n(\tilde{t}) + B_n(t), \\ \tilde{t} = \tilde{t}(t), \quad \widetilde{n} = n, \quad A_{n,t} = 0, \quad \tilde{t}_t \neq 0, \quad A_n \neq 0. \end{gathered} \tag{22}$$

The symmetry algebra will be realized by vector fields of the form

$$\widehat{v} = \tau(t, u_n)\partial_t + \phi_n(t, u_n)\partial_{u_n}. \tag{23}$$

Applying the general algorithm (4), (5), (6) we find that for any interaction F_n we have

$$\tau = \tau(t), \quad \phi_n = \left(\frac{1}{2}\dot{\tau} + a_n\right) u_n + \beta_n(t), \quad \dot{a}_n = 0 \tag{24}$$

where $\tau(t)$, $\beta_n(t)$ and a_n satisfy the determining equation:

$$\frac{1}{2}\dddot{\tau}u_n + \ddot{\beta}_n + \left(a_n - \frac{3}{2}\dot{\tau}\right)F_n - \tau F_{n,t} - \sum_\alpha \left[\left(\frac{1}{2}\dot{\tau} + a_\alpha\right)u_\alpha + \beta_\alpha\right]F_{n,u_\alpha} = 0. \tag{25}$$

Allowed transformations will transform F_n and also take the vector field (23) into a new one with

$$\begin{gathered}\tilde{\tau}(\tilde{t}) = \tau(t(\tilde{t}))\tilde{t}_t, \quad \tilde{a}_n = a_n \\ \widetilde{B}_n(\tilde{t}) = \frac{\sqrt{\tilde{t}_t}}{A_n}\left[\left(\frac{1}{2}\dot{\tau} + a_n\right)B_n + \beta_n - \tau\dot{B}_n\right].\end{gathered} \tag{26}$$

Let us briefly sum up the classification strategy that we follow:

1. First find all interactions F_n that allow at least one-dimensional symmetry algebras. We assume that eq. (25) is satisfied for a vector field. We use the allowed transformations to simplify the vector field and then view eq. (25) as an equation for F_n, to be solved by the method of characteristics. This leads to exactly three possibilities:

(i) $\tau(t) \neq 0$:

$$\hat{v} = \partial_t + a_n u_n \partial_{u_n}, \quad F_n = f_n(\xi_{n-1}, \xi_n, \xi_{n+1})e^{a_n t}, \quad \xi_k = u_k e^{-a_k t} \tag{27}$$

(ii) $\tau(t) = 0$, $a_n \neq 0$:

$$\hat{v} = a_n u_n \partial_{u_n}, \quad F_n = f_n(t, \xi_{n-1}, \xi_{n+1}), \quad \xi_k = u_k^{a_n} u_n^{-a_k} \tag{28}$$

(iii) $\tau(t) = 0$, $a_n = 0$, $\beta_n \neq 0$:

$$\hat{v} = \beta_n(t)\partial_{u_n}, \quad F_n = \frac{\ddot{\beta}_n}{\beta_n}u_n + f_n(t, \xi_{n-1}, \xi_{n+1}), \quad \xi_k = \beta_n u_k - \beta_k u_n. \tag{29}$$

2. Find all interactions allowing Abelian symmetry algebras. We proceed by dimension. Starting from a basis element X_1 in standard form, i.e. in form (27), (28) or (29) we find X_2 such that $[X_1, X_2] = 0$ and again use eq. (25) to restrict the form of F_n further. Adding further operators successively, we find that interactions exist, allowing for Abelian symmetry algebras L with $1 \leq \dim L \leq 4$.

3. Find all interactions allowing for nilpotent nonabelian symmetry algebras. These exist for $3 \leq \dim L \leq 5$ and can be found by extending the previously found abelian symmetry algebras.

4. Find all solvable nonnilpotent symmetry algebras. They can be constructed starting from the already known abelian and nilpotent ones, using these as nilradicals. Solvable symmetry algebras exist for $2 \leq \dim L \leq 7$.

5. For nonsolvable Lie algebras we use the Levi decomposition. It is easy to show that any nonsolvable symmetry algebra of eq. (20) must contain $\mathrm{sl}(2,\mathbb{R})$ as a subalgebra. The algebra $\mathrm{sl}(2,\mathbb{R})$ can, up to allowed transformations, be realized in one way only, namely as

$$X_1 = \partial_t, \quad X_2 = t\partial_t + \frac{1}{2}u_n\partial_{u_n}, \quad X_3 = t^2\partial_t + tu_n\partial_{u_n}. \tag{30}$$

The invariant interaction is

$$F_n = \frac{1}{u_n^3} f_n(\xi_{n-1}, \xi_{n+1}), \quad \xi_k = \frac{u_k}{u_n}. \tag{31}$$

Up to four further symmetry operators can be added as a radical of the symmetry algebra (implying further restrictions on f_n in eq. (31)).

The complete classification will be presented elsewhere [25], here we just give three further examples.

(i) Four-dimensional solvable Lie algebra with a nonabelian nilradical

$$\begin{gathered} X_1 = \partial_{u_n}, \quad X_2 = t\partial_{u_n}, \quad X_3 = \partial_t, \quad X_4 = t\partial_t + \gamma_n\partial_{u_n}, \\ \gamma_n \neq \gamma_{n+1}, \quad \dot{\gamma}_n = 0. \end{gathered} \tag{32}$$

The invariant interaction is

$$\begin{gathered} F_n = e^{-2(u_n - u_{n+1})/(\gamma_n - \gamma_{n+1})} f_n(\xi), \\ \xi = (\gamma_{n+1} - \gamma_n)u_{n-1} + (\gamma_{n-1} - \gamma_{n+1})u_n + (\gamma_n - \gamma_{n-1})u_{n+1}. \end{gathered} \tag{33}$$

As a special case, we obtain the Toda lattice with its symmetry algebra (14) by putting

$$\gamma_n = 2n, \quad f_n(\xi) = (-1 + e^{\xi/2}). \tag{34}$$

It would be interesting to establish whether the Toda lattice is the only integrable member of the family (33).

(ii) Seven-dimensional nonsolvable symmetry algebra:

$$\begin{gathered} X_1 = \partial_t, \quad X_2 = t\partial_t + \frac{1}{2}u_n\partial_{u_n}, \quad X_3 = t^2\partial_t + tu_n\partial_{u_n} \\ X_4 = \partial_{u_n}, \quad U_5 = t\partial_{u_n}, \quad X_6 = \gamma_n\partial_{u_n}, \quad X_7 = \gamma_n t\partial_{u_n} \\ \dot{\gamma}_n = 0, \quad \gamma_{n+1} \neq \gamma_n. \end{gathered} \tag{35}$$

The invariant interaction is

$$F_n = c_n \xi^{-3}, \quad \dot{c}_n = 0 \tag{36}$$

with ξ as in eq. (33).

(iii) Six-dimensional solvable symmetry algebra with five-dimensional non-abelian nilradical

$$X_1 \equiv \partial_{u_n}, \quad X_2 = (-1)^n \partial_{u_n}, \quad X_3 = t\partial_{u_n}, \quad X_4 = (-1)^n t\partial_{u_n}$$
$$X_5 = \partial_t, \quad X_6 = [a + (-1)^n] u_n \partial_{u_n}, \quad \dot{a} = 0, \quad |a| \neq 1.$$

The interaction is

$$F_n = c_n (u_{n-1} - u_{n+1})^{(a+(-1)^n)/(a-(-1)^n)}. \tag{37}$$

5. Conclusions

The main conclusion is that Lie group theory works just as well for differential-difference equations as it does for differential ones. The construction of the symmetry group is algorithmic and can be computerized. Symmetry reduction can be performed. The group classification of differential-difference equations is feasible and useful. Contact and higher symmetries exist and can be found.

Many questions remain open. Among them is that of finding all discrete symmetries of an equation and of symmetries acting simultaneously on discrete and continuous variables. Progress in this direction has been made using the machinery of quantum groups [26], so far for linear equations. The challenge is to extend the approach to nonlinear ones.

A somewhat related problem is that of discretizing differential equations in a manner that preserves all Lie point symmetries [27]. This can be done, but it usually leads to discrete equations on irregular lattices [28].

6. Acknowledgements

The authors thank G. Rideau for many interesting discussions and profound comments on the topic of this article and many others.

The research of P.W. was partially supported by research grants from NSERC of Canada and FCAR du Québec.

References

1. P.J. Olver (1986) *Applications of Lie Groups to Differential Equations*, Springer, New York.

2. P. Winternitz (1993) Lie Groups and Solutions of Nonlinear Partial Differential Equations, *Integrable systems, Quantum Groups and Quantum Field Theories*, Kluwer, Dordrecht, (L.A. Ibort and M.A. Rodriguez, eds)., pp. 429–495.
3. D. Levi and P. Winternitz (1991) Continuous Symmetries of Discrete Equations, *Phys. Lett. A* **152**, pp. 335–338 .
4. D. Levi and P. Winternitz (1993) Symmetries and Conditional Symmetries of Differential Difference Equations, *J. Math. Phys.* **34**, pp. 3713–3730.
5. S. Maeda (1980) Canonical Structure and Symmetries for Discrete Systems, *Math. Japon.* **25**, pp. 405–420
6. S. Maeda (1987) The Similarity Method for Discrete Equations, *IMA J. Appl. Math*, **38**, pp. 129–134.
7. G.R.W. Quispel, H.W. Capel and R. Sahadevan (1992) Continuous Symmetries of Difference Equations; the Kac-van Moerbeke Equation and Painlevé Reduction, *Phys. Lett.* **170A**, pp. 379–383.
8. G.W. Bluman and J.D. Cole (1969) The General Similarity Solution of the Heat Equation, *J. Math. Mech.* **18**, pp. 1025–1042.
9. P.J. Olver and Ph. Rosenau (1987) Group Invariant Solutions of Differential Equations, *SIAM J. Appl. Math.* **47**, pp. 263–278.
10. D. Levi and P. Winternitz (1989) Nonclassical Symmetry Reduction: Example of the Boussinesq Equation, *J. Phys. A: Math. Gen.*, **22**, pp. 2915–2924.
11. D. David, D. Levi, and P. Winternitz (1988) Equations Invariant under the Symmetry Group of the Kadomtsev-Petviashvili Equation, *Phys. Lett. A* **129**, pp. 161–164.
12. G. Rideau and P. Winternitz (1990) Nonlinear Equations Invariant under Poincaré, Similitude and Conformal Groups in Two-Dimensional Space-Time, *J. Math. Phys* **31**, pp. 1095–1105.
13. G. Rideau and P. Winternitz (1993) Invariant Nonlinear Schrödinger Equations in Two-Dimensional Space-Time, *J. Math. Phys.* **34**, pp. 558–569.
14. J.P. Gazeau and P. Winternitz (1992) Symmetries of Variable Coefficient Korteweg-de Vries Equations, *J. Math. Phys.* **33**, pp. 4087–4102.
15. A.V. Mikhailov (1979) Integrability of a Two Dimensional Generalization of the Toda Chain, *JETP Letters* **30** pp. 414–418.
16. A.P. Fordy and J. Gibbons (1980) Integrable Nonlinear Klein-Gordon Equations and Toda Lattices, *Commun. Math. Phys.* **77**, pp. 21–30.
17. D. David, N. Kamran, D. Levi, and P. Winternitz (1985) Subalgebras of Loop Algebras and Symmetries of the Kadomtsev-Petviashvili Equation, *Phys. Rev. Lett.* **55**, pp. 2111–2113.
18. B. Champagne and P. Winternitz (1988) On the Infinite Dimensional Symmetry Group of the Davey Stewartson Equations, *J. Math. Phys.* **29**, pp. 1–8.
19. L. Martina and P. Winternitz (1989) Analysis and Applications of the Symmetry Group of the Multidimensional Three Wave Resonant Interaction Problem, *Ann. Phys.* **196**, pp. 231–277.
20. A. Yu. Orlov and P. Winternitz (1994) Loop Algebra Symmetries and Commuting Flows for the Kadomtsev-Petviashvili Hierarchy, *Preprint CRM-1936*, Montréal.
21. G. Gasper and M. Rahman (1990) *Basic Hypergeometric Series*, Cambridge Univ. Press, Cambridge.
22. R. Floreanini and L. Vinet (1993) Quantum Algebras and q-Special Functions, *Ann. Phys.* **221**, pp. 53–79.
23. E.G. Kalnins, H.L. Manocha and W. Miller Jr. (1992) Models of q-algebra Representations: Tensor Products of Special Unitary and Oscillator Algebras, *J. Math. Phys.* **33** pp. 2365–2383.
24. G. Rideau and P. Winternitz (1993) Representations of the Quantum Algebra $su_q(2)$ on a Real Two-Dimensional Sphere, *J. Math. Phys.* **34**, pp 6030–6044.
25. D. Levi and P. Winternitz (1995) Symmetry Classification of Discrete Dynamical Systems, *Preprint CRM-2256*, Montréal.
26. R. Floreanini and L. Vinet (1994) Symmetries of q-Difference Heat Equation, *Lett.*

Math. Phys. **32**, pp. 37–44.

27. V.A. Dorodnitsyn (1994) Symmetry of Finite Difference Equations, *CRC Handbook of Lie Group Analysis of Differential Equations*, **Vol. 1** CRC Press, Boca Raton (N.H. Ibragimov, ed).
28. V.A. Dorodnitsyn, M. Sanielevici, and P. Winternitz (1995) Symmetry Preserving Discretization of Variable Coefficient KdV Equations, *Preprint CRM-2230*, Montréal.

PROCESSUS LIMITES ET IRRÉVERSIBILITÉ EN MÉCANIQUE STATISTIQUE HORS D'ÉQUILIBRE

RAYMOND JANCEL
Laboratoire de Physique Théorique et Mathématique,
Université Denis Diderot - Paris VII,
Tour Centrale 3e étage, 2 place Jussieu,
F-75251 Paris Cedex 05, France.

A mon ami Guy Rideau,
en souvenir de notre très longue collaboration et de nos nombreux et enrichissants échanges intellectuels.

Sommaire. On se propose dans cet article d'étudier la démarche qui conduit à la démonstration de résultats rigoureux en Mécanique statistique hors d'équilibre, en montrant qu'elle est rendue possible par l'existence, à certaines limites, de propriétés asymptotiques caractéristiques des systèmes à un très grand nombre de particules. Après avoir défini les importantes notions de description réduite autonome et de processus limite, on examine le passage du niveau microscopique au niveau cinétique qui fournit un exemple significatif d'une telle démarche. Dans le cas d'un fluide idéal, ce passage est bien décrit par la limite de Boltzmann-Grad et par le théorème de Lanford, dont la démonstration permet de *déduire* le régime cinétique de Boltzmann à partir de la description microscopique et de déterminer clairement dans quelles conditions peuvent être obtenues l'irréversibilité de l'évolution et la réduction de la description de l'état du système.

1. Introduction.

Dès ses débuts, la théorie atomistique de la matière s'est trouvée confrontée au problème de la compatibilité entre deux descriptions de la même réalité physique, à deux niveaux d'observation très différents:
La description macroscopique, qui est celle de notre expérience sensible, à notre échelle; elle utilise un petit nombre de grandeurs continues, telles que les variables locales de l'hydrodynamique, dont l'évolution est régie

J. Bertrand et al. (eds.), Modern Group Theoretical Methods in Physics, 283–314.

par des lois essentiellement irréversibles (lois de Fick, de Fourier, etc...,), impliquant dissipativité et croissance de l'entropie. Ces lois macroscopiques sont généralement asymétriques par rapport au temps, la "flèche du temps" coïncidant avec la direction de l'évolution et celle de l'entropie croissante pour un système isolé.

La description microscopique, où la matière est conçue comme discontinue et constituée d'un très grand nombre $N(\simeq 10^{23})$ de particules, en agitation perpétuelle et désordonnée, dont le mouvement obéit aux lois réversibles de la Mécanique classique ou quantique. Celles-ci n'impliquant pas de direction privilégiée du temps, il y a totale symétrie entre passé et futur, de sorte que le rôle du temps se réduit à celui d'un simple paramètre entrant dans la prédiction (ou la "rétrodiction") des états successifs d'un système mécanique conservatif.

Pour que la compatibilité entre ces deux descriptions puisse être établie, il faut donc être en mesure de répondre aux questions suivantes: comment expliquer l'émergence à notre échelle de processus physiques réguliers, décrits par quelques variables continues, à partir de l'extrême complexité des mouvements d'un nombre énorme de particules? Pourquoi cette émergence s'accompagne-t-elle de la brisure de symétrie que comporte nécessairement le passage de l'évolution microscopique réversible à l'irréversibilité de fait des phénomènes macroscopiques?

On sait que c'est précisément l'objet de la Mécanique statistique d'établir un pont entre ces deux descriptions, en mettant en œuvre des méthodes fondées sur l'introduction des notions de probabilité et d'ensembles statistiques de systèmes. Les problèmes qui lui sont posés par le passage de la description microscopique à la description macroscopique peuvent alors se ramener aux deux questions fondamentales:

A - Comment s'opère la *réduction de la description* de l'état du système au "niveau" macroscopique?

B - Comment expliquer la *brisure de symétrie* qui accompagne ce processus de réduction?

Ce sont certains aspects de ces problèmes que nous nous proposons de traiter ici, essentiellement ceux qui se rapportent à la recherche de *résultats rigoureux* en Mécanique statistique hors d'équilibre (MSHE), pour des systèmes classiques. En effet, bien que les profondes intuitions des pères fondateurs de ces disciplines (Boltzmann, Gibbs, ...), si magistralement exposées par P. et T. Ehrenfest [1], n'aient pas laissé de doute sur le bien fondé des méthodes de la physique statistique, il n'en demeure pas moins que la nature qualitative de leurs arguments, et de ceux de leurs successeurs, ne permet pas d'apporter des réponses claires et définitives aux questions A et B. Cela tient pour l'essentiel à ce qu'il est souvent difficile de distinguer dans les raisonnements la part qui revient aux arguments de nature probabiliste (loi des grands nombres), et à ceux qui résultent du comportement dynamique du système (réversibilité, propriétés ergodiques,...). Dans bien des cas, les conclusions semblent s'appuyer sur des hypothèses ou des approximations plus ou moins bien justifiées, toujours susceptibles d'entrer en conflit avec les lois dynamiques (cas de l'hypothèse du chaos moléculaire).

C'est ce qui fait l'intérêt de la recherche de résultats rigoureux en MSHE, dans l'esprit de ceux obtenus pour les systèmes à l'équilibre avec le passage à la *limite thermodynamique*, qui est à la base de toute la thermodynamique statistique. Les premiers travaux dans ce domaine ne sont apparus qu'assez tardivement dans la littérature, avec les travaux de pionnier de Grad (1958) sur l'équation de Boltzmann [2] et le comportement qualitatif des systèmes à un grand nombre de degrés de liberté (1967) [3], puis avec le théorème de Lanford (1975) [4] concernant l'évolution limite d'un fluide de N sphères dures.

Ces premiers résultats peuvent en fait s'inscrire dans une démarche plus générale visant à développer le point de vue suivant: puisque les deux descriptions microscopique et macroscopique d'un même système sont l'une et l'autre considérées comme valables, chacune dans leur domaine propre, leur compatibilité exige qu'il soit possible d'élaborer une méthode permettant une *déduction* rigoureuse des équations du domaine macroscopique à partir des lois dynamiques exactes du "niveau" microscopique. Les méthodes mises en œuvre dans cette démarche s'appuient essentiellement sur l'argument selon lequel les réponses aux questions A et B sont à rechercher dans le comportement asymptotique, pour certaines *limites*, des systèmes macroscopiques, ce comportement étant étroitement lié aux propriétés géométriques spécifiques des espaces à un très grand nombre de dimensions.

Ce sont les implications de cet argument en MSHE que nous traitons dans cet exposé, en les plaçant dans la perspective du développement des concepts fondamentaux de la Mécanique statistique. C'est pourquoi nous partons de la théorie des systèmes à l'équilibre, dont les idées essentielles (limite thermodynamique) sont brièvement rappelées dans la section 2, pour examiner ensuite dans la section 3 comment ces idées peuvent être adaptées aux problèmes spécifiques de la MSHE. Nous verrons ainsi se dégager les importantes notions de "niveaux de description", associés aux diverses échelles caractéristiques du système, puis de " *descriptions réduites autonomes*" et de *processus limites* qui décrivent des régimes particuliers de l'évolution macroscopique. Ces différents "niveaux" correspondent à des descriptions de plus en plus "macroscopiques" de l'état du système, chaque changement dans ce sens s'accompagnant d'un certain "processus de réduction" de la description où se trouvent impliquées des questions du type A et B.

Etant donné la grande variété des phénomènes physiques hors d'équilibre, on conçoit sans peine que l'application de ces méthodes conduit en général à des solutions très différentes selon la nature du problème étudié. Comme il est impossible d'être exhaustif dans le cadre d'un tel exposé, nous avons choisi de traiter dans les sections 4-6 les problèmes posés par la réduction de la description au niveau cinétique, pour le cas particulier d'un fluide idéal satisfaisant à l'équation de Boltzmann. Il y a plusieurs raisons à ce choix :

(i) d'abord, le régime cinétique d'un fluide idéal est parfaitement défini par la limite de Boltzmann-Grad;

(ii) ensuite, le formalisme de la hiérarchie BBGKY fournit un cadre naturel pour traiter le problème de la réduction de la description à cette limite;

(iii) de plus, la comparaison (effectuée dans les sections 4 et 5) entre la méthode de la hiérarchie et celle de la théorie cinétique des gaz offre, au-delà de son intérêt historique, la possibilité de mieux cerner la signification physique des opérations de passage à la limite;

(iv) enfin, on dispose dans ce cas d'un résultat rigoureux, le théorème de Lanford, qui établit, sous certaines conditions, la convergence à la limite de Boltzmann-Grad des solutions (réversibles) de la hiérarchie, vers celles de l'équation (irréversible) de Boltzmann; en outre, chaque étape de cette démonstration se prête à une certaine interprétation physique et apporte un élément de réponse aux questions A et B.

Ainsi, l'étude du régime cinétique de Boltzmann-Grad fournit un exemple significatif de la démarche générale évoquée ci-dessus et permet notamment, comme nous le verrons dans la section 6, de clarifier les mécanismes qui sont à l'origine du comportement irréversible du système et de la réduction de la description de son état.

2. Les Principes généraux de la Mécanique statistique - Cas de l'équilibre.

L'introduction de la notion de probabilité en Physique statistique trouve sa justification dans deux propriétés remarquables de l'observation macroscopique fine: d'une part, la mesure d'une grandeur macroscopique est en réalité un événement *non-reproductible*, en raison de l'existence de fluctuations imprévisibles et totalement irrégulières des valeurs de cette grandeur; d'autre part, c'est une observation hautement *incomplète* de l'état du système, en ce sens qu'il existe toujours un *ensemble* d'états microscopiques possibles, compatibles avec le *même* état macroscopique. Il est ainsi naturel de considérer des ensembles d'échantillons (E), identiques au système considéré, et de les munir d'une certaine loi de probabilité; ce sont les ensembles de Gibbs de la Mécanique statistique. En théorie classique, où l'état d'un système est représenté par un point P de l'espace des phases Γ, ils s'identifient à des ensembles de l'espace Γ sur lequel on a défini une certaine densité de probabilité $\rho(P)$. Dans ce qui suit, nous considérerons toujours un système *classique*, constitué de N particules de coordonnées $\vec{X}_i = (\vec{r}_i, \vec{p}_i)$, contenues dans une enceinte Λ de volume $V(\Lambda)$, de sorte que l'on aura:

$$\begin{aligned} P &= (\vec{X}_1, \ldots, \vec{X}_N) \in \Gamma_N = (\Lambda \times R^3)^N, \\ \text{et:} \quad \rho(P) &= \rho_N(\vec{X}_1, \ldots, \vec{X}_N) \equiv \rho_N(P). \end{aligned} \tag{2.1}$$

Il importe de remarquer que le point de vue adopté revient à considérer la donnée d'une densité en phase $\rho_N(P)$ comme une "notion première" de la théorie.

Le postulat fondamental. Ceci étant, la méthode générale de la Mécanique statistique consiste à faire correspondre, à toute grandeur macroscopique A, une *variable aléatoire* $A(E;t)$, définie sur l'échantillon (E), telle que sa valeur moyenne $\overline{A}^E(E;t)$ soit égale à la valeur de A effectivement observée, et que sa dispersion relative soit très petite et tende vers zéro à la limite des "*grands systèmes*" (pour laquelle $N \longrightarrow \infty$ d'une manière qui est à préciser). C'est le *postulat fondamental* de la Mécanique statistique, qui s'exprime par les deux conditions:

$$\overline{A(E;t)}^E (\equiv \overline{A}^E(t)) = \text{ grandeur macroscopique observée;} \tag{2.2}$$

$$\begin{gathered} \overline{(A - \bar{A}^E(t))^2}^E / (\overline{A}^E(t))^2 \equiv \overline{(\delta A)^2}^E / (\bar{A}^E(t))^2 \longrightarrow 0, \\ \text{à la limite des "grands systèmes" } (N \longrightarrow \infty). \end{gathered} \tag{2.3}$$

Mécanique statistique à l'équilibre (MSE). L'évolution dynamique n'étant pas en jeu dans ce cas, l'application des principes fondamentaux ne rencontre pas de difficultés majeures pour deux raisons:

(i) D'une part, les ensembles représentatifs de l'état d'équilibre sont bien connus (ce sont les ensembles micro-canonique, canonique et grand-canonique de la MSE) et les grandeurs physiques sont bien définies (variables thermodynamiques de densité n, de pression p, de température T...).

(ii) D'autre part, l'état d'équilibre étant généralement unique, un seul passage à la limite des "grands systèmes" est bien approprié à cette situation: c'est la *limite thermodynamique*, définie par:

$$N, V, E \longrightarrow \infty, \text{ avec } N/V \longrightarrow n, E/N \longrightarrow e, \tag{2.4}$$

où N, V et E sont respectivement le nombre total de particules, le volume et l'énergie totale du système, et où n et e sont des constantes (finies) représentant les limites de la densité numérique et de la densité d'énergie.

Le rôle de ce passage à la limite est bien illustré par le comportement d'une grandeur extensive, qui croît proportionnellement à N; on sait en effet, d'après la théorie des fluctuations à l'équilibre, que la dispersion $\overline{(\delta A)^2}$ d'une telle grandeur est aussi proportionnelle à N [5], de sorte que l'on a pour sa dispersion relative: $\overline{(\delta A)^2}/(\overline{A})^2 \simeq 0(1/N)$. On constate ainsi que la valeur relative des fluctuations de A est négligeable dès que N est très grand et devient rigoureusement nulle lorsque $N \longrightarrow \infty$; ce n'est donc que lorsque cette limite est atteinte que la valeur moyenne $\overline{A}$ (si elle existe!) peut être *rigoureusement* identifiée à la grandeur macroscopique observée.

Ces considérations, qui permettent d'établir le statut de la MSE sur une base *rigoureuse* [6], fournissent une bonne illustration de la démarche générale suivie en Mécanique statistique. Elles mettent en évidence le rôle crucial du passage à la limite thermodynamique, qui conduit à une extension, aux ensembles de systèmes hamiltoniens, des résultats du théorème limite central du calcul des probabilités. Comme la loi des grands nombres dont ils sont une autre expression, ces résultats reposent en dernière analyse sur les propriétés géométriques asymptotiques des espaces à un très grand nombre de dimensions $N(\simeq 10^{23})$. Celles-ci ont une conséquence capitale pour notre propos, à savoir que la mesure (dans Γ_N) de l'ensemble correspondant à l'état maxwellien est considérablement plus grande que celle de tous les autres états; cette énorme *prédominance de l'état maxwellien* confère à certains types de fonctions de phase un comportement caractéristique de "*plateau*" au voisinage de leur valeur pour cet état (cf. section 5).

On voit ainsi comment le recours à la limite thermodynamique permet de répondre à la question A concernant la réduction de la description de l'état du système dans le cas de l'équilibre. On constate aussi que ce processus de réduction s'accompagne d'un changement *qualitatif* de l'état du système, qui se manifeste dans le fait que les grandeurs thermodynamiques ont à la limite thermodynamique, et *seulement* à cette limite, un comportement non-analytique caractérisé par l'existence de discontinuités au voisinage de points critiques.

3. Problèmes propres à la Mécanique statistique hors d'équilibre.

Nous étudions dans cette section l'application des principes généraux à la description des systèmes hors d'équilibre. Pour ce faire, nous passons brièvement en revue quelques-unes des difficultés les plus importantes rencontrées dans ce domaine, afin d'être en mesure de préciser la nature et les caractères spécifiques des passages à la limite de la MSHE (pour un exposé plus détaillé, voir par exemple [7]).

Equation de Liouville. Conformément aux principes généraux de la section 2, l'état d'un système macroscopique hors d'équilibre doit être représenté par un ensemble statistique dépendant du temps défini par une densité en phase à N particules, $\rho_N(P;t)$. Cette densité en phase correspond à la " description maximale", c'est à dire la plus complète possible de l'état du système (au sens de la Mécanique statistique), et son évolution instantanée est gouvernée par l'équation de Liouville. Celle-ci est une conséquence directe des équations du mouvement et peut se mettre sous la forme:

$$\frac{\partial \rho_N}{\partial t} + [\rho_N, H_N] = \frac{\partial \rho_N}{\partial t} - L_N \rho_N = 0, \tag{3.1}$$

où $L_N \equiv [H_N, \quad]$ est l'opérateur de Liouville à N particules associé au flot hamiltonien ($P \longrightarrow P_t = T_t P$) engendré par le mouvement du système dans Γ_N. Mise sous la forme $\frac{d\rho_N}{dt} = 0$, où $\frac{d}{dt}$ est la dérivée totale "suivant le mouvement", on voit qu'elle n'est autre que l'équation de continuité d'un fluide incompressible dans Γ_N, dont la densité ρ_N est conservée au cours du mouvement: c'est le *théorème de Liouville*, qui implique en particulier l'invariance de la mesure élémentaire $d\Gamma_N$.

Tout comme les équations de la Mécanique classique, dont elle est une conséquence directe, l'équation de Liouville (3.1) est réversible et ne comporte aucune direction privilégiée du temps; elle ne peut donc, à elle seule, décrire le comportement irréversible d'un système macroscopique. Comme elle est du premier ordre en t, sa solution peut s'écrire formellement:

$$\rho_N(P;t) = U_N(t)\, \rho_N(P;0), \tag{3.2}$$

où $U_N(t) \equiv \exp[Lt]$ est l'opérateur d'évolution induit dans l'espace fonctionnel des ρ_N par le flot hamiltonien , et où $\rho_N(P;0) \equiv \rho_N(P)$ est la densité en phase à l'instant initial $t = 0$. On vérifie ainsi que l'équation de Liouville conduit à une évolution *rigoureusement déterministe*, puisque $\rho_N(P;t)$ est (en principe) bien déterminée, d'après (3.2), dès que $\rho_N(P;0)$ est connue; *tout* le contenu statistique de la théorie se trouve donc inclus dans la densité initiale $\rho_N(P;0)$.

Le choix de cette densité initiale comporte en fait une grande part d'arbitraire, à l'exception de quelques conditions très générales auxquelles doivent satisfaire les ensembles statistiques hors d'équilibre "*physiquement possibles*". La plus importante pour notre propos, soulignée particulièrement par Penrose [8], consiste dans le fait que de tels ensembles doivent être en général *asymétriques* par renversement du temps, cette asymétrie ayant pour fonction d'éliminer les évolutions inverses théoriquement possibles, mais physiquement irréalisables et inobservées. En bref, le choix d'un certain $\rho_N(P;0)$ "physiquement possible" vaut à titre de condition initiale mais *non* comme condition finale; cette propriété apparaît en fait comme une simple conséquence du *principe de causalité* [8, 9], qui a pour effet d'introduire en Mécanique statistique la distinction entre le passé et le futur, c'est-à-dire de fixer le sens de la "flèche du temps".

Ceci étant, le choix d'une densité initiale reste encore largement indéterminé, de sorte que l'on ne dispose en MSHE, contrairement au cas de l'équilibre, d'aucune prescription générale permettant de construire sans ambiguïté les états statistiques correspondant à des situations physiques (hors d'équilibre) données. La détermination de tels ensembles demande en fait un examen approfondi des conditions mêmes de l'observation macroscopique, et elle rencontre deux difficultés essentielles, l'une qui est liée au problème de l'irréversibilité, l'autre à la définition de variables macroscopiques.

Irréversibilité et coarse-graining. Comme l'on s'intéresse maintenant à l'évolution instantanée du système, on se trouve confronté au problème fondamental posé par l'interprétation de l'irréversibilité macroscopique (question B), dont les difficultés ont été illustrées par deux paradoxes bien connus: celui de Loschmidt (lié à la réversibilité des systèmes mécaniques), et celui de Zermelo (résultant de leurs propriétés de récurrence).

Pour tenter de surmonter ces difficultés, on a recours au procédé bien connu de "*coarse-graining*" (ou "granulation") qui joue un rôle majeur en MSHE; il consiste à considérer une certaine partition de l'espace des phases (ou de l'ensemble des valeurs possibles d'une grandeur) en "cellules" macroscopiques, dont les dimensions sont liées aux imprécisions inhérentes à toute observation macroscopique. On admet alors que l'observateur peut seulement distinguer les états du système appartenant à *différentes* cellules, d'où l'introduction de quantités "*semi-fines*" (coarse-grained) obtenues en prenant la moyenne des quantités "fines" correspondantes sur tous les états d'une même cellule. Il est alors aisé de voir que ces quantités semi-fines ne satisfont pas à l'équation de Liouville et que leur évolution est différente de celle des quantités fines correspondantes; il apparaît donc possible, sous certaines conditions, d'éviter les paradoxes évoqués ci-dessus et d'aboutir à une évolution irréversible pour ces quantités semi-fines.

En réalité, l'application de ce procédé rencontre deux difficultés importantes; l'une provient de la grande part d'arbitraire que comporte la définition d'un certain système de cellules; l'autre est liée au fait que la connaissance des quantités semi-fines à un instant donné ne suffit pas pour déterminer leur valeur à un instant ultérieur. Il s'ensuit que le "coarse-graining" ne permet pas à lui seul de répondre aux questions A et B (en particulier, on ne peut ainsi établir l'approche monotone de l'équilibre), et que ces difficultés ne peuvent être surmontées que par une étude plus approfondie de la nature des variables macroscopiques.

Variables macroscopiques et "niveaux de description". La seconde difficulté de principe évoquée plus haut concerne précisément ce problème; elle tient au fait qu'il existe une certaine indétermination dans le choix et la définition des variables macroscopiques pertinentes pour décrire un certain état hors d'équilibre. A cette indétermination, s'ajoute celle qui est due à l'existence de différents "*niveaux de description*" possibles de l'état macroscopique d'un système. Ces niveaux, auxquels correspondent des régimes particuliers de l'évolution macroscopique, sont eux-mêmes liés à l'existence des différentes échelles (d'espace et de temps) qui sont associées aux longueurs et aux temps caractéristiques du système.

Naturellement, les propriétés du système sont très différentes selon l'ordre de grandeur respectif de ces grandeurs; un cas particulièrement intéressant pour notre propos est celui des *fluides dilués*, pour lesquels on a:

$$r_0 << d << l << L, \quad \text{avec } nr_0^3(\simeq r_0^3/d^3) \equiv \epsilon << 1, \tag{3.3}$$

où r_0 est la portée du potentiel interparticulaire, $d \simeq n^{-1/3}$ la distance moyenne entre les particules, $l \simeq 1/nr_0^2$ le libre parcours moyen (l.p.m.) et L une longueur macroscopique caractérisrique, et où la condition vérifiée par la densité "sans dimension" ϵ résulte des deux premières inégalités de (3.3). Selon les valeurs de ϵ, on distingue : les fluides *modérément denses* où ϵ, bien que petit, *reste fini*, et le *fluide idéal* pour lequel $\epsilon \longrightarrow 0$; c'est le cas que nous considérons dans cet article. Quant à la dernière inégalité de (3.3), elle caractérise le *régime hydrodynamique* du fluide (avec un paramètre de Knudsen $l/L \simeq \epsilon_K << 1$), dans lequel s'est établi un certain équilibre local sous l'effet de très nombreuses collisions. On est finalement conduit à considérer, dans le cas des fluides dilués, les trois niveaux de description suivants:

- *La description fine ou "maximalc"*, dans laquelle l'état du système est défini par la densité en phase $\rho_N(\vec{X}_1, \ldots \vec{X}_N; t)$ gouvernée par l'équation de Liouville (3.1); c'est le niveau le plus "profond", avec un temps caractérisrique de l'ordre de $\tau_0 \simeq r_0/\overline{v}$ ($\overline{v}$ vitesse thermique).

- *Le niveau cinétique*, qui procède d'une première réduction de la description maximale, de telle manière que l'état du système est maintenant décrit par une fonction de distribution à *une* particule $f(\vec{r}, \vec{p}; t)$ définie sur l'espace $\mu(\ni \vec{r}, \vec{p})$; cette équation obéit à une équation de type cinétique, telle que l'équation de Boltzmann dans le cas d'un fluide idéal, avec un temps caractérisrique de l'ordre du temps de l.p.m. : $\tau_{\text{l.p.m.}} \simeq l/\overline{v}$.

- *Le niveau hydrodynamique*, qui est celui de l'expérience usuelle où l'état du système est défini (dans le cas le plus simple) par les cinq variables hydrodynamiques locales $n(\vec{r}, t)$, $\vec{u}(\vec{r}, t)$ et $T(\vec{r}, t)$; celles-ci satisfont aux équations de l'hydrodynamique, avec des temps caractéristiques qui dépendent de ϵ_K. Notons que ce dernier niveau peut être conçu comme résultant, soit d'une seconde réduction de la description à partir du niveau cinétique, soit d'une réduction globale effectuée directement à partir de la description fine. On comprend dès lors l'importance du niveau cinétique:

il représente en effet une étape intermédiaire entre la description fine et le niveau hydrodynamique, et se prête aisément aux discussions relatives aux questions A et B.

Descriptions réduites autonomes et processus limites. Les niveaux de description qui viennent d'être définis sont caractérisés par différents régimes d'évolution, gouvernés par des équations qui doivent être déduites des principes généraux. Bien que les méthodes mises en œuvre à cet effet soient très variées, elles présentent cependant un point commun important: ce sont toutes des " descriptions réduites" de l'état du système, formulées en termes de variables "semi-fines", dans lesquelles n'est utilisée qu'une très faible partie de l'information "maximale" contenue dans $\rho_N(P;t)$. Ce sont donc des descriptions très "incomplètes", en ce sens qu'à une situation macroscopique donnée, correspond toujours un très grand nombre d'états microscopiques possibles, compatibles avec cette même situation. Comme l'évolution dynamique du système dépend de toutes les variables microscopiques, la connaissance de l'état macroscopique initial ne suffit pas pour déterminer l'état macroscopique ultérieur: l'évolution des variables macroscopiques n'est donc *pas autonome* [10].

Pour obtenir une *description réduite autonome*, où ne figurent que les seules variables macroscopiques, il faut que l'effet des variables microscopiques non observées puisse s'exprimer en fonction des variables macroscopiques elles-mêmes. Compte-tenu de l'extrême complexité des interactions élémentaires, ceci implique nécessairement une *hypothèse de nature statistique* portant sur la contribution macroscopique de ces interactions; c'est naturellement par de telles hypothèses que l'irréversibilité peut être introduite dans la description. Mais, comme elles mettent en jeu les effets conjugués de la statistique et des lois mécaniques déterministes (cas de l'hypothèse du chaos moléculaire), elles peuvent devenir incompatibles avec l'évolution dynamique sous-jacente; en réalité, ces hypothèses se présentent le plus souvent comme des approximations plus ou moins bien définies qui demandent à être fondées sur des bases rigoureuses. En conclusion, deux ingrédients apparaissent nécessaires au développement de la MSHE:

(i) un certain processus de réduction du nombre des variables, qui aboutit à une description réduite de l'état du système;

(ii) la formulation d'une hypothèse de nature statistique, nécessaire pour obtenir une description réduite autonome, généralement irréversible, en termes des seules variables macroscopiques.

La justification rigoureuse de ces hypothèses est donc le problème central de la MSHE; comme dans le cas de l'équilibre, on y parvient en ayant recours à des passages à la limite qui doivent être appropriés à la nature des processus macroscopiques étudiés. C'est pourquoi la limite des "grands systèmes" (ou limite thermodynamique) de la section 2 n'est pas adaptée aux systèmes hors d'équilibre, où l'on a affaire à des états non-uniformes qui évoluent différemment selon la description réduite considérée et la nature des interactions élémentaires. En fait, les passages à la limite de la MSHE doivent dépendre, à la fois, des paramètres caractéristiques du potentiel d'interaction entre les particules (par exemple la portée r_0 et l'ordre de grandeur λ du potentiel), et des grandeurs qui définissent le niveau de la description. Ils impliquent en général des quantités qui croissent indéfiniment (telle que N) et d'autres qui tendent vers zéro, de telle manière que soit conservée une certaine grandeur caractérisrique du régime étudié. Il en est ainsi, par exemple, de la *limite de Boltzmann-Grad*, qui concerne le régime cinétique dans lequel le l.p.m. l doit demeurer *fini*: lorsque $N \longrightarrow \infty$, on a aussi, pour la densité numérique, $n = N/V(\Lambda) \longrightarrow \infty$, (puisque $V(\Lambda)$ reste fini), de sorte que $l \simeq 1/n\sigma^2$ reste fini, si $\sigma \longrightarrow 0$ de telle manière que $n\sigma^2 = cte$; on vérifie alors que l'on a aussi, à cette limite, $n\sigma^3 = 0$, ce qui correspond au cas d'un fluide idéal.

Ayant ainsi précisé les caractères spécifiques des passages à la limite en MSHE, il reste à déterminer de quelle manière peuvent être formulées les hypothèses nécessaires à la déduction rigoureuse des équations relatives à une description réduite autonome. Or, on ne dispose en MSHE que de l'équation de Liouville qui décrit l'évolution déterministe de $\rho_N(P;t)$; il s'ensuit que les seules hypothèses de nature statistique, compatibles avec la dynamique sous-jacente, ne peuvent porter que sur la condition initiale $\rho_N(P;0)$. Le problème central de la MSHE se trouve ainsi ramené à la recherche des conditions nécessaires et suffisantes que doivent remplir les *densités en phase initiales*, pour qu'il soit possible, à une certaine limite, de déduire rigoureusement, de l'équation de Liouville, les équations décrivant le comportement du système au niveau macroscopique correspondant. Ces équations portent alors sur des variables macroscopiques, dont la dispersion est nulle à cette limite (conformément à la condition 2.2), cette propriété étant *conservée* au cours du temps (c'est par exemple le cas de la "propagation" du chaos moléculaire, cf. section 6).

En conclusion, on voit que la démonstration de résultats rigoureux en MSHE implique la démarche suivante: d'abord rechercher s'il est possible de caractériser une situation physique donnée par un passage à la limite approprié; ensuite, établir les conditions nécessaires et suffisantes qui doivent être satisfaites à cette limite par la densité initiale $\rho_N(P;0)$, pour qu'une description réduite autonome puisse être obtenue. Le comportement macroscopique du système est ainsi décrit exactement par un *certain processus limite*, dont les équations, rigoureusement *déduites* de l'équation de Liouville, présentent en général les propriétés requises d'irréversibilité et de dissipativité.

Hiérarchie B.B.G.K.Y..[1] Pour traiter le problème de la réduction de la description d'un système à partir de la description "maximale" à N particules, on est conduit à développer le formalisme de la hiérarchie B.B.G.K.Y., qui repose sur la définition de fonctions de distribution réduites. Celles-ci s'introduisent naturellement si l'on remarque que les fonctions de phase usuelles ont généralement une structure telle que leurs valeurs moyennes ne dépendent que des fonctions de distribution à une ou deux particules. Pour tenir compte de l'indiscernabilité (au sens classique) des particules de même espèce, nous considérons dans ce qui suit les *fonctions de distribution réduites génériques* à s particules, $\widetilde{\rho}_s(\vec{X}_1,\ldots\vec{X}_s;t)$, symétriques en $\vec{X}_1,\ldots\vec{X}_s$, définies par:

$$\widetilde{\rho}_s(\vec{X}_1,\ldots\vec{X}_s;t)=\frac{N!}{(N-s)!}\int_{(\Lambda\times R^3)^{N-s}}\rho_N(\vec{X}_1,\ldots\vec{X}_N;t)d\vec{X}_{s+1}\ldots d\vec{X}_N, \tag{3.4}$$

avec $\int_{(\Lambda\times R^3)^s}\widetilde{\rho}_s d\vec{X}_1\ldots d\vec{X}_s=\dfrac{N!}{(N-s)!}$. Il est clair que la représentation de l'état du système par l'ensemble des fonctions $\widetilde{\rho}_s,(1\leq s\leq N-1)$, est équivalente à celle définie par la densité en phase $\rho_N(\vec{X}_1,\ldots\vec{X}_N;t)$.

En portant ces définitions dans l'équation de Liouville (3.1) et en intégrant sur $(\vec{X}_{s+1},\ldots\vec{X}_N)$, on démontre[2] que les fonctions $\rho_s(\vec{X}_1,\ldots\vec{X}_s;t)$ satisfont au système d'équations couplées:

$$\frac{\partial\widetilde{\rho}_s}{\partial t}=L_s\widetilde{\rho}_s+C_{s,s+1}\widetilde{\rho}_{s+1},\quad(1\leq s\leq N-1), \tag{3.5}$$

où L_s est l'opérateur de Liouville relatif à un "essaim"de s particules, et où l'opérateur de collisions $C_{s,s+1}$ est défini par:

$$C_{s,s+1}\widetilde{\rho}_{s+1}=\sum_{i=1}^{s}\int_{(\Lambda\times R^3)}d\vec{X}_{s+1}\frac{\partial\phi(r_{i,s+1})}{\partial\vec{r}_i}.\frac{\partial}{\partial\vec{p}_i}\widetilde{\rho}_{s+1}. \tag{3.6}$$

On voit d'après (3.6) que l'évolution de $\widetilde{\rho}_s(\ldots;t)$ résulte de la somme de deux termes: le terme "liouvillien", $L_s\widetilde{\rho}_s$, qui décrit le mouvement des s particules en interaction entre elles (avec le potentiel $\phi(r_{ij})$); le terme de "collisions", $C_{s,s+1}\widetilde{\rho}_{s+1}$, qui représente la contribution due aux interactions des s particules de l'essaim avec les $(N-s)$ particules restantes (en vertu de la symétrie des $\widetilde{\rho}_s$), et qui a pour effet de coupler de proche en proche les équations de (3.5).

Les équations (3.5) sont établies pour le cas général d'un système constitué de N particules supposées ponctuelles, dont l'interaction est décrite par

[1]Bogolioubov, Born et Green, Kirkwood, Yvon.

[2]Nous n'indiquons pas la démonstration qui figure dans tous les ouvrages classiques.

des forces centrales dérivées d'un potentiel $\phi(r_{ij})$. Cependant, pour l'étude de la limite de Boltzmann-Grad et du théorème de Lanford (où $r_0 \longrightarrow 0$), il est commode de se placer dans le cas où l'interaction entre particules est représentée par un modèle *de sphères dures* de diamètre σ. Comme le formalisme de la hiérarchie suppose une évolution et des forces d'interaction continues, l'utilisation de ce modèle rencontre des difficultés d'ordre technique, en raison du caractère manifestement discontinue de l'évolution.

On peut néanmoins adapter le formalisme général à cette situation particulière, en modifiant la définition de l'espace des phases de manière à retrouver dans le nouvel espace une évolution continue. Au terme d'une analyse non-triviale que nous ne pouvons reproduire ici [11], on montre que l'évolution du système est encore décrite par une équation de Liouville d'où l'on peut déduire les *équations de la hiérarchie pour le modèle des sphères dures*: celles-ci ont la même forme que (3.5), avec un terme liouvillien $L_s\widetilde{\rho}_s$, qui représente le mouvement de s sphères dures (y compris leurs collisions élastiques binaires), et un terme de "collisions" $C_{s,s+1}\widetilde{\rho}_{s+1}$ qui est maintenant défini par:

$$\begin{aligned} C_{s,s+1}\widetilde{\rho}_{s+1} = \sigma^2 \textstyle\sum_{i=1}^{s} \int_{S^2\times R^3} d\vec{p}_{s+1} d\hat{w}\ \hat{w}. \left[\dfrac{\vec{p}_{s+1}-\vec{p}_i}{m}\right] \times \\ \widetilde{\rho}_{s+1}(\vec{X}_1,\ldots,\vec{X}_s,\vec{r}_i+\sigma\hat{w},\vec{p}_{s+1}), \end{aligned} \tag{3.7}$$

où $\hat{w}$ représente le vecteur-unité porté par la ligne des centres des deux particules (i) et $(s+1)$ à l'instant de leur collision (et pointant du centre de (i) vers $(s+1)$), et où $d\hat{w}$ est l'élément d'aire de la sphère-unité S^2.

4. Le niveau cinétique et l'équation de Boltzmann.

Nous abordons maintenant l'étude du niveau cinétique dont on peut proposer deux descriptions: l'une part de la notion d'ensemble de Gibbs et passe par le procédé de fermeture de la hiérarchie; l'autre, qui est celle de la théorie cinétique des gaz, est fondée sur une certaine définition de l'état macroscopique du système.

Problème de la fermeture de la hiérarchie. Pour étudier le comportement du système au niveau cinétique, on doit considérer la première équation de la hiérarchie (3.5) relative à la fonction de distribution à une particule $\widetilde{\rho}_1(\vec{X}_1;t)$. Comme l'on a, pour $s=1, L_1 = -(\vec{p}_1/m).\partial/\partial\vec{r}_1$, cette équation peut s'écrire dans le cas du modèle des *sphères dures*:

$$\begin{aligned} \frac{\partial\widetilde{\rho}_1}{\partial t} + \frac{\vec{p}_1}{m}\cdot\frac{\partial\widetilde{\rho}_1}{\partial\vec{r}_1} = \sigma^2 \int_{S^+\times R^3} d\hat{w} d\vec{p}_2\ \hat{w}\cdot\left[\frac{\vec{p}_1-\vec{p}_2}{m}\right] \times \\ \left\{\widetilde{\rho}_2(\vec{r}_1,\vec{p'}_1,\vec{r}_1-\sigma\hat{w},\vec{p'}_2;t) - \widetilde{\rho}_2(\vec{r}_1,\vec{p}_1,\vec{r}_1+\sigma\hat{w},\vec{p}_2;t)\right\}, \end{aligned} \tag{4.1}$$

où l'on a donné au terme de collisions la forme habituelle d'un "bilan de gains et pertes"; on obtient ce terme en décomposant l'intégrale sur la sphère-unité S^2 dans (3.7), en deux intégrales prises sur les deux demi-sphères S^- et S^+, définies respectivement par $\hat{w} \cdot (\vec{p}_2 - \vec{p}_1) > 0$ (cas où les deux particules se séparent après une collision), et $\hat{w} \cdot (\vec{p}_2 - \vec{p}_1) < 0$ (cas où les deux particules entrent en collision), et en remplaçant l'intégrale sur S^- par une intégrale sur S^+ par le changement de $\hat{w}$ en $-\hat{w}$.

Ceci étant, il est important pour notre propos de remarquer que l'équation (4.1) ne permet *pas* de définir une description réduite autonome de l'état du système à ce "niveau". En effet, on constate immédiatement que l'équation relative à $\widetilde{\rho}_1$ dépend de $\widetilde{\rho}_2$, que celle relative à $\widetilde{\rho}_2$ dépend de $\widetilde{\rho}_3$, ... et ainsi de suite jusqu'à ρ_N, qui obéit elle-même à l'équation de Liouville; on en conclut que la connaissance de $\widetilde{\rho}_1$ à chaque instant t implique l'intégration de l'équation de Liouville, c'est-à-dire le recours à la description maximale.

Pour obtenir une description réduite autonome, il faut donc être en mesure de s'arrêter dans la hiérarchie (3.5) à un certain rang, ce qui n'est possible que si l'on peut exprimer le terme d'interaction (3.6) au moyen des seules fonctions de distribution $\widetilde{\rho}_s$ de rang inférieur à $s+1$. C'est le problème de la *fermeture de la hiérarchie*: il consiste à chercher à établir, dans une situation physique donnée, une relation fonctionnelle adéquate entre $\widetilde{\rho}_{s+1}$ et les $\widetilde{\rho}_s$ de rang inférieur. Dans le cas $s = 1$, une telle relation prend la forme $\widetilde{\rho}_2 = \mathcal{F}[\widetilde{\rho}_1]$: c'est le cas par exemple du "chaos moléculaire" défini par (6.1); en portant cette expression dans le second membre de (4.1), on obtient alors une équation fermée en $\widetilde{\rho}_1$ qui représente une certaine description réduite autonome au niveau cinétique.

Pour formuler de telles relations de fermeture, on recherche habituellement des solutions approchées de la hiérarchie sous la forme de développements de perturbation suivant l'un des paramètres caractéristiques (supposé petit) de l'état du système. Un exposé de ces méthodes sortant du cadre de cet article, il suffira pour notre propos de souligner qu'elles rencontrent à la fois des difficultés d'ordre technique et des problèmes qui concernent les fondements mêmes de la théorie.

Si l'on considère par exemple le cas des fluides neutres dilués , où l'on développe les fonctions $\tilde{\rho}_s$ suivant les puissances de $\epsilon = nr_0^3$, l'application de cette méthode conduit à une équation de Boltzmann "généralisée", qui décrit en principe le régime cinétique d'un fluide modérément dense (avec $\epsilon > 0$). Toutefois la portée de ce résultat se trouve singulièrement limitée en raison de difficultés bien connues liées à l'existence des divergences (pour $t \longrightarrow \infty$) qui apparaissent dans le calcul des termes d'ordre supérieur à deux [12]. De plus, la mise en œuvre de cette méthode repose sur des hypothèses qui portent à la fois sur les conditions initiales et sur *l'atténuation des corrélations*, dans lesquelles se trouvent étroitement impliqués le comportement dynamique et l'état statistique du fluide, et dont la justification s'avère de ce fait particulièrement délicate. En réalité, il semble bien que ces difficultés de la théorie des fluides modérément denses proviennent de ce que l'on ne sait pas clairement définir un régime cinétique, et la limite correspondante, pour de tels systèmes [7,13,14].

Théorie cinétique des gaz et équation de Boltzmann. Pour préciser la signification physique du passage à la limite de Boltzmann-Grad, il est utile de rappeler les idées essentielles qui sont à la base d'une autre " description réduite" de l'état d'un fluide au niveau cinétique. Il s'agit de la théorie cinétique des gaz au sens de Boltzmann qui est fondée sur l'introduction directe de variables "coarse-grained" ou "semi-fines" relatives au seul système considéré, et qui diffère de ce fait de la description cinétique obtenue à partir de la hiérarchie et de l'équation (4.1) concernant un certain ensemble statistique de systèmes.

Pour définir ces nouvelles variables, on considère l'espace des phases à *une* particule $\mu = \Lambda \times R^3$, dans lequel l'état dynamique d'un système à N particules est représenté par un ensemble de N points (ou de N sphères dures). Pour tenir compte de l'imprécision des observations macroscopiques, on introduit une partition de l'espace μ en cellules $w_i(\mu = \cup_i w_i)$, supposées petites à l'échelle macroscopique mais suffisamment grandes pour que chacune d'entre elles contienne un grand nombre n_{w_i} de particules. Ceci étant, l'état "macroscopique" d'*un* système est alors défini en théorie cinétique des gaz par l'ensemble des *nombres d'occupation* $\{n_{w_i}\}$ associés aux cellules $\{w_i\}$, avec $\sum_i n_{w_i} = N$.

A chaque état dynamique du système, représenté par un point P de l'espace des phases Γ_N, correspond un ensemble de nombres d'occupation bien déterminés $\{n_{w_i}(P)\}$. Au cours du mouvement, le point P décrit une trajectoire dans Γ, $P \longrightarrow P_t$, entraînant une variation des nombres n_i qui deviennent des fonctions de P_t : $n_{w_i}(P) \longrightarrow n_{w_i}(P_t)$; l'ensemble de ces nombres $\{n_{w_i}(P_t)\} \equiv \{n_i(t)\}$ n'est donc bien déterminé au cours du temps que si l'on connaît la trajectoire exacte de P_t.

Il est alors facile de montrer, par l'argumentation habituelle du "coarse-graining" (cf. section 3 et [7]), que cette description "réduite" n'est *pas* autonome, c'est-à-dire que la donnée initiale des nombres $n_{w_i}(P)$ ne permet pas d'en déduire les $n_{w_i}(P_t)$. Toutefois, comme ces nombres sont supposés très grands, il est *vraisemblable* que l'ensemble des valeurs possibles des $n_{w_i}(P_t)$ aura tendance à se grouper, avec une faible dispersion, autour d'une certaine valeur moyenne, conformément aux considérations de P. et T. Ehrenfest concernant la "courbe" H (cf. section 5).

Ces considérations permettent d'introduire la notion de *fonction de distribution des vitesses dans l'espace* μ. C'est la notion centrale de la théorie cinétique des gaz qui repose sur les hypothèses suivantes:

(a) On admet d'abord que les cellules w_i, bien que contenant un très grand nombre de particules, peuvent être supposées *infiniment petites* à l'échelle macroscopique, d'où la notation $w_i \simeq \triangle\vec{X} = \triangle\vec{r}\triangle\vec{p}$; de notre point de vue, il importe de remarquer que ceci implique en fait un passage à la limite très particulier, où l'on a, d'une part, $N \longrightarrow \infty$, et où, d'autre part, les dimensions des w_i, ainsi que celles des particules, tendent vers zéro, mais de telle manière que leurs nombres d'occupation restent très grands.

(b) Compte tenu de l'hypothèse précédente, on admet ensuite qu'il existe une fonction $f(\vec{r},\vec{p})$, lentement variable à l'échelle des $\triangle\vec{X}$ et telle que $n_{\triangle\vec{X}}(P) \simeq Nf(\vec{r},\vec{p})\triangle\vec{r}\triangle\vec{p}$.

(c) On suppose enfin qu'au cours de l'évolution $f(\vec{r},\vec{p}) \longrightarrow f(\vec{r},\vec{p};t)$ de telle manière que $n_{\triangle\vec{X}}(P_t) \simeq Nf(\vec{r},\vec{p};t)\triangle\vec{r}\triangle\vec{p}$, où $f(\vec{r},\vec{p};t)$ est la solution d'une certaine équation cinétique correspondant à la condition initiale $f(\vec{r},\vec{p})$.

Ces hypothèses, qui constituent le contenu essentiel de la théorie, suffisent à préciser la nature de la fonction de distribution des vitesses et à déterminer son évolution. D'après (a) et (b), on voit en effet que $f(\vec{r},\vec{p})$ représente la *densité* des particules du système au voisinage du point $(\vec{r},\vec{p})$ de l'espace μ, et qu'elle a le sens d'une quantité "coarse-grained", puisqu'elle n'est définie que pour des cellules $\triangle\vec{X}$ de dimensions finies.

Ce caractère apparaît encore plus clairement si l'on compare $f(\vec{r},\vec{p})$ à la densité de probabilité "fine" $\widetilde{\rho}_1(\vec{X}_1)$ qui est bien définie en tout point $\vec{X}_1 \in \mu$; en revenant à leur définition respective, on voit que ces deux fonctions sont liées par la relation:

$$\frac{n\triangle\vec{X}(P)}{N} \simeq f(\vec{r},\vec{p})\triangle\vec{X} \simeq \frac{1}{N}\int_{\triangle\vec{r}\triangle\vec{p}} \widetilde{\rho}_1(\vec{X}_1)d\vec{X}_1, \tag{4.2}$$

avec $\int_{\Lambda\times R^3} f(\vec{r},\vec{p})d\vec{X} = 1$, d'où il ressort que f est une quantité "semi-fine" (coarse-grained) qui peut être considérée comme la moyenne de $\widetilde{\rho}_1$ prise sur la cellule $\triangle\vec{r}\triangle\vec{p}$.

D'autre part, l'hypothèse (c) montre que la fonction $f(\vec{r},\vec{p};t)$ définit une certaine " description réduite autonome" de l'évolution du système au niveau cinétique, qui reste à compléter par l'établissement de l'équation à laquelle cette fonction doit satisfaire. Dans le cas des gaz neutres dilués, un raisonnement bien connu conduit à la célèbre équation de Boltzmann qui joue un rôle essentiel en théorie cinétique. Il suffira pour notre propos de rappeler que cette équation repose sur la célèbre *hypothèse du chaos moléculaire*, selon laquelle deux particules sur le point d'entrer en collision ne sont *pas* corrélées, juste *avant* la collision. Cette hypothèse donne lieu à deux remarques importantes: (a) d'une part, elle introduit manifestement une dissymétrie dans l'évolution temporelle du système, puisque l'indépendance statistique entre les deux particules concernées est postulée *avant* la collision et non après; (b) d'autre part, comme elle doit être valable à tout instant t, elle peut entrer en conflit avec les lois du mouvement et ne sera en fait justifiée qu'à une certaine limite.

Ceci étant, l'équation de Boltzmann pour un fluide idéal constitué de N sphères dures de diamètre σ, s'écrit (avec les notations habituelles):

$$\frac{\partial f(\vec{r}_1,\vec{p}_1;t)}{\partial t} + \frac{\vec{p}_1}{m} \cdot \frac{\partial f(\vec{r}_1,\vec{p}_1;t)}{\partial \vec{r}_1} = N\sigma^2 \int_{S^+ \times R^3} d\hat{w} d\vec{p}_2 \, \hat{w} \cdot \left[\frac{\vec{p}_1 - \vec{p}_2}{m}\right] \times$$

$$\left\{f(\vec{r}_1,\vec{p'}_1;t)f(\vec{r}_1,\vec{p'}_2;t) - f(\vec{r}_1,\vec{p}_1;t)f(\vec{r}_1,\vec{p}_2;t)\right\}, \qquad (4.3)$$

où les impulsions $\vec{p'}_1, \vec{p'}_2$ s'expriment en fonction des $\vec{p}_1, \vec{p}_2$ d'après les lois gouvernant le processus de collision. On remarque la présence, dans le terme de collisions, du facteur $N\sigma^2$ caractérisrique de l'état granulaire de la matière puisqu'il est inversement proportionnel au l.p.m. des particules. On sait que cette équation permet d'obtenir des résultats remarquables dans la théorie des gaz neutres dilués, même pour les problèmes qui sortent de son strict domaine de validité. Mais, pour notre propos, son importance tient à son caractère irréversible qui est illustré par le célèbre

Théorème H de Boltzmann : Soit $H(t)$ la fonction du temps défini par

$$H(t) = \int_{\Lambda \times R^3} d\vec{r}_1 d\vec{p}_1 \, f(\vec{r}_1,\vec{p}_1;t) \log f(\vec{r}_1,\vec{p}_1;t); \qquad (4.4)$$

si $f(\vec{r}_1,\vec{p}_1;t)$ est solution de l'équation de Boltzmann (4.3), on a $\frac{dH(t)}{dt} \leq 0$, l'égalité n'ayant lieu que dans le cas où f est maxwellienne.

$H(t)$ est donc une fonction monotone décroissante du temps; on peut en conclure que, pour des conditions assez générales, la fonction de distribution $f(\vec{r}_1\vec{p}_1;t)$ tend vers la distribution d'équilibre de Maxwell-Boltzmann lorsque $t \longrightarrow \infty$, avec un temps caractérisrique de l'ordre de $\tau_{\text{l.p.m.}}$. Ainsi se trouve mise en évidence l'irréversibilité de l'évolution au niveau cinétique, tel qu'il est défini en théorie cinétique des gaz. Comme il est bien connu, c'est l'incompatibilité entre cette évolution et celle de la description microscopique sous-jacente qui conduisit Boltzmann à élaborer une interprétation statistique de (4.3), selon laquelle cette équation décrit en fait le comportement "le plus probable" du système.

5. Retour à l'interprétation statistique selon Boltzmann.

Avant d'en venir à l'étude de la limite de Boltzmann-Grad et de la déduction rigoureuse de l'équation (4.3) à partir des principes généraux de la Mécanique statistique, il est utile d'approfondir la nature des rapports existant entre la théorie cinétique des gaz et la théorie des ensembles de Gibbs, en revenant sur l'interprétation statistique proposée par Boltzmann et reprise par P. et T. Ehrenfest [1]. Dans ce qui suit, nous nous bornons à esquisser les idées essentielles de l'analyse de Boltzmann qui furent à l'origine de l'interprétation statistique de l'entropie.

Fonction et "courbe" $H[Z(t)]$. On reprend à cet effet la définition de l'état macroscopique d'un système par l'ensemble des nombres d'occupation $n_i(t)$ associés à un certain système de cellules w_i, avec $w_i \simeq \triangle r \triangle p$. A l'état macroscopique Z_i du système représenté par l'ensemble des nombres $\{n_i(t)\}$, avec $\sum_i n_i(t) = N$, il correspond dans l'espace des phases total Γ_N une "macro-cellule" dont l'extension en phase a pour volume:

$$W(Z_i) = \frac{N!}{\Pi_i n_i!} \Pi_i w_i^{n_i}. \tag{5.1}$$

L'entropie au sens de Boltzmann est alors définie par:

$$S[Z_i(t)] = k \log W[Z_i(t)] = K - k \sum_i n_i \log n_i, \tag{5.2}$$

à laquelle on fait correspondre la célèbre fonction $H[Z(t)]$, en posant:

$$H[Z_i(t)] = -\frac{S[Z_i(t)]}{k} = \sum_i n_i \log n_i + K', k = \mathrm{C^{te}} \text{ de Boltzmann.} \tag{5.3}$$

Ceci étant, l'argumentation de Boltzmann s'appuie sur une propriété capitale de la distribution de Maxwell-Boltzmann caractérisant l'état d'équilibre: parmi tous les états Z ayant une énergie macroscopique donnée $E_0 = E(Z) = \sum_i n_i \epsilon_i$ (ϵ_i étant l'énergie associée à la cellule w_i), l'état $Z_{\max}$ défini par la distribution maxwellienne des $n_i(n_i^{\max} = \alpha e^{-\beta \epsilon_i})$ est celui dont la "macro-cellule" associée a le volume $W_{\max} \equiv W(Z_{\max})$ le plus grand dans Γ_N. Par suite, l'entropie de l'état maxwellien $S_{\max} = k \log W_{\max}$ est maximum, alors que la fonction $H(Z_{\max})$ correspondante est minimum, soit $H_{\min} = -k \log W_{\max} \equiv H_0$; comme l'entropie thermodynamique peut être identifiée à $S_{\max}$, la relation (5.2) permet ainsi d'étendre la définition de l'entropie d'un système à des états hors d'équilibre.

En fait, compte tenu du très grand nombre de dimensions de l'espace Γ_N, on vérifie que la fonction $W(Z)$, définie sur l'hypersurface d'énergie constante E_0, présente un pic très aigu au voisinage de $Z = Z_{\max}$, de sorte que l'état maxwellien a sur cette hypersurface une extension considérablement plus grande que celle de tous les autres états. Cette *énorme prédominance* de l'état maxwellien montre une fois de plus le rôle central joué par les propriétés géométriques spécifiques des espaces à un très grand nombre de dimensions (cf section 2). Celle-ci a une conséquence immédiate sur le comportement de la fonction $H[Z(t)]$: au cours de l'évolution du système, la fonction $H[Z(t)]$ restera, sauf cas "très exceptionnels", au voisinage de la valeur $H = H_0$ pour des durées extrêmement longues, de l'ordre du temps de récurrence prévu par le théorème du retour de Poincaré. Cette "propriété de plateau" des valeurs de $H[Z(t)]$, qui n'exclut pas la possibilité d'une grande fluctuation au terme d'un temps très long, est partagée par toutes les fonctions de l'état macroscopique Z, et permet de concevoir le comportement macroscopique du système comme résultant de l'évolution "la plus probable " de son état.

A l'appui de cette conception, il suffit en effet d'observer que le point représentatif du système dans l'espace Γ_N tendra, au cours de son mouvement, à passer "*presque* toujours" par des macro-cellules dont le volume est de plus en plus grand, jusqu'à ce que la "macro-cellule" Z_{max} soit atteinte. En interprétant le volume d'une macro-cellule comme la "probabilité" (relative) de l'état correspondant, on est ainsi fondé à concevoir l'évolution macroscopique comme se déroulant des états les moins probables vers les états les plus probables, pour aboutir finalement à l'état d'équilibre où le système demeurera pour une durée extrêmement longue.

Ce comportement est bien illustré par la représentation graphique dans le plan (H, t) de la fonction $H[Z(t)]$ pour un système supposé isolé. Comme cette représentation est extrêmement complexe et irrégulière – elle est en effet constituée d'une suite discrète de "sauts" intervenant à chaque changement des valeurs des n_i – il est commode de lui substituer la "*courbe*" $H[Z(t)]$ obtenue par un certain procédé de lissage : celui-ci consiste à ne retenir des valeurs de $H[Z]$ que celles observées sur une suite d'intervalles finis $\triangle t$, petits à l'échelle macroscopique mais suffisamment grands pour contenir un grand nombre de sauts de la fonction $H[Z]$. Cette "courbe" $H[Z(t)]$, qui est en fait constituée d'une suite de segments de droite, a des propriétés singulières analogues à celles de la fonction $H[Z]$, que les analyses bien connues de Boltzmann et P. et T. Ehrenfest ont permis de préciser de la manière suivante: (a) comme la fonction $H[Z]$, elle reste dans le voisinage de H_0 durant des intervalles de temps extrêmement longs; (b) elle décroît presque toujours à partir de tout point $H_1 > H_0$; (c) elle est telle que presque tous ses points sont des maximums; (d) ces énoncés sont valables *quel que soit le signe de t* (réversibilité du mouvement), de sorte que la "courbe" H d'un système isolé doit être conçue comme se prolongeant indéfiniment à la fois dans le sens des t positifs et des t négatifs.

Ces propriétés, en apparence paradoxales, sont en fait directement liées à celles des espaces à un très grand nombre de degrés de liberté et elles entraînent notamment deux conséquences importantes: (i) lorsque le système est observé dans un certain état hors d'équilibre pour lequel $H_1 > H_0$, la propriété (b) permet d'affirmer que l'évolution "la plus probable" de ce système est un retour rapide vers l'état d'équilibre, conformément à ce qui est effectivement constaté à l'échelle macroscopique; (ii) comme, d'après (d), le modèle est totalement réversible, il s'ensuit que "la flèche du temps" de notre expérience doit être identifiée au sens de l'évolution "la plus probable".

Faisceaux de "courbes" $H[Z(t)]$. La courbe de concentration et la conjecture de Boltzmann-Ehrenfest. Bien que les résultats précédents soient très suggestifs concernant le comportement d'un système macroscopique, ils n'en demeurent pas moins des arguments de nature qualitative, où les expressions "la plus probable", "presque toujours", ..., demandent à recevoir une définition précise. A cet effet, on ne peut que revenir aux principes fondamentaux de la Mécanique statistique en vue d'approfondir les liens existant entre la description cinétique de Boltzmann et la théorie des ensembles de Gibbs. Pour ce faire, la méthode adéquate consiste à introduire, avec P. et T. Ehrenfest, la notion de faisceau de "courbes" associé à un certain état macroscopique.

Considérons donc un système macroscopique qui a été observé, à un instant initial t_0, dans un état hors d'équilibre $Z_A(t_0)$, avec par conséquent $H(Z_A) > H_0$. A cet état macroscopique se trouve associé un ensemble d'états microscopiques possibles, représentés dans Γ_N par tous les points de la macro-cellule $Z_A(t_0)$. A chacun de ces points $P \in Z_A(t_0)$, il correspond dans le plan (H, t) une certaine "courbe" $H[Z_A(t)]$, de sorte qu'à l'état Z_A se trouve associé un *faisceau de "courbes"* H, toutes issues du *même* point $H[Z_A(t_0)]$, qui représente en fait l'évolution d'un certain ensemble statistique hors d'équilibre. Après un intervalle de temps $n\triangle t$, on obtient ainsi dans le plan (H, t) un "nuage de points" $H[Z_A(t_0 + n\triangle t)]$ correspondant au point $H[Z_A(t_0)]$. Pour établir un lien entre ce faisceau de "courbes" et la description cinétique, on est alors conduit à supposer que cet ensemble de points a les propriétés suivantes (ce sont les hypothèses VI a-c de P. et T. Ehrenfest [1]):

Propriété 1: L'ensemble des points $H[Z_A(t_o + n\triangle t]$ se concentre avec une *très petite dispersion* autour d'une certaine valeur $\mathcal{H}_n$. La suite discrète de valeurs $\mathcal{H}_1, \mathcal{H}_2, \ldots, \mathcal{H}_n$ ainsi obtenue définit la *courbe de concentration* $\mathcal{H}$ du faisceau.

Propriété 2: La courbe de concentration $\mathcal{H}$ décroît de manière monotone à partir de $H[Z_A(t_0)]$, et tend vers sa valeur minimum H_0, dont elle ne s'écarte plus jamais ensuite.

Propriété 3: A partir de t_0, pour une durée très longue, l'immense majorité des "courbes" $H[Z_A(t)]$ resteront très voisines de la courbe de concentration $\mathcal{H}$.

Naturellement, ces propriétés ne sont que des hypothèses qui demandent à être justifiées pour chacune des situations physiques considérées; mais, une fois obtenue une telle justification, on est alors fondé à décrire le comportement macroscopique du système par la "courbe de concentration" $\mathcal{H}$. Il est toutefois important de souligner que le remplacement d'une "courbe" quelconque du faisceau par la courbe de concentration implique un changement *radical* dans le comportement du système, puisque l'on passe d'une évolution réversible (dans le cas de la "courbe" H) à l'évolution manifestement irréversible représentée par la courbe $\mathcal{H}$ (propriétés 2 et 3); ce changement est donc équivalent à la brisure de symétrie de l'état du système au niveau macroscopique, évoquée dans la question B. On remarquera de plus qu'un tel changement se trouve nécessairement lié à une certaine situation limite, dans laquelle a été prise en compte la nature "physiquement" infinitésimale[3] de cellules contenant un très grand nombre de particules.

Ceci étant, il est facile d'énoncer la conjecture de Boltzmann-Ehrenfest qui est à la base de l'interprétation statistique de la théorie cinétique des gaz:

Conjecture: *La courbe du théorème H associée à l'équation de Boltzmann est identique à la courbe de concentration $\mathcal{H}$ du faisceau de "courbes" $H[Z_A(t)]$, telle qu'est définie par les propriétés 1-3* (c'est l'hypothèse VII de P. et T. Ehrenfest [1]).

D'après les remarques précédentes, la justification de cette conjecture passe par la définition d'une limite appropriée au cas d'un fluide idéal; c'est précisément l'objet du passage à la limite de Boltzmann-Grad et du théorème de Lanford qui en découle.

6. Limite de Boltzmann-Grad et théorème de Lanford.

Cette section est essentiellement consacrée à la démonstration du théorème de Lanford qui repose sur la mise en œuvre du passage à la limite de Boltzmann-Grad dans la hiérarchie B.B.G.K.Y.. Au cours de cette étude, notre intention n'est pas d'entrer dans tous les détails techniques du sujet, mais seulement de mettre en évidence les étapes successives de la démonstration, en insistant sur la contribution que chacune d'entre-elles apporte à la solution des problèmes relatifs aux questions A et B.

Limite de Boltzmann-Grad. C'est à Grad que revient le mérite d'avoir montré que la signification précise de l'équation de Boltzmann était directement liée à cette situation limite particulière [2]. Il suffit pour le voir de comparer l'équation de Boltzmann (4.3) à la première équation de la hiérarchie (4.1) pour le modèle des sphères dures, et de remarquer que l'on peut passer *formellement* de l'une à l'autre, à condition :

(a) d'exprimer, dans le second membre de (4.1), la distribution réduite $\widetilde{\rho}_2$ sous la forme d'un produit de $\widetilde{\rho}_1$, en posant:

$$\widetilde{\rho}_2(\vec{X}_1, \vec{X}_2; t) = \widetilde{\rho}_1(\vec{X}_1; t)\, \widetilde{\rho}_1(\vec{X}_2; t); \tag{6.1}$$

[3]Signalons que Klimontovich a fait un usage systématique de cette notion, voir notamment [15].

(b) de remplacer $\frac{1}{N}\widetilde{\rho}_1(\vec{X}_1;t)$ par $f(\vec{r}_1,\vec{p}_1;t)$, ce qui fait apparaître le facteur N au second membre de (4.1);
(c) de faire tendre $\sigma \longrightarrow 0$ dans f, de sorte que $f(\vec{r}_1 \pm \sigma\hat{w},\vec{p}_2;t) \longrightarrow f(\vec{r}_1,\vec{p}_2;t)$.

Ces trois conditions permettent de déduire formellement l'équation de Boltzmann de la première équation de la hiérarchie en opérant le passage à la *limite de Boltzmann-Grad*, définie par:

$$\sigma \longrightarrow 0; N \longrightarrow \infty, \text{ avec } N\sigma^2 = \text{cte} \quad \left(\simeq \frac{V(\Lambda)}{l}\right), \tag{6.2}$$

pour laquelle le terme de collisions demeure fini, et proportionnel à $1/l$; il s'ensuit donc que la condition (c) conduit à effectuer le passage à la limite (6.2) et que l'équation de Boltzmann n'a de sens que pour cette situation limite particulière. La condition (b) permet d'autre part d'identifier, à cette limite, $(1/N)\widetilde{\rho}_1(\vec{X}_1;t)$ avec la fonction $f(\vec{r}_1,\vec{p}_1;t)$ de la théorie cinétique, ce qui implique en fait que la cellule $\triangle\vec{r}\triangle\vec{p}$ de la section (4.2) peut être considérée comme "physiquement" infinitésimale, soit $\triangle\vec{r}\triangle\vec{p} \longrightarrow 0$. Quant à la condition de factorisation (6.1), elle joue évidemment le même rôle que l'hypothèse du chaos moléculaire; pour le vérifier, il suffit de remarquer que l'intégrale de collisions de (4.1) est définie sur la sphère-unité S^+ (correspondant au cas où les particules 1 et 2 sont sur le point d'entrer en collisions), et que la condition $\widetilde{\rho}_2 = \widetilde{\rho}_1\widetilde{\rho}_1$ exprime de ce fait l'absence de corrélations entre ces particules *avant* la collision. Elle donne donc lieu aux mêmes remarques que précédemment (cf. section 4); en particulier, elle n'est pas compatible avec le modèle dynamique sous-jacent, puisqu'elle doit être valable quel que soit t, alors que les collisions créent des corrélations au cours du mouvement.

Signalons encore une autre conséquence de la condition de factorisation (6.1), importante pour notre propos. Elle concerne les nombres d'occupation $n_w(t)(=n_w(P_t))$ de la théorie cinétique, dont la dispersion, compte tenu de son expression à partir des fonctions $\widetilde{\rho}_1$ et $\widetilde{\rho}_2$, satisfait en vertu de (6.1) à la relation:

$$\frac{\overline{n_w^2(t)} - (\overline{n_w}(t))^2}{(\overline{n_w}(t))^2} \simeq O(\frac{1}{\overline{n_w}(t)}) \longrightarrow 0, \tag{6.3}$$

où le second membre tend vers zéro dès que $\overline{n_w} \simeq n$ est suffisamment grand. On vérifie ainsi que la condition (6.1) exprime aussi le fait que les variables $n_w(t)/\overline{n_w}(t)$ deviennent,à la limite considérée, de "bonnes" variables macroscopiques qui rendent possible la réduction de la description de l'état du système au niveau cinétique.

Théorème de Lanford. Les raisonnements précédents montrent que le passage à la limite (6.2) permet de déduire formellement l'équation de

Boltzmann de la première équation de la hiérarchie. Mais pour établir cette analyse formelle sur une base rigoureuse (et donner notamment un sens à des expressions telles que "presque tous les points P", "très petite" dispersion, etc...), il faut prouver que les solutions de (4.1) convergent effectivement à cette limite vers une solution de l'équation de Boltzmann (4.3). Comme nous savons que l'équation (4.1) implique *toute* la hiérarchie (3.5), on est ainsi amené à étudier le comportement de cette hiérarchie à la limite de Boltzmann-Grad (6.2). C'est précisément l'objet du théorème démontré par Lanford dans le cas des sphères dures [4].

Conformément à l'argumentation de ce théorème, on considère un système de N sphères dures de diamètre σ (contenues dans une enceinte Λ de volume $V(\Lambda)$ fini), dont l'état est décrit par un ensemble statistique $\rho_N(P;t)$, défini dans l'espace des phases $\Gamma_N = (\Lambda \times R^3)^N$, et par les fonctions de distribution associées $\widetilde{\rho}_s(\vec{X}_1, \ldots, \vec{X}_s; t)$. Pour procéder au passage à la limite (6.2), on considère alors une suite infinie de tels systèmes, indexés par le nombre σ, dont l'état est défini par les ensembles statistiques $\rho_N^{(\sigma)}(P;t)$ et par la suite correspondante des fonctions de distribution réduites $\widetilde{\rho}_s^{(\sigma)}(\vec{X}_1, \ldots, \vec{X}_s; t)$. Comme chacune de ces fonctions $\widetilde{\rho}_s^{(\sigma)}$ est une solution de la hiérarchie (3.5) pour la valeur de σ considérée, le problème posé revient à étudier le comportement de ces solutions à la limite (6.2) où $\sigma \longrightarrow 0$.

Ceci étant, la démonstration des résultats obtenus par Lanford passe par les trois étapes suivantes:

- 1ère *étape*. Elle consiste en un changement d'échelle des fonctions de distribution $\widetilde{\rho}_s^{(\sigma)}$; celui-ci est en effet rendu nécessaire en raison de la divergence de ces fonctions à la limite $\sigma = 0$. (On le vérifie en remarquant qu'elles se comportent, d'après la relation de normalisation de (3.4), comme $\lim_{N \to \infty} N(N-1)..(N-s+1) \simeq N^s$). On est ainsi conduit à introduire de nouvelles fonctions de distribution réduites $f_s^{(\sigma)}$ définies par:

$$\begin{aligned} f_s^{(\sigma)}(\vec{X}_1, \ldots, \vec{X}_s; t) &= N^{-s} \widetilde{\rho}_s^{(\sigma)}(\vec{X}_1, \ldots, \vec{X}_s; t) \simeq \sigma^{2s} \widetilde{\rho}_s^{(\sigma)}, \qquad (6.4) \\ \text{avec } \lim_{\sigma \to 0} \int_{(\Lambda \times R^3)^s} & f_s^{(\sigma)}(\vec{X}_1, \ldots, \vec{X}_s; t)\, d\vec{X}_1 \ldots d\vec{X}_s = 1. \end{aligned}$$

Ces fonctions $f_s^{(\sigma)}$ satisfont également aux équations de la hiérarchie (3.5), qui prennent la forme:

$$\frac{\partial f_s^{(\sigma)}}{\partial t} = L_s^{(\sigma)} f_s^{\sigma} + C_{s,s+1}^{(\sigma)} f_{s+1}^{(\sigma)}, \qquad (6.5)$$

où les opérateurs de Liouville $L_s^{(\sigma)}$ sont identiques à ceux de (3.5), et où les opérateurs de collisions $C_{s,s+1}^{(\sigma)}$ sont obtenus en remplaçant σ^2 par $N\sigma^2$ dans l'opérateur de collisions (3.7).

- 2ème *étape. La hiérarchie de Boltzmann.* Elle s'introduit naturellement lorsque l'on considère la limite $\sigma \longrightarrow 0$ du système d'équations (6.5). En procédant *formellement*, on obtient ainsi le nouveau système d'équations linéaires couplées:

$$\frac{\partial f_s^{(0)}}{\partial t} = L_s^{(0)} f_s^{(0)} + C_{s,s+1}^{(0)} f_{s+1}^{(0)}, \tag{6.6}$$

où $L_s^{(0)} \equiv \sum_{i=1}^{s} (\vec{p}_i/m_i) \cdot \partial/\partial\vec{r}_i$ est maintenant l'opérateur de Liouville correspondant au mouvement libre de s sphères dures (y compris les réflexions sur la paroi $\partial\Lambda$), et où l'opérateur de collisions $C_{s,s+1}^{(0)}$ peut se mettre, comme en (4.1), sous la forme d'un "bilan de gains et pertes", soit:

$$C_{s,s+1}^{(0)} f_{s+1}^{(0)} = N\sigma^2 \sum_{i=1}^{s} \int_{S^+ \times R^3} d\vec{p}_{s+1} d\hat{w}\ \hat{w} \cdot (\frac{\vec{p}_i - \vec{p}_{s+1}}{m}) \times \tag{6.7}$$

$$\left[f_{s+1}^{(0)}(\vec{X}_1, \ldots, \vec{r}_i, \vec{p'}_i, \ldots, \vec{r}_i, \vec{p'}_{s+1}; t) \ - \ f_{s+1}^{(0)}(\vec{X}_1, \ldots, \vec{r}_i, \vec{p}_i, \ldots, \vec{r}_i, \vec{p}_{s+1}; t) \right];$$

c'est la "hiérarchie de Boltzmann" qui a été étudiée pour la première fois par Cercignani [11].

De notre point de vue, cette nouvelle hiérarchie possède deux propriétés capitales, qui la distingue fondamentalement de la hiérarchie (6.5):

1) Par le changement de t en $-t$, le terme de collisions $C_{s,s+1}^{(0)}$ change de signe. En effet, si l'on examine le calcul conduisant à l'expression de $C_{s,s+1}^{(\sigma)}$ (avec $\sigma \neq 0$!) sous la forme d'un bilan de "gains et pertes" analogue à celle de $C_{1,2}$ dans (4.1), on vérifie que le remplacement de t par $-t$ revient à échanger le rôle des impulsions $\vec{p}_i, \vec{p}_{s+1}$ et $\vec{p'}_i, \vec{p'}_{s+1}$. Il s'ensuit qu'à la limite $\sigma = 0$ (et seulement à cette limite !), $C_{s,s+1}^{(0)}$ est alors remplacé par $-C_{s,s+1}^{(0)}$. La hiérarchie de Boltzmann est donc *irréversible*, cette irréversibilité étant due au passage à la limite $\sigma = 0$ dans la hiérarchie B.B.G.K.Y. (6.5).

2) Si les fonctions de distribution $f_s^{(0)}$ satisfont *initialement* à la condition de factorisation:

$$f_s^{(0)}(\vec{X}_1, \ldots, \vec{X}_s) = \Pi_{i=1}^{s} f_1^{(0)}(\vec{X}_i), \tag{6.8}$$

il existe alors une solution $f_s^{(0)}(\vec{X}_1, \ldots \vec{X}_s; t)$ de la hiérarchie (6.6) correspondant à la condition initiale (6.8), avec $f_s^{(0)}(\ldots, 0) = f_s^{(0)}$, et cette solution *reste* factorisée au cours du temps, de telle sorte que l'on a:

$$f_s^{(0)}(\vec{X}_1, \ldots, \vec{X}_s; t) = \Pi_{i=1}^{s} f_1^{(0)}(\vec{X}_i; t), \tag{6.9}$$

où $f_1^{(0)}(\vec{X};t)$ est la solution de l'équation non linéaire de Boltzmann (4.3) correspondant à la condition initiale $f_1^{(0)}(\vec{X};0) = f_1^{(0)}(\vec{X})$. Cette seconde propriété montre que la hiérarchie de Boltzmann assure la *conservation au cours du temps du chaos moléculaire*; si celui-ci est réalisé à l'instant initial, il "*se propage*" à un instant t quelconque, sans qu'il y ait contradiction avec les lois du mouvement.

- 3ème *étape. Démonstration du théorème de Lanford.* Ayant ainsi établi la convergence *formelle* de la hiérarchie B.B.G.K.Y. (6.5) vers celle de Boltzmann (6.6), il reste à prouver que les solutions $f_s^{(\sigma)}(\ldots;t)$ de (6.5) convergent effectivement en un certain sens vers les solutions $f_s^{(0)}(\ldots;t)$ de (6.6): c'est l'objet même du théorème de Lanford, dont la démonstration se déroule en deux temps.

(i) On exprime d'abord les solutions respectives de (6.5) et de (6.6) sous la forme de développements en séries de perturbations dépendant du temps (Dyson), dans lesquelles ce sont les termes de collisions qui jouent le rôle des perturbations. Pour assurer la convergence uniforme en σ de ces développements sur un certain intervalle de temps, on doit imposer aux fonctions de distribution initiales $f_s^{(\sigma)}(\ldots,0)$ d'être uniformément bornées en σ, leur borne supérieure étant définie par la condition:

(C_1). Il existe, pour tout $\sigma < \sigma_0$, une paire de nombres $z, \beta > 0$ et une constante positive M, indépendante de σ et de s, tels que l'on a sur $(\Lambda \times R^3)^s$: $|f_s^{(\sigma)}(\vec{X}_1,\ldots\vec{X}_s)| \leq M z^s \Pi_{i=1}^s \phi_\beta(\vec{p}_i)$, où les $\phi_\beta(\vec{p}_i)$ sont des distributions maxwelliennes normalisées à la température $1/k\beta$.

En portant cette condition dans les séries de perturbation, on peut alors établir leur convergence uniforme pour des intervalles de temps $|t| \leq t_0(z,\beta)$, avec $t_0(z,\beta) \simeq \tau_{\text{l.p.m.}}/A$, où $\tau_{\text{l.p.m.}}$ est le temps de libre parcours moyen pour des particules à la température $1/k\beta$, et où A est une constante, indépendante de z, β et σ, qui intervient dans les majorations des termes du développement.

(ii) Ayant ainsi obtenu une expression formelle des solutions $f_s^{(\sigma)}(\ldots,t)$ et $f_s^{(0)}(\ldots,t)$ pour $0 \leq t \leq t_0(z,\beta)$, il reste à déterminer sous quelle condition on peut assurer la convergence terme à terme des séries de perturbation définissant respectivement les $f_s^{(\sigma)}(\ldots;t)$ et les $f_s^{(0)}(\ldots;t)$; selon la méthode suivie, il est clair que cette nouvelle condition ne peut porter que sur le mode de convergence des fonctions de distribution initiales $f_s^{(\sigma)}(\ldots,0)$ et $f_s^{(0)}(\ldots,0)$. Avant de la formuler, on est conduit à introduire une restriction de l'espace des phases $\Gamma_s = (\Lambda \times R^3)^s$ définie par: $\Gamma_s^{(0)} = \{(\vec{X}_1\ldots\vec{X}_s) \in (\Lambda \times R^3)^s \mid (\vec{r}_i \neq \vec{r}_j)\}$, qui est obtenue en excluant de l'espace Γ_s tous les points tels que $\vec{r}_i = \vec{r}_j$, qui donnent lieu à une collision binaire à l'instant initial. Soulignons que. par sa définition même, $\Gamma_s^{(0)}$ ne diffère de Γ_s que par un ensemble de mesure de Lebesgue *nulle*, puisque les points exclus sont tous situés sur certaines hypersurfaces de Γ_s.

Ceci étant, le mode de convergence auquel doivent satisfaire les fonctions de distribution initiales $f_s^{(\sigma)}(\ldots;0)$ est défini par la condition:

(C_2). Il existe une fonction continue $f_s^{(0)}(\vec{X}_1,\ldots\vec{X}_s)$ sur $\Gamma_s^{(0)}$, telle que l'on a uniformément, sur tous les ensembles compacts de $\Gamma_s^{(0)}$,

$$\lim_{\sigma\to 0} f_s^{(\sigma)}(\vec{X}_1,\ldots\vec{X}_s;0) = f_s^{(0)}(\vec{X}_1,\ldots\vec{X}_s). \tag{6.10}$$

Les deux conditions (C_1) et (C_2) permettent alors d'achever la démonstration du théorème de Lanford [4], qui s'énonce:

Théorème: Soit $f_s^{(\sigma)}(\vec{X}_1,\ldots\vec{X}_s)$ une suite, indexée par σ, de fonctions de distribution représentant l'état statistique initial d'un fluide de N sphères dures de diamètre σ, et satisfaisant aux conditions (C_1) et (C_2). Si $f_s^{(\sigma)}(\vec{X}_1,\ldots\vec{X}_s;t)$ est la solution de la hiérarchie B.B.G.K.Y. (6.5) pour la condition initiale $f_s^{(\sigma)}(\vec{X}_1,\ldots\vec{X}_s)$, et si $f_s^{(0)}(\vec{X}_1,\ldots\vec{X}_s;t)$ est celle de la hiérarchie de Boltzmann (6.6) pour la condition initiale $f_s^{(0)}(\vec{X}_1,\ldots\vec{X}_s)$ figurant dans (C_2), il existe alors un temps $t_0(z,\beta)$ tel que l'on a pour $0 \le t \le t_0(z,\beta)$:

$$\lim_{\sigma\to 0} f_s^{(\sigma)}(\vec{X}_1,\ldots\vec{X}_s;t) = f_s^{(0)}(\vec{X}_1,\ldots\vec{X}_s;t) \tag{6.11}$$

presque partout sur $\Gamma_s^{(0)}$, où les fonctions $f_s^{(\sigma)}(\ldots;t)$ admettent une borne du type (C_1) avec des constantes $z' > z$ et $\beta' < \beta$.

Comme la hiérarchie de Boltzmann a la propriété de conserver le "chaos moléculaire" au cours du temps, ce théorème comporte l'important corollaire suivant:

Corollaire: Si les conditions du théorème précédent sont remplies et si, de plus, la fonction continue $f_s^{(0)}(\vec{X}_1,\ldots\vec{X}_s)$ de (C_2)satisfait à la condition de factorisation (6.8), on a alors presque partout sur $\Gamma_s^{(0)}$, pour $0 \le t \le t_0(z,\beta)$,

$$\lim_{\sigma\to 0} f_s^{(\sigma)}(\vec{X}_1,\ldots\vec{X}_s;t) = \Pi_{i=1}^{s} f_1^{(0)}(\vec{X}_i;t), \tag{6.12}$$

où $f_1^{(0)}(\vec{X};t)$ est la solution de l'équation de l'équation non linéaire de Boltzmann (4.3) pour la condition initiale $f_1^{(0)}(\vec{X})$.

En d'autres termes, ce résultat permet d'établir rigoureusement, pour un fluide de sphères dures, l'existence à la limite de Boltzmann-Grad d'un régime cinétique décrit par l'équation de Boltzmann (4.3), si les fonctions de distribution initiales satisfont à la condition de convergence (C_2) et à la condition de chaos moléculaire (6.8). Ainsi se trouvent clairement précisées les conditions qui rendent possible le processus de réduction de la description de l'état d'un fluide idéal aboutissant au régime cinétique de Boltzmann-Grad.

Il reste à faire quelques remarques concernant les limitations de la validité de ce théorème et de son corollaire:

(i) Celles qui proviennent des deux conditions (C_1) et (C_2) (relatives à la norme et à la convergence des fonctions de distribution initiales) sont peu restrictives du point de vue physique.

(ii) Il n'en est pas de même pour celle relative à la valeur de $t_0(z,\beta)$ qui borne supérieurement l'intervalle de temps sur lequel est assurée la convergence des séries de perturbation définissant $f_s^{(\sigma)}(\ldots;t)$ et $f_s^{(0)}(\ldots;t)$. Elle limite en effet la validité des résultats à une durée de l'ordre d'une fraction du temps de l.p.m. (on le vérifie en optimisant la valeur de la constante A de manière à avoir le domaine de convergence le plus grand possible, ce qui conduit à la valeur approximative $A^{-1} \simeq 1/5$ [4]); il s'ensuit que le théorème précédent ne peut pas s'appliquer à l'étude du comportement du système sur de longues durées.

Toutefois, la portée de cette dernière limitation est atténuée par les deux remarques suivantes: d'abord, elle apparaît essentiellement d'ordre technique, puisqu'elle résulte de l'utilisation de développements de perturbation et du fait que la convergence des séries correspondantes ne peut être maîtrisée que pour de petites valeurs de t; de plus, le résultat obtenu apporte néanmoins une information capitale concernant l'établissement du régime cinétique, puisque le temps de l.m.p. est précisément de l'ordre du temps de relaxation de la fonction de distribution $f_1^{(0)}(\vec{X};t)$ vers une maxwellienne locale.

(iii) Enfin, une autre limitation de ce théorème pourrait provenir du modèle particulier (sphères dures) pour lequel il est démontré; en fait, il a été établi que la plupart des résultats précédents étaient également valables dans le cas de potentiels d'interaction positifs et de portée finie, satisfaisant à certaines conditions de régularité [16].

Pour conclure cette section, il convient de revenir sur la signification des résultats précédents en se plaçant au point de vue des questions A et B, ce qui conduit aux importantes remarques suivantes:

(i) L'irréversibilité résulte du *passage à la limite* $\sigma \longrightarrow 0$ (avec l constant), pour lequel les solutions de la hiérarchie (réversible) B.B.G.K.Y. tendent vers les solutions de la hiérarchie (irréversible) de Boltzmann. On constate ainsi que l'irréversibilité n'est *pas* due à une quelconque hypothèse de nature statistique (telle que le "chaos moléculaire"), mais qu'elle apparaît comme une propriété de *l'évolution limite* du système (décrite par le hiérarchie de Boltzmann) lorsque $\sigma \longrightarrow 0$.

(ii) C'est l'hypothèse de chaos moléculaire à *l'instant initial*, et la *conservation* de cette propriété au cours du temps, qui rendent possible la réduction, à la limite $\sigma \longrightarrow 0$, des solutions de *toute* la hiérarchie B.B.G.K.Y. aux *seules* solutions de l'équation non-linéaire de Boltzmann. Ce processus de réduction est donc lié au choix de conditions initiales particulières pour lesquelles le système considéré (un fluide idéal) évolue suivant le régime cinétique de Boltzmann-Grad; dans ce cas, l'état du système est rigoureusement représenté, à la limite $\sigma \longrightarrow 0$, par son état moyen dont l'évolution est décrite par l'équation de Boltzmann.

Ces deux remarques intéressent directement les réponses qui peuvent être ainsi apportées aux questions A et B; elles demandent encore à être complétées par des observations concernant le problème de l'irréversibilité:

(iii) La réversibilité dynamique et l'irréversibilité de l'évolution limite sont rendues compatibles, dans le théorème de Lanford, par la différence existant entre les modes de convergence à l'instant initial et à l'instant t. D'après C_2, on impose en effet aux fonctions de distribution initiales de satisfaire à une condition de convergence uniforme sur $\Gamma_s^{(0)}$ qui n'est *pas* conservée à l'instant t, puisque le résultat (6.11) est seulement valable *presque partout* sur $\Gamma_s^{(0)}$. Cette restriction est due à l'existence d'un ensemble de points de phase exceptionnels, dont la mesure devient rigoureusement nulle lorsque $\sigma \longrightarrow 0$, pour lesquels le théorème de Lanford ne s'applique pas; cet ensemble est en fait défini par les hypersurfaces associées aux collisions binaires qui se sont produites sur l'intervalle $(0, -t)$ précédant l'instant initial, et qui créent des corrélations ne satisfaisant pas à C_2[4]. Notons d'ailleurs que c'est l'existence même de telles corrélations qui est la cause du retour du système à son état initial, après renversement des impulsions. Il s'ensuit que les $f_s^{(\sigma)}(\ldots;t)$ ne peuvent pas être choisies comme nouvelles conditions initiales; il en est de même pour les fonctions de distribution obtenues par renversement de toutes les impulsions à l'instant t, ce qui permet d'éviter les contradictions liées à la réversibilité du modèle dynamique. Il apparaît ainsi que l'irréversibilité est associée à un affaiblissement progressif, dans la *direction des t croissants*, des conditions de convergence des $f_s^{(\sigma)}(\ldots;t)$ vers les solutions limites $f_s^{(0)}(\ldots;t)$ [4,17].

(iv) Le théorème précédent s'applique également dans le cas où $t \leq 0$, puisque la démonstration de la convergence des séries de perturbation vaut également pour les t négatifs, tels que $-t_0(z,\beta) \leq t \leq 0$. Il s'ensuit que les $f_s^{(\sigma)}(\ldots;t)$ convergent alors vers une solution de la hiérarchie de Botzmann avec un terme de collisions *changé de signe* (c'est l'évolution "anti-Boltzmann"), ce qui s'accorde avec la réversibilité des courbes $H[Z(t)]$ évoquée dans la section 5.

[4]Ces corrélations sont à l'origine des "recollisions" entre particules, qui sont responsables des "divergences" de la théorie des fluides modérément denses (cf. section 4); ce sont précisément ces "recollisions" que le théorème de Lanford a pour effet d'éliminer.

(v) Le signe de t, qui fixe la "flèche du temps" est déterminé par le fait que la condition C_2 doit être considérée comme une condition *initiale* et non finale. Ceci s'accorde d'ailleurs avec les remarques de la section 3 concernant la propriété d'asymétrie par rapport au temps de la densité en phase initiale. Finalement, on voit que l'irréversibilité est la conséquence des propriétés asymptotiques (6.11) du système *et* du choix de la "flèche du temps".

7. Conclusion.

Malgré ses limitations, le théorème de Lanford est un résultat rigoureux de grande importance pour notre propos , car il montre qu'il est possible de concilier la description mécaniste microscopique avec l'irréversibilité de fait à l'échelle macroscopique. Il montre aussi que cette possibilité repose sur l'existence de régimes – ou processus – limites qui résultent des propriétés asymptotiques, pour certaines limites, de systèmes mécaniques à un très grand nombre de degrés de liberté. On voit ainsi que la définition d'un passage à la limite approprié (s'il existe!) joue un rôle central dans la réponse aux questions A et B: d'une part, il aboutit à la détermination d'un processus limite dont l'évolution est gouvernée par des équations généralement[5] irréversibles portant sur des variables "macroscopiques" sans dispersion; d'autre part, il permet de rendre compte du changement qualitatif qui se produit dans l'état du système lors du passage à la limite.

Il convient encore de souligner que ce changement qualitatif, en particulier l'irréversibilité, ne concerne *que* le processus limite ainsi obtenu, et que le comportement du système, au *voisinage* de la limite, conserve les caractères propres aux systèmes mécaniques, et notamment la réversibilité de l'évolution. On peut donc dire que l'évolution du système, au voisinage de la limite, se présente comme une bonne "*simulation mécaniste*" du comportement macroscopique irréversible. Cette discontinuité dans la description de l'état du système, qui correspond à la brisure de symétrie évoquée dans la question B, est bien illustrée en théorie cinétique des gaz par le passage de la "courbe" $H[Z(t)]$, relative à un seul système, à la courbe de concentration $\mathcal{H}$ de la conjecture de Boltzmann-Ehrenfest. Il n'est d'ailleurs pas sans intérêt de remarquer que le théorème de Lanford apporte, pour le cas d'un fluide idéal, une justification "a posteriori" de cette conjecture [18], et qu'il permet ainsi d'établir un "pont" entre deux définitions de l'état macroscopique du système - celle qui est fondée sur la théorie des ensembles de Gibbs et celle qui s'appuie, comme la théorie cinétique des gaz, sur l'introduction de variables "coarse-grained" - dont la comparaison s'avère utile pour approfondir la signification physique des passages à la limite.

[5]On doit mettre à part le cas de l'équation de Vlasov, qui est réversible.

On notera enfin que les résultats de la section 6 conduisent à clarifier les conditions qui sont à l'origine de l'irréversibilité de l'évolution, et à rectifier notamment l'opinion habituelle selon laquelle celle-ci serait due à des hypothèses de nature statistique toujours susceptibles d'entrer en conflit avec les lois dynamiques.

Le théorème de Lanford fournit ainsi un bon exemple de la démarche générale esquissée dans la section 3; pour la développer, il faut naturellement l'appliquer à d'autres situations limites, telles que celles signalées dans les sections 3 et 4. Un certain nombre de résultats ont été ainsi obtenus dans le cas du *régime cinétique*; ils concernent essentiellement: la limite du couplage faible, où la validité d'une équation cinétique du type Landau a pu être établie dans la cadre restreint du modèle de Lorentz [19]; la limite du champ moyen, où le théorème remarquable de Braun et Hepp [20] permet une déduction rigoureuse de l'équation de Vlasov, qui décrit le "régime de Debye" caractéristique des milieux ionisés. Mais la situation s'avère beaucoup moins favorable pour le régime hydrodynamique, sauf si le fluide est suffisamment dilué pour qu'il existe un régime cinétique intermédiaire. Dans le cas général où la densité sans dimension reste finie, on est amené à définir une *limite hydrodynamique* (telle que $l \longrightarrow 0$, avec $nr_0{}^3 = cte$), et à considérer directement le passage du niveau microscopique au niveau hydrodynamique accessible à notre expérience. On doit alors tenter d'élaborer une théorie des phénoménes de transport, dont les lois phénoménologiques bien connues (lois de Fick, Fourier, ...) seraient rigoureusement déduites, à cette limite, à partir de l'évolution microscopique du système [14]. On conçoit que la réalisation de ce programme très ambitieux rencontre des difficultés considérables, et qu'elle en reste le plus souvent au stade des conjectures; seules sont à signaler quelques avancées fragmentaires dans le problème le plus simple, celui de la self-diffusion [21].

Si les résultats ainsi obtenus constituent un progrès réel, en ce sens qu'il apparaît possible de concevoir un programme visant à *déduire* le comportement irréversible d'un fluide à partir de sa description microscopique, on voit que sa mise en œuvre rencontre de grandes difficultés, au moins d'ordre technique. Une autre limitation de cette démarche tient au point de vue même qui a été adopté, qui a pour effet de privilégier les arguments fondés sur les propriétés asymptotiques du système (loi des grands nombres), alors que le rôle de la dynamique se trouve réduit au minimum; c'est ainsi par exemple que les propriétés ergodiques d'un fluide de N sphères dures [22] n'interviennent pas dans la démonstration du théorème de Lanford. En fait, la dynamique n'entre dans les raisonnements que par le "jeu" entre les paramètres qui caractérisent le potentiel d'interaction, et ceux relatifs à l'état du fluide. Ces remarques permettent de mieux comprendre pourquoi la démarche décrite dans cet article s'applique bien au cas du régime cinétique d'un fluide idéal, et quelle est l'origine des difficultés propres aux fluides modérément denses, pour lesquels l'existence problématique d'un régime cinétique apparaît liée à l'impossibilité de définir un passage à la limite approprié (les grandeurs l et nr_0^3 devant rester finies); dans ce dernier cas, il semble donc nécessaire, pour aboutir à une théorie satisfaisante, de prendre en compte les effets dynamiques des interactions élémentaires, d'une manière qui reste à déterminer.

En conclusion, l'interrogation fondamentale que comportent les questions A et B demeure un problème ouvert, qui ne peut que continuer à susciter de nouveaux travaux, tant dans le domaine cinétique que dans celui de l'hydrodynamique. Soulignons d'ailleurs qu'il ne s'agit pas d'un simple problème académique, dont l'objet serait de justifier et d'établir sur des bases rigoureuses des résultats déjà bien acquis, mais que son étude peut au contraire contribuer à ouvrir des voies nouvelles ou à en infléchir d'autres, qui se sont révélées moins fructueuses. Enfin, ce problème comporte aussi d'autres prolongements, et non des moindres, qui intéressent deux questions fondamentales que nous ne pouvons que mentionner ici. La première porte sur le problème du "statut" des lois macroscopiques qui peut se formuler ainsi: celles-ci sont-elles des lois "secondaires"qui doivent être *dérivées* de lois plus fondamentales, considérées comme "primaires" (au sens d'Eddington [23]), ou bien sont-elles également des lois "primaires" au même titre que les autres lois fondamentales? La seconde question, qui n'est d'ailleurs pas indépendante de la précédente, concerne l'universalité bien connue des phénomènes irréversibles, que la démarche exposée ici ne suffit pas à expliquer, puisqu'elle revient à étudier l'évolution d'un système à partir d'une situation initiale hors d'équilibre résultant d'une certaine "préparation" de son état. Les résultats ainsi obtenus conduisent en fait à l'interrogation suivante : pour quelles raisons, les conditions initiales qui permettent de répondre aux questions A et B sont-elles toujours pratiquement réalisées dans la nature? Cette dernière question renvoie presque inévitablement aux problèmes de la cosmologie, qui ont été profondément renouvelés à la suite des développements rapides intervenus notamment avec la description fine du "big-bang" et la théorie des univers "inflationnaires".

References

1. P. and T. Ehrenfest: The Conceptual Foundations of the Statistical Approach in Mechanics, (English translation), Cornell Univ. Press, 1959.
2. H. Grad: Principles of the Kinetic Theory of Gases, in Handbuch der Physik, S. Flugge ed, Springer-Verlag, Berlin, 1958, p.205.
3. H. Grad: Levels of description in Statistical Mechanics, in Delaware Seminar in the Foundations of Physics (M. Bunge ed.), Springer, Berlin, 1967, Vol.1, p.49.
4. O.E. Lanford: Time Evolution of Large Classical Systems, in Dynamical Systems, Theory and Applications, J. Moser ed., Lecture Notes in Physics n°38, Springer-Verlag,

Berlin, 1975, p.1.
5. A. I. Khintchine: Mathematical Foundations of Statistical Mechanics, Dover, New-York, 1949.
6. (a) D. Ruelle: Statistical Mechanics, Rigorous Results, Benjamin, New-York, 1969; (b) R. B. Griffiths: Rigorous Results and Theorems, in Phase Transitions and Critical Phenomena, C. Domb. and M.S. Green eds, Vol.1, New-York Acad. Press, 1972, p.7.
7. R. Jancel: Ann. Fondation L. de Broglie, **8**, 1983, p. 83 et 189.
8. O. Penrose: Rep. Prog. Physics, **42**, 1979, p. 1937.
9. (a) R. Peierls: Surprises in Theoretical Physics, Princeton Univ. Press, Princeton 1979, p. 73-84.
(b) R.P.Feynman: The Character of Physical Law., B.B.C. Publications, 1965, Lecture 5; trad. fr. dans La Nature de la Physique, ed. du Seuil, Paris, 1980, p. 129-150.
10. O.E. Lanford: Soc. Math. de France, Astérisque, **40**, 1976, p. 117.
11. C. Cercignani: Theory and Application of the Boltzmann Equation, Scottish Acad. Press, Edinburgh and London, 1975.
12. (a) E.G.D. Cohen: The Kinetic Theory of Dense Gases, in Fundamental Problems in Statistical Mechanics, Vol. II, (E.G.D. Cohen ed.), North Holland, Amsterdam, 1968.
(b) J.R. Dorfman et H. Van Beijeren: The Kinetic Theory of Gases, in Statistical Mechanics, Part B, (B.J. Berne ed.), Plenum Press, New-York, 1977.
13. N.N. Bogolioubov: Problems of Dynamical Theory in Statistical Physics, (Engl. trans.), in Studies in Statistical Mechanics, J. de Boer and G.E. Uhlenbeck eds., North-Holland, Amsterdam, 1962.
14. J.L. Lebowitz et H. Spohn: J. Stat. Phys., **28**, 1982, p. 539; Comm. Pure and Appl. Math., **36**, 1983, p. 595.
15. Y.L. Klimontovich: Statistical Physics, (Engl. trans.), Harwood Acad. Publishers, Chur, London, 1986.
16. F.G. Kiang: Ph. D. Thesis, Univ. of California, Berkeley, 1975.
17. R. Jancel: Ann. Fondation L. de Broglie, **12**, 1987, p. 221.
18. O.E. Lanford, Physica, **106A**, 1981, p.70.
19. H. Spohn: Comm. Math. Phys., **60**, 1978, p. 277; Rev. Mod. Phys., **52**, 1980, p. 569.
20. W. Braun et K. Hepp: Comm. Math. Phys., **56**, 1977, p. 101.
21. L.A. Bunimovich et Ya. G. Sinai: Comm. Math. Phys., **78**, 1980, p. 247; **78**, 1981, p.49.
22. Ya. G. Sinai: (a) in Statistical Mechanics: Foundations and Applications (Proc. IUAP Meeting, 1966, T.A. Back ed., New-York, Benjamin, 1967; p. 559.
(b) Development of Krylov's Ideas, in Works on the Foundations of Statistical Physics by N.G. Krylov, (Engl. Transl.), Princeton Univ. Press, 1979.
23. A. Eddington: The Nature of the Physical World, Ann. Arbor Paperbacks, Michigan Press, 1958.

ALLOCUTION DE CLÔTURE

GUY RIDEAU

D'abord, et avant tout, merci à tous qui êtes venus ici me faire l'honneur et m'apporter la joie de participer à ces deux journées qui ont été, grâce à vous, le plus beau cadeau qui pouvait m'être fait à ce tournant de ma vie. Je le dois à l'initiative de Moshé Flato et Daniel Sternheimer, des amis de plus de trente ans, et de Jean-Pierre Gazeau dont l'amitié a sa source dans les bouillonnements soixante huitards. Je suis heureux de leur exprimer ici toute ma gratitude. Elle va aussi à celles qui se sont dépensées sans compter pour assurer la réussite matérielle de ce colloque, Jacqueline Bertrand et Michèle Irac, collègues, collaboratrices et néanmoins amies, et à Irène Mazzella, la dévouée secrétaire du laboratoire dont tous ceux qui la connaissent peuvent apprécier la bonne humeur, la gentillesse et la compétence. Tous ont longtemps travaillé dans le secret avant de me révéler cette surprise si bien préparée.

Mais il n'y a pas de fête parfaite. Certains avec qui j'ai partagé longtemps discussions, et parfois controverses, ne sont plus. Je voudrais associer à ce colloque mes deux amis prématurément disparus, Bernard Jouvet et Michel Sirugue. J'y associerai aussi le souvenir d'Asim Barut, dont nous avons appris le décès tout récent, et celui de ma femme, si longtemps témoin de mes doutes et de mes espoirs et dont j'aurais tant aimé la présence aujourd'hui.

J'ai donc pendant quarante cinq ans été un chercheur en physique mathématique. Sans doute ma vie ne se résume pas à cette seule activité, mais elle aura été tout au long, au moins comme une basse obstinée d'une partition, parfois même un peu plus. Certains ici savent combien le travail m'aura aidé à faire face à une douloureuse épreuve.

Je ne reviendrai pas sur le bilan qu'on peut en faire ; il me suffira de dire que je serai beaucoup moins louangeur que Moshé Flato, ayant toujours plus profondément ressenti mes échecs que mes réussites. Mais il me paraît intéressant d'essayer de trouver avec vous la racine, le point de

J. Bertrand et al. (eds.), Modern Group Theoretical Methods in Physics, 315–318.

départ, l'impulsion initiale qui m'ont si fortement marqué que toute ma vie professionnelle en a été décidée.

Il faut, pour cela, que je me reporte cinquante ans en arrière. J'arrivais alors à Paris pour poursuivre mes études dans les classes préparatoires du lycée Louis-le-Grand. La seconde guerre mondiale s'achevait, apportant, en même temps, la révélation de ses horreurs et des difficultés quotidiennes de tous ordres. Ce sombre arrière plan aggravait le mal que j'avais à vivre l'enfermement et la discipline arbitraire qui caractérisaient les internats de l'époque. J'essayais d'échapper à cette morosité en gagnant, le plus souvent clandestinement, à travers le passage de la Sorbonne, la librairie des Presses Universitaires, non pour y faire un quelconque achat, dont je n'avais pas les moyens, mais pour m'y détendre dans l'environnement du papier imprimé. Les Presses Universitaires n'étaient pas la ruche bourdonnante et surchargée qu'elle est devenue. Avec la complicité bienveillante des vendeurs, il était possible de s'isoler dans un coin pour y bouquiner paisiblement. J'ai ainsi lu des livres entiers en plusieurs épisodes. Je revenais souvent aux rayons consacrés aux ouvrages scientifiques. On y trouvait la collection des cours de de Broglie sur la Mécanique Quantique et quelques autres volumes présentant les principaux travaux et idées des pères fondateurs. Etant donnée la formation que l'on recevait alors dans les terminales scientifiques des lycées qui allait difficilement jusqu'à la notion de dérivée et d'intégrale simple, je feuilletais ces livres sans y comprendre rien. Des symboles apparemment simples, des carrés, des triangles m'étaient aussi hermétiques que les idéogrammes d'une langue inconnue. Certains objets possédaient apparemment l'étrange propriété d'anticommuter. D'autres, dont le nom de fonction d'onde paraissait évocateur, vivaient dans un espace inconcevable, sinon pour Hilbert, dont je sentais presque l'infinité de dimensions m'environner de son bruissement. J'avais l'impression d'être devant l'accès d'un monde surréel et poétique, plus vrai que le monde borné à trois dimensions, rendu si pénible par la dureté des temps. Je ne pensais même pas qu'il pouvait m'être possible de pénétrer ces mystères. Ce regard jeté à la dérobée au-delà des parois de la caverne suffisait à me rasséréner pour tout un moment. Avec des ruses multiples, je regagnais alors subrepticement mon lycée pour me plonger dans les peu enthousiasmants problèmes de résolution des équations linéaires. D'ailleurs, ni durant le temps passé à Louis-le-Grand, ni pendant mes études de licence, il ne fut question de protons et d'électrons sinon allusivement et encore moins d'équation de Schrödinger. La physique alors dénommée générale était résolument classique. Mais le mal était déjà fait et quand se présenta l'occasion de passer du rêve à la réalité, je m'obstinais à accomplir la vocation qui était née dans un coin retiré des Presses Universitaires.

Si j'ai raconté un peu longuement cette anecdote, c'est sans doute en partie pour le plaisir de revivre un peu de ma jeunesse, mais c'est surtout pour en tirer une leçon qui me paraît importante. Je suis convaincu qu'il n'y a pas de recherche valable sans ces deux composantes : la vocation et le rêve.

La première permet de trouver un certain bonheur dans sa réalisation même et de poursuivre son travail en surmontant amertumes et déceptions.

Quant au rêve, il me paraît être un moteur essentiel de la découverte. Deux rêves au moins traversent toute l'histoire de la physique mathématique et n'ont pas atteint sans doute l'extrémité de leurs trajectoires, celui de la géométrisation de l'Univers depuis les tétraèdres du Timée jusqu'à la théorie de la relativité générale et l'introduction des groupes de symétrie, celui, encore plus ancien, de la simplicité de l'univers, depuis le principe élémentaire des philosophes ioniens jusqu'à la très contemporaine théorie de la grande unification ; ces deux rêves sont d'ailleurs de plus en plus inextricablement mêlés dans la recherche actuelle, plusieurs communications à ce colloque en sont la parfaite illustration.

Malheureusement, il ne suffit pas de rêver pour trouver et l'histoire de la physique mathématique est aussi un vaste cimetière de visions fausses. Les exemples en sont moins facilement présents à l'esprit parce qu'évacués peu à peu hors de la mémoire collective, quoique les atomes crochus de Descartes soient encore bien connus de tous.

Mais il arrive que même une vision fausse conduise à des résultats positifs. L'exemple bien connu est celui de Képler qui découvrit ses trois lois faute de pouvoir enfermer dans des polyèdres les orbites des planètes. Plus près de nous, ce sont les tentatives avortées pour rendre compte de la contraction de Lorentz par des moyens purement physiques qui permirent à Einstein de fonder la relativité restreinte. Le cimetière des visions fausses est en quelque sorte le terreau dont se nourrissent les vraies découvertes. Il faut accepter cette accumulation d'échecs sous peine de bloquer tout progrès. Plus brièvement, je dirai qu'on ne peut entreprendre l'exploration de nouveaux territoires sans courir de risques, ou plus poétiquement, comme l'énonçait un de nos grands ancêtres, il y a 2500 ans, dans un de ces distiques lapidaires qui l'on fait surnommer Héraclite l'obscur

Si tu n'espères pas l'inespéré
Tu ne le trouveras pas

Je pense que cette situation ambiguë entre rêves et échecs est spécifiquement la nôtre et qu'elle constitue la meilleure justification de nos efforts, de nos espoirs et de notre persévérance.

Elle ne facilite pas nos rapports avec nos autorités de tutelle. Par essence, elles ont horreur du risque et leur devoir leur semble être de ne distribuer leur manne parcimonieuse qu'avec l'assurance de n'en rien perdre. On nous a donc demandé d'établir des projets individuels puis collectifs, de définir des programmes pour un an, puis deux ans, puis quatre ans. On nous a aussi demandé de contribuer à l'élaboration de rapports de prospective à plus ou moins long terme dont devaient résulter des synthèses tendant à fixer le devenir d'une discipline sur plusieurs années. Par un ironique retournement dialectique, nous avons trouvé là une autre façon de nous abandonner aux rêves et à l'imaginaire, dans un jeu distordu par la nécessaire supputation de ce qui pourrait plaire et par l'indispensable gymnastique pour rattacher nos désirs à des objectifs prioritaires établis parfois loin de notre contrôle. Il y a sans doute eu quelques aspects positifs dans cet encadrement paperassier, ne serait-ce que nous inciter à un certain regard critique sur nos propres travaux. Mais les aspects négatifs n'ont pas été négligeables : prolifération de publications, parfois insuffisamment préparées ou répétitives, effets de mode ravageurs, création de cercles restreints de chercheurs ne se citant qu'entre eux, sans parler de quelques règlements de compte. Comme on aurait dit, en d'autres temps, nos conditions objectives de production s'en sont trouvées changées. A vrai dire, cette évolution a été quelque peu tempérée par le simple fait que les administrations concernées ont dû confier à certains chercheurs le soin d'évaluer les projets ou programmes que d'autres soumettaient. Les évaluateurs risquant, à leur tour, d'être évalués, il s'est établi une certaine connivence, une sorte de langage codé, à l'abri de quoi nous avons pu sauvegarder jusqu'à maintenant une part de liberté et de nos droits à l'imaginaire. Mais nous devons rester vigilants et réaffirmer notre conviction qu'une discipline dont le développement serait effectivement prévisible est une discipline morte, qui n'aurait plus qu'à tirer ses dernières cartouches.

A en juger par la variété et le haut niveau des communications que nous avons entendues durant ces deux jours, tel ne paraît pas être le cas de la nôtre. On sent un frémissement, une agitation de la pensée, une remise en cause d'acquis hier encore intangibles, qui autorisent tous les espoirs de renouvellement et promettent de nouveaux départs.

De quoi demain sera-t-il fait ? Je ne me risquerai pas à prophétiser, mais au moment où je quitte la scène, je me réjouis de la voir si encombrée, si active et comme exemplaire de cet autre fragment héraclitéen, sur quoi j'aimerai conclure

"Qui cherche la vérité doit être prêt à l'inattendu,
Car elle est difficile à trouver
Et déconcertante quand on la rencontre ".

Encore une fois, merci à tous, merci d'être venus si nombreux me témoigner votre amitié, et bonne chance pour vos travaux futurs.